INTRODUCTION TO STATISTICAL REASONING

Introduction to Statistical Reasoning

PHILIP J. McCARTHY

Professor of Statistics, New York State School of Industrial and Labor Relations, Cornell University

ROBERT E. KRIEGER PUBLISHING COMPANY
HUNTINGTON, NEW YORK
1978

Original Edition 1957
Reprint 1978 with corrections

Printed and Published by
ROBERT E. KRIEGER PUBLISHING CO., INC.
645 NEW YORK AVENUE
HUNTINGTON, NEW YORK 11743

Printed in the United States of America

Library of Congress Cataloging in Publication Data

McCarthy, Philip John, 1918-
Introduction to statistical reasoning.

Reprint of the edition published by McGraw-Hill, New York; with corrections.
Includes index.
1. Statistics. I. Title.
[HA29.M16 1978] 519.5 78-1044
ISBN 0-88275-661-3

To
Mary Ann

PREFACE

This book is intended for use in a one-semester, nonmathematical course in statistics in which the instructor wishes to present a careful introduction to statistical reasoning. In writing this book, I was guided by the following considerations: (1) Statistics, as that branch of scientific method which provides the approach and the tools for dealing with the collection, analysis, interpretation, presentation, and use of quantitative data derived from either counting or measurement, deserves serious study in its own right. (2) The great majority of students who do or should take courses in statistics are not equipped to pursue a completely mathematical approach. (3) The use of a nonmathematical approach does not absolve one from exercising great care in stating the reasoning and techniques involved in statistics. (4) A first course should emphasize the concepts of statistical reasoning rather than attempt to cover a wide variety of techniques. (5) The concepts of statistical reasoning are novel and difficult for most students in a first course and therefore need to be introduced gradually and to be repeated time after time. (6) Although the elements of statistical reasoning are the same no matter what may be the subject-matter field in which they are to be used, different fields tend to require different types of methods. (7) Illustrative material should be drawn from investigations that are as significant as possible, and should not merely consist of sets of data upon which to perform computations.

Because of the interests of my students and other individuals with whom I have had close professional contacts, the illustrative material in this book has been chosen to range broadly over the social sciences. A very brief account of research problems usually accompanies each illustrative example, and any student who chooses to read some or all of the cited references may expect to improve his insight into the problems of research methodology in the social sciences.

This selection of illustrative material from the social sciences has also influenced to some extent the topics discussed in the book. Thus Chapter 10 gives an elementary and precise account of the elements of sample design as applied to the sampling of human populations. If I had been writing for an audience in the biological sciences, this chapter would undoubtedly have been devoted to the elements of experimental design. It is true that the biological sciences do employ sample designs and that the social sciences do find use for experimental designs. Yet there is a limit to the number of topics that can be carefully covered in a one-semester course.

The early chapters of the book follow the more or less traditional line of approach, discussing populations, variables, distributions, and measures of location and variability. The fundamental ideas and techniques provided by this approach are, I believe, required by the beginning student before he can begin to appreciate the need for modern statistical arguments. In dealing with these fundamentals, however, I have tried to lay the groundwork for the later work on probability and inference by repeatedly emphasizing the difference between populations and samples and by stressing the usefulness of conceptual models. The need for models is brought to a focus in Chapter 6 on simple random sampling and in Chapter 7 on probability. The remaining topics in the book are then treated in terms of models. The reader is first introduced to confidence-interval estimation, tests of statistical hypotheses, and the ideas of statistical decision theory in Chapter 8, which contains a fairly extensive treatment of the binomial probability model. This emphasis on the binomial model was chosen for several reasons: (1) the binomial model can frequently be employed in describing qualitative data that arise in the social sciences; (2) one can actually derive the binomial probability model from basic assumptions, at the mathematical level of maturity assumed of readers, and this would not be possible for any other theoretical probability distribution; and (3) the development of confidence intervals and hypothesis-testing procedures can be carried out in almost complete detail for the binomial model. Chapter 9 treats large sample theory for means, percentages, and differences between means and percentages, continues the discussion of confidence intervals and tests of hypotheses, and introduces the operating characteristic of a test procedure to expand on the treatment of Type II errors in Chapter 8; Chapter 10 discusses the elements of sample design; Chapter 11 presents a careful discussion of χ^2 procedures for qualitative data; and Chapter 12 describes the models used in linear regression and correlation. Not only are the appropriate models and their consequences dealt with in each case, but some indication is always given of the common situations in which the requirements of the model are violated. Chapter 10, on the elements of sample design, may be omitted from a first course if so desired. The remaining portions of the book are highly interrelated and will ordinarily have to be covered in the order in which they are presented.

A number of topics are conspicuous by their absence, namely: (1)

small-sample procedures based on Student's t distribution, the χ^2 distribution, and the F distribution; (2) the analysis of variance; and (3) nonparametric, or distribution-free, procedures. This is not a reflection on their importance. Since it is possible to cover carefully only a limited number of topics in a first course, it was necessary to exercise personal judgment in selecting those topics which would best introduce the student to the various aspects of statistical reasoning and which would be encountered most frequently in the social sciences. In presenting this material, every effort has been expended in attempting to provide the student with the basic ideas and tools that will be required in studying such omitted topics, and reference is made to them when the occasion warrants. Thus footnote references to Student's t distribution are given in Chapters 9 and 12 and to the parametric use of the χ^2 distribution in Chapter 11. Also, several examples of nonparametric tests are given, including the Spearman rank correlation coefficient in Chapter 12. It is my personal feeling that detailed discussion of any or all of these topics can best be originated in a second course after the basic concepts of statistical method have been assimilated.

I wish to express my appreciation to the many individuals and organizations who kindly gave me permission to use quotations, data, tables, and diagrams appearing in previously published work. Appropriate reference is made in the text as to the actual source of each of these items. I owe special debts to S. S. Wilks, from whom I received my original training in statistics; to Frederick F. Stephan, whose penetrating insight into the use of quantitative methods in the social sciences has always been a source of inspiration to me; to J. E. Morton, who was involved with me in setting up and teaching a general introductory course in statistics; and to Isadore Blumen, who has shared with me the teaching responsibility for this course and who offered many helpful suggestions during the preparation of the manuscript. Needless to say, all errors and faults in the presentation are mine alone.

I am also indebted to the New York State School of Industrial and Labor Relations, which provided the physical facilities for carrying out this work, and to the secretaries who labored over various drafts of the manuscript, particularly Helen Resnick and Agnes Murray, who typed the final version of the manuscript.

PHILIP J. MCCARTHY

CONTENTS

CHAPTER 1

INTRODUCTION

Quantitative data derived from either counting or measurement play an extremely important role in all types of human activity, and it is only necessary to scan the pages of a large daily newspaper in order to appreciate the full significance of this fact. For example, in a single issue of *The New York Times* we might find items like the following: the predicted number of bushels of wheat for the coming harvest, the estimated number of unemployed workers in the United States as of a certain date, an index of the stock-market level, the fraction of adults in a sample who say they will vote for a particular candidate in an approaching election, the results of an experiment to determine the efficacy of a vaccine for poliomyelitis, the level of the Consumer Price Index of the U.S. Bureau of Labor Statistics for a specified month, data which attempt to relate smoking and the incidence of lung cancer, and advertisements presenting data on the size of audience reached by various periodicals. This list could be multiplied many times without exhausting the kinds of numerical information which are presented to the public through the medium of the press—information with respect to which any well-informed individual should be able to react with some degree of critical judgment.

The results of a large number and variety of quantitative investigations find their way into widely circulated newspapers, periodicals, and books, but these investigations constitute only a minor portion of all such work that is done. The remainder are aimed at specific audiences and appear only in government reports, trade journals, technical publications, or reports which are confidential within a company or other organization. Thus a sociologist attempts to discover what motivates the members of a community to "pull together"; a plant breeder tries to determine the type of corn which is best suited to a given geographical region; a public health official wishes to trace the spread of a particular disease among members of a family or among families; the management of a company conducts an employee-attitude survey in order to discover ways in which its relations with employees can be improved; a company hires a market research organization to survey its customers and potential customers concerning their attitudes toward the company's products; or labor economists study the manner in which workers move between occupations, industries, and geographic areas in order that more realistic plans can be made for emergency mobilization.

Each of the foregoing examples represents an instance in which the final results depend upon the collection, analysis, interpretation, and presentation of quantitative data. Moreover, these and similar investigations usually have a more serious motivation than the mere satisfaction of idle curiosity. This implies that decisions and actions will be based, at least in part, upon these data, e.g., to stimulate further research in a particular direction, to guide a government or company in developing a program of action, or to advance knowledge in a subject-matter field. Under such circumstances, both the producer and the consumer of the data—and these may be the same individuals or organizations—have an important stake in knowing that the data have been correctly collected, analyzed, interpreted, and presented. *Statistics* is that branch of scientific method which attempts to provide the approach and the tools for dealing with these problems. Since any set of tools or method of approach may be correctly or incorrectly applied, it is vital that both producers and consumers of quantitative data have a thorough understanding of the methods of statistics.

This book will present a body of fundamental concepts and elementary methods which are common to all statistical investigations, regardless of the field in which they are applied. Naturally some special emphases will be developed through the choice of illustrative material, and the material of this book will be directed toward the social sciences rather than toward the physical or biological sciences.

In order to provide an over-all view and to demonstrate the wide applicability of statistical methods, the broad outlines of several illustrative examples will now be presented. These examples will serve to introduce some of the concepts which underlie every statistical investigation, concepts which will be examined in greater detail throughout the remaining chapters of this book.

Sample Surveys. We are all familiar to some extent with the work of the Bureau of the Census, particularly that which relates to the decennial censuses. Every 10 years the Census Bureau enumerates, as of a given date, all individuals living in the United States and presents data showing the geographical distribution of the total population (by regions, states, counties, cities, and the like), the age distribution, the sex distribution, distributions for many other characteristics of the individual members of the population, and the relationships between these distributions. This complete census is a very expensive undertaking, costing millions of dollars, and many people do not realize that a large part of the 1940 and

1950 censuses were carried out on an incomplete or sample basis. For certain purposes a part of the total population was selected; studies were made on this sample; and inferences were then drawn concerning the relationship between the sample and the total population.

One may well ask why sampling was used in conjunction with a complete census. The answer, as is so frequently the case, involves considerations of economy with respect to time and money. In this instance the primary motivating factors were as follows: (1) The feasible census schedule, to be used for all individuals, could not meet the public need for population statistics. However, by adding a set of "supplementary questions" to be asked of a sample of only one person in twenty,* the Bureau of the Census was able to go further in meeting these needs than it could have otherwise done. (2) By asking these questions on a sample basis, fieldwork was speeded up and carried out at reduced cost. (3) Tabulations based on the sample were completed months ahead of the regular tabulations prepared for the full count. (4) Because the sample tabulations were made at greatly reduced cost, more of them could be prepared and published than would otherwise have been possible. (5) The sample cards could be stored for subsequent tabulation as the need arose, whereas storage space would not have been available for all census cards. One of the statistical problems arising in this undertaking is that of selecting and analyzing the sample in such a manner that one can measure how much the tabulations carried out on the sample differ from those which would have been obtained if the entire population had been studied. This can be done by the use of appropriate methods [1].

Many phases of the data collecting carried out by the Census Bureau between the decennial censuses are also based on samples. This includes the reports on employment and unemployment and on current population estimates. All of these involve the above-stated problem of measuring the agreement between the sample results and the results which would have been obtained if the entire population had been studied.

A Statistical Appraisal of Worker Attitudes and Industrial Absenteeism. During World War II, the problem of industrial absenteeism became a very important one. Absenteeism cut into the efficient use of available manpower, and many studies were carried out to see what was causing the absenteeism and how it could be reduced [2].

The procedure used in the cited study was as follows. One plant was chosen in the central New York area, and a questionnaire was distributed to all workers in the plant. Questions were asked concerning absenteeism, job satisfaction, opinion of the efficiency of management, opinion

* The 1950 census used a sampling rate of one person in five.

of the attitude of management, home situation, comfort at job, life organization (morale), community situation, conception of absenteeism, education, age, number of dependents, similarity of present job to former jobs, amount of vacation taken, average length of former jobs, and average number of annual visits to the doctor. The determination of these questions was not a statistical problem in the ordinary sense. It depended primarily on the investigator's ability to analyze the problem of absenteeism and to forecast those factors which would be most likely to affect absenteeism. Nevertheless, the exact form of the questions to be asked and the manner in which the responses were to be recorded were influenced and to some extent determined by the statistical techniques that were to be used in the analysis and interpretation of the completed questionnaires.

The main statistical problems in this study arose after the questionnaires had been filled out. Obviously the response of an individual worker could indicate very little about absenteeism in general. Thus it would be easy to find one worker who was absent frequently because of his health and another who was absent frequently because he did not need the money and felt no responsibility toward the plant. The fact that such variability exists between individuals is one of the principal reasons why statistical methods are needed. The more pertinent problems in this instance were concerned with the determination of those causes which arose most frequently and were most directly related to absenteeism. The determination of the most frequent causes is a purely descriptive matter, and it is only necessary to count the number of workers who mention ill health, for example, and then perhaps express this number as a percentage of the total number. The determination of the relative importance of factors is a more difficult problem. Once the more important factors are discovered and their interrelationships described, then it may be that the plant management can introduce means to correct the situations leading to absenteeism.

Because of the fact that questionnaires were distributed to all workers in the plant one may at first glance think that there are no sampling problems associated with this example. Actually, there are many sampling problems. For example:

1. No investigator would be able to study all conceivable factors which might influence absenteeism. Furthermore, any set of questions which he writes down concerning a given set of factors represents only a portion of those questions that might have been used. Thus we have a sample of factors and a sample of questions.

2. Such a study will be carried out at a given point in time. Conditions change through time, both with respect to an individual worker (his health, his financial responsibilities, etc.) and with respect to his social

and work environment. Therefore a single study represents a sample in time.

3. Not all the workers in the plant filled out the questionnaire. Only 54 per cent returned usable ones. Consequently, one must ask whether or not the workers who filled out questionnaires are "representative" of all workers in the plant. We speak of those workers who filled out a questionnaire as a sample from the population of all workers in the plant. This question of "representativeness" can be answered with appropriate statistical procedures, provided that the sample is chosen in accordance with certain principles. If each worker drawn into the sample can decide for himself whether or not to answer the questionnaire, then additional difficulties are encountered in analysis.

4. War production depended not upon this one plant alone, but on many plants—plants of larger or smaller size, plants in other regions of the country, plants manufacturing different products, etc. Therefore, we can think of a population consisting of all industrial plants in the United States, and we have a sample of one plant. Little can be said about this population on the basis of one plant. It would be somewhat like saying a restaurant has good food after drinking one glass of milk in the restaurant. Actually, the cited article does not attempt to draw any such inferences and stresses mainly the statistical techniques employed in finding out the most important causes of absenteeism *in this one plant.*

In closing this account, it might be noted that absenteeism raises some very challenging conceptual problems which have counterparts in studies of sickness, industrial or occupational mobility, and industrial accidents. For example, one may ask whether there exist "absence-prone" individuals and whether an individual who has a past record of absences is more apt to be absent in the future than are individuals with no such past record. Answers to these questions can be obtained only by devoting special attention to the manner in which data are collected, analyzed, and interpreted.

Determination of the Amount of Light Which the Human Eye Can See. Since the preceding two illustrations have come from the field of social science it seems desirable to present an application of statistical methodology to a problem which the uninitiated might think of as being a question not of statistics or variation, but as a matter of exact measurement. It is concerned with the amount of energy which is needed to produce a sensation of light in the human eye [3]. The problem of the amount of energy needed for vision was first attacked about 1890, and energy was measured in quanta, or discrete packets of energy. Each unit of light contains a quantum of energy. The first measurements showed that about 1,000 quanta was needed foı vision. Subsequent experiments gave figures of 17–30, 34–68, 40–90. Since there was so much variation between these measurements, Hecht and his collaborators again attacked the problem.

One of the procedures used by Hecht was the following. Flashes of light were presented to an observer, and the observer simply stated that he did or did not see each flash. Because of physical difficulties, it is impossible to devise a light flash which will always give the same number of quanta. The best that can be done is to present a series of flashes which, on the average, will give a specified number of quanta, say 10. Sometimes the flash will give a smaller number of quanta and sometimes a larger, and these numbers occur in a random fashion, like the appearance of heads and tails in the tosses of a coin. Although one can never know exactly how many quanta are emitted by a given flash, it is possible through the use of statistical methods to predict the fraction of times that a given number of quanta will be emitted. Experimentally, this means that when flashes of light are repeatedly presented to the eye, there will be a certain fraction of them for which the flash will be seen as light. If the average energy is high, the flash will be seen nearly every time. If the average energy is low, the flash will not be seen every time but only that fraction of the times when it delivers at least the requisite number of quanta. The basic data are simply the fractions of the time that flashes of a number of different intensities are seen. Comparison with the results predicted by statistical theory then provides an estimate of the number of quanta which the eye requires in order to see.

This procedure was carried through by Hecht, and he obtained the result that between 5 and 8 quanta of energy were needed for minimum visual effect. Other independent approaches to the same problem gave values of between 5 and 14 quanta. It must be recognized that the foregoing account has been overly simplified and that the interested reader should refer to Hecht's paper.

In this example, just as in the preceding ones, we are concerned with describing a population. However, in this instance, it is a population consisting of an infinite or indefinitely large number of elements, namely, all possible flashes of light that can be emitted by a light of given average intensity. Each flash has an exact but unknown number of quanta, this number varying from flash to flash. The best that can be done is to state, through the use of statistical theory, the fraction of flashes which will have a given number of quanta.

REFERENCES CITED

1. Stephan, Frederick F., W. Edwards Deming, and Morris Hansen: "The Sampling Procedure of the 1940 Population Census," *Jour. Amer. Stat. Assoc.*, **35**:615–630 (1940).

2. Noland, E. William: "Worker Attitudes and Industrial Absenteeism: A Statistical Appraisal," *Amer. Sociol. Rev.*, **10**:503–510 (1945).
3. Hecht, Selig: "Energy and Vision," *Amer. Scientist*, **32**:159–177 (1944).

OTHER READING

Stephan, Frederick F.: "History of the Uses of Modern Sampling Procedures," *Jour. Amer. Stat. Assoc.*, **43**:12–39 (1948). This paper gives quickly and clearly a general idea of the wide applicability of modern sampling procedures.

Tippett, L. H. C.: *Statistics*, Oxford University Press, New York, 1943. This short book is worth reading in its entirety. Written in an easy style with no formulas, it gives a bird's-eye view of elementary statistics.

CHAPTER 2

THE COMPONENTS OF A STATISTICAL INVESTIGATION

Although the applications of statistical method cited in the preceding chapter were extremely diverse in nature, they would, if examined in sufficient detail, exhibit a large number of common features. The purpose of this chapter is to set forth the most basic of these common features. Some of these will be carefully examined in the succeeding portions of this book; others are not entirely appropriate for inclusion in a treatment of statistical method as such and must be studied in the context of a particular field of application; and still others require a more advanced treatment than is possible here. Moreover, the coverage of this chapter will not extend to all aspects of every possible form of statistical investigation.

A variety of illustrations will be given under each topic, but in order to maintain some degree of continuity a single application will be carried through the entire chapter. The particular application that will be used is the Survey of Consumer Finances (SCF) of the U.S. Federal Reserve Board. This survey has been conducted on an annual basis since 1946 by the Survey Research Center of the University of Michigan, and its methods and results have been carefully described in the literature [1]. The purpose of the SCF has been set forth [1, July, 1950, pp. 795–809] as follows:

> The Survey of Consumer Finances seeks to extend our knowledge of the economic position and attitudes of consumers in the United States. It gathers information about their current financial status, their recent spending and saving behavior, their attitudes toward their own financial situation and prospects as well as those of the country as a whole, and their spending and saving plans for the near future. These are important factors in the ebb and flow of economic activity and information concerning them should be collected periodically if it is to make its greatest contribution to our knowledge of economic change.

2.1. The Population, or Universe

The SCF is devoted to the study of consumers in the United States, although the individual consumer is not the unit in terms of which most of the analyses are carried out. Rather, the survey is concerned primarily with the *spending unit*, defined as "a group of persons living in the same dwelling and related by blood, marriage, or adoption, who pool their incomes for their major items of expense." Therefore the SCF wishes to describe the *population*, or *universe*, composed of all spending units in the United States. For practical and theoretical reasons, however, certain exclusions are made from this population of *all* spending units in order to arrive at the *working population* of the survey. This working population does not include "members of the armed forces or other persons living on military reservations, residents in hospitals or other institutions (penal, educational, religious, etc.), or residents in hotels, large boarding houses, and tourist camps."

As in the SCF, any statistical investigation is directed toward the description of a particular class of elements, and this class of elements is referred to as the *population*, or *universe*, for the study. Each investigator has the responsibility for delimiting the particular group of elements which he choses to study, and for doing this in a manner such that neither he nor the users of his results can question the meaning of the definition. There are many conceptual and practical difficulties which arise in this connection, and we shall now examine a few of these.

The conceptual difficulties will be kept to a minimum if we first restrict ourselves to *finite populations*, that is, populations where we can think of gathering together and counting all elements in the population. Thus the students registered in a college course form a finite population, and the fact that each student has registered for the course serves to define the population; similarly, the population of spending units in the United States is finite, although very large. Other illustrations of finite populations are individuals 21 years of age and over living within a particular city, county, state, or country; workers employed by a company in a stated capacity; farms which are located in a given state and which have 10 or more dairy cows; firms manufacturing a certain product; members of the American Medical Association; members of a fraternal organization; subscribers to a magazine; and so on. Of course it must be clearly recognized that such populations as these are changing in composition through time (elements may enter or leave most populations) and that time is therefore an important factor in the definition. As a matter of fact, many statistical investigations have as their main purpose the study of changes in a population through time, both with respect to composition and size and also with respect to character.

The practical difficulties arise from the fact that great care must be taken in defining a finite population in order that no ambiguity will arise as to the inclusion or exclusion of any given element, either during the course of the investigation or in the minds of those who read the report of the study. Thus the population of voters in the United States might mean citizens 21 years of age and over to one person, citizens 21 years of age and over who are registered to vote in a specific national election to another, and individuals who actually have voted in a specific election to a third. An exact statement dispels all doubt. Related situations can and do occur frequently. The U.S. Bureau of the Census must define a dwelling unit, or a farm, or a member of the labor force; a social psychologist must define what he means by a unit of interaction between two persons; and a sociologist must define what is meant by a "community leader." A reader may quarrel with the population used for a particular purpose, but at least he will know what it was if the investigator defines it with proper care.

Finite populations by no means exhaust the types of population with which statistical method must deal. It may happen that the population we wish to describe exists only in our imaginations and has an indefinitely large number of elements. We shall refer to such populations as *infinite populations.* For these populations it is physically impossible to examine every element since they exist only conceptually until some act or sequence of acts is performed to bring them into reality. Perhaps the simplest illustration is provided by the tossing of a coin. In this situation, we say that our population consists of all tosses which could be made with a coin, *under stated conditions.* We can clearly conceive of this population, and we can clearly obtain elements from it. However, the population cannot be exhausted by considering the elements one by one. This does not mean that an infinite population cannot be described, and many of the ideas and techniques in this book will be aimed at this very problem.

Infinite populations may need to be considered in almost any type of statistical investigation, and a few examples will now be described to illustrate this point.

Repeated Physical Measurements. Suppose that an individual is given a line marked on a sheet of paper and a ruler graduated in centimeters (no finer subdivisions are marked on the ruler) and is asked to measure the length of the line to the nearest tenth of a centimeter, i.e., to the nearest millimeter. He will mark down his estimate of the length of the line. It is possible to consider asking this same individual to perform the same measurement at some future time, the conditions of the measurement remaining the same (length of line, lighting, ruler, etc.), but with the assumption that the individual has forgotten the answer he obtained the first time. Under such circumstances we cannot necessarily expect the two measurements to be the same. The alignment of the ruler with the line may be different, and the judgment of the individual between two centimeter marks may be different. Generalizing from this, we may now think of the infinite population of all measurements which this individual might make *under these stated conditions.* Although this is clearly an oversimplified example, it is exactly the type of situation that arises whenever a physical measurement is being performed. One can always think of repeating such a measurement an indefinitely large number of times, and a statement of the conditions under which the measurement is to be conducted serves to define the population. Sometimes this statement of conditions will allow for the repeated measurements to be made by different individuals or by different kinds of measuring instruments.

Repeated Responses of an Individual to a Survey Question. When an interviewer asks an individual a question concerning a fact, an opinion, a like or a dislike, an attitude, or a want, it may seem like stretching the imagination to an unwarranted degree to think of this in terms of an infinite population. Nevertheless such an approach is frequently helpful and necessary. For example:

1. Any individual almost certainly changes through time. He is conditioned by the people with whom he associates, by local, national, and international affairs, and the like. Since he could conceivably be questioned at any one of an infinite number of points in time, we may wish to speak of an infinite population of "questioning points" or of responses. This population is defined by the individual and the conditions under which the question is asked. An element of this population comes into existence (or can be observed) only when a point or points in time are chosen at which to ask the question. One can, of course, also regard an entire finite population as changing through time and thus arrive at an infinite population, each element of which is itself a finite population. This approach has been discussed in some detail by Deming and Stephan [2], and often it is this change through time which is of principal interest in a study. This very point was emphasized to some extent in the quotation relating to the SCF which was given in the opening portion of this chapter.

2. As is well known (and this will be illustrated in Sec. 2.5), the answer which an individual gives to a fixed question at a fixed point in time may be conditioned by the personality and other characteristics of the person asking the question. Again we may think of an infinite population of possible interviewer types and thus obtain an infinite population of responses to the given question.

3. Finally, not only may the interviewer have an effect on the response to a given question, but there may also be an effect due to the manner in which the question is phrased or the exact aspect of an attitude or opinion toward which it is directed. Thus it may be desirable to speak of an infinite population of questions, any one of which might be used in a particular set of circumstances. Guttman [3, p. 80] has made use of this concept in connection with his work on scaling and refers to it in the following terms:

> . . . Any question asked in an attitude or opinion survey is ordinarily but a single sample of indefinitely many ways the question could be put. . . . It is, therefore, essential to inquire into the nature of the *universe of all possible questions of the same content*. . . .

As a concluding footnote to this illustration, we note that an investigator may wish to consider various finite and infinite populations in conjunction with one another. That is, although the ultimate goal may be to describe a finite population of individuals, the response which any element in this population gives must be interpreted in terms of the point in time at which the question was asked, the interviewer who asked the question, the form of the question, and so on. All these factors must be taken into account in defining the setting within which statistical methods are to be applied.

2.2. The Variables

We have seen that a purpose of a statistical investigation is to provide a (numerical) description of a defined population. When one speaks of describing any object, be it a man or an experiment or a population, it is necessary to specify those characteristics of the object in terms of which the description is to be made. In this section we shall be concerned with the description of a single element in a population, while in the following section we shall discuss the manner in which these separate descriptions can be combined to provide a description of the entire population.

The elements of a population will always be identical in certain respects. At the very least they will have in common those characteristics which serve to define the population. Thus the spending units of the SCF are all located in the continental United States, and their members all live in private households (see exclusions noted in the preceding section). In addition, however, there are many other characteristics on which these spending units may differ. For example, any two spending units may or may not be the same with respect to such items as the following: number of members of the spending unit, age and occupation of the head of the spending unit, money income before taxes for the calendar year preceding the survey, opinion concerning expected change in consumer prices for the coming year, whether or not an automobile was purchased during the preceding year, plans for purchasing an automobile during the coming year, and whether or not the unit owns its home. The existence of characteristics that are different, or that vary, among spending units makes it desirable to conduct surveys of consumer finances and necessitates the use of statistical method.

Characteristics for which some population elements differ from others, as in the examples given above, are referred to as *variables*, or *variates*. The study of populations with respect to variables has motivated most of the present body of statistical method. If all elements of a population were identical, then it would be necessary only to examine and describe a single element, although one might still have difficulty in determining the *number* of elements in certain specified finite populations and would need statistical tools to cope with this problem. This latter situation is particularly well illustrated in the field of fisheries biology where much effort has been devoted to the development of sound procedures for estimating the total number of a certain species of fish in a pond or lake.

There are several major types of variables which arise in practice. These will now be briefly described and illustrated in preparation for the detailed treatment which will follow in the succeeding portions of this book.

Qualitative Variables. When differences between population elements (with reference to a stated characteristic) must be stated in qualitative terms rather than measured in quantitative units, we speak of this characteristic as a *qualitative variable*, or as an *attribute*. Formally, then, a qualitative variable is defined by specifying a set of two or more classes or categories. An individual population element is "measured" on the variable by being assigned to one and only one of these classes. Since we require that it be possible for *all* population elements to be assigned to classes and that a single element be assigned to only one class, we speak of these categories as being exhaustive and mutually exclusive.

There are many situations in which the categories of a qualitative variable are "natural" and in which the classification of a population element into a category is relatively straightforward. For example, the result of a coin toss is a qualitative variable with two categories, heads or tails; the sex of the head of a spending unit in the SCF, male or female; the current marital status of an individual, single, married, separated, divorced, or widowed; and the geographic location of the residence of a spending unit is a qualitative variable with categories defined in terms of geographic region, state, county, city, and so on. Opposed to these "simple" qualitative variables are those for which either the definition of the categories of the variable or the classification of individuals into these categories, or both, may raise difficult problems. Thus the variable "employment status," considered as having the two categories "employed" and "unemployed," has raised innumerable difficulties for investigators interested in labor-force measurement; the qualitative variable occupation is difficult to treat since there is a practical limit to the number of occupational categories that can be handled meaningfully in analysis, and this inevitably necessitates the grouping of "like" occupations into a single category; a study of community organization must define the categories of community leadership and must devise methods of classifying individuals into these categories; and a person working with racial prejudice must not only define different degrees of prejudice but must also be able to differentiate among individuals according to these defined degrees or classes of prejudice.

Although the difficulties mentioned above are primarily difficulties of measurement, and will be considered again in Sec. 2.5, it is clear that they will have an effect on the data gathered in a study. The formal procedure which we shall later discuss for dealing with qualitative variables will not distinguish between "simple" and "difficult" situations—and yet the differences are there. The investigator must recognize and allow for this in his choice of statistical techniques and methods of analysis.

Quantitative Variables. Whenever it is possible to assign to each population element a numerical value of the variable, we speak of a *quantitative variable.* The SCF, for example, makes use of such quantitative variables as the following: number of members of a spending unit 21 years of age and over at the time of the survey; money income (before taxes) in dollars for the calendar year preceding the survey; age in years of the head of a spending unit; and age in months of any automobile possessed by a spending unit. Many other illustrations of quantitative variables immediately come to mind, a few of which are the following: finite population of hourly-paid workers in a specified plant at a specified time—a variable is the *hourly rate* in dollars and cents; infinite population of all possible throws that can be made with a die under stated conditions—a variable is the *number of dots* appearing on the upper face of the die; finite population of students in a school—a variable is the *height* of a student in inches; infinite population of all possible tosses that can be made with two coins (i.e., population element is *one* toss of *two* coins)—a variable is the *number of heads* that appear; and finite population of families in a city—a variable is the *number of times* that a physician made a professional visit to a family's home during the preceding year.

Discrete and Continuous Quantitative Variables. One of the distinguishing features of all the preceding examples of quantitative variables is that the different values assumed by any one of them are separated from each other by at least some stated amount. Thus the uppermost face on a die can show one of the six possible values 1, 2, . . . , 6, and these are separated by at least one; similarly, if a spending unit's annual income is measured in dollars, then the annual incomes of two different units differ, if at all, by at least one dollar; and, if heights are measured in inches, then two heights differ, if at all, by at least one inch. When the possible values of a variable are separated from one another by at least some stated amount, *no matter how small this may be,* we refer to the variable as a *discrete* quantitative variable.

Separations between the different possible values of a quantitative variable arise for one of two reasons. These are:

1. The very nature of the variable makes it impossible to consider anything other than separated values. That is, there exists a fundamental unit which cannot be subdivided, and assigning a value of the variable to a population element consists merely of *counting* the number of such units that the element possesses. Elements must therefore differ, if at all, by at least one unit, and we have a "naturally" discrete variable. Thus we have number of children in a family (unit being a child), number of "serious" accidents in a plant during a year (unit being a "serious" accident), or number of social contacts that an individual has during a stated period of time with a member of some racial minority group (unit being a social contact).

2. Separations between different values of the variable are imposed by considerations such as custom, convenience, or limitations of the measuring instrument, even though it is possible to think of subdividing any unit of measurement that is being used. For example, we frequently choose to measure age in years. Yet it would also be possible to measure age in months, or days, or hours, or minutes, if the need arose and if the appropriate data were recorded. Similar comments could be made with respect to such quantitative variables as height, weight, and temperature.

Whenever the unit of measurement for a quantitative variable can be divided into arbitrarily small pieces, we speak of that variable as being *continuous*. That is, no matter how close together two values of the variable are, other possible values of the variable can be inserted between them. Thus time measurements can be refined successively through minutes, seconds, tenths of a second, hundredths of a second, thousandths of a second, and so on. There is, of course, a limit imposed on how far these subdividing processes can be continued, a limit which is determined by our ability to construct more and more accurate timing devices, more and more accurate weighing balances, or more and more accurate micrometers. This means that observational data will always be discrete in character—the values of the variable that we can distinguish from one another being separated by some noticeable amount, no matter how small this may be. Nevertheless, we can imagine that this subdivision process is continued indefinitely and arrive at the *concept* of a *continuous variable.* Although we can never observe all possible values of a continuous variable, the concept is still an extremely useful one and is drawn upon repeatedly throughout the theory of statistics.

Further Relationships between Qualitative and Quantitative Variables. In order to emphasize the major points, the preceding portions of this discussion have been presented as though all variables were either qualitative (no numerical description of the categories is possible) or quantitative (there exists a basic unit of measurement which allows one to speak meaningfully of quantitative differences between population elements with respect to the variable). Actually, there are many gradations between these two extremes, and problems in social research must frequently deal with intermediate situations.

The foregoing point can be illustrated very simply by considering the following question, which is similar to those asked on employee-attitude surveys:

"How closely does your foreman check up on his men?"

________ Much too closely
________ Somewhat too closely
________ Just about right
________ Not quite closely enough
________ Not nearly closely enough

The categories or classes of possible answers to this question are not entirely unrelated from a quantitative point of view—as would be the head and tail categories for a coin toss—since there is an implied *order* relationship among them. That is, the "closeness" of supervision is greater for the "much too closely" category than it is for the "not quite closely enough" category. However, there is no unit of measurement in terms of which we can assign a quantitative measure to the difference between these two categories. In addition there is no *absolute zero point* such that categories on one side represent "too close supervision" and categories on the other side represent "too loose supervision." Every individual answering the question may have such a personal zero point, but this is not the same as having, for example, a zero length which has the same meaning to everyone.

Considerations similar to the foregoing arise repeatedly in social research. We may be able to state that individual A is more prejudiced than individual B, A is better informed on national affairs than B, A is a better supervisor than B, and still not be able to measure the difference in quantitative terms. The choice of statistical methods is conditioned by the type of variable that is being studied, and it therefore behooves every investigator to become familiar to some extent with the theory of measurement. One brief introduction to this has been provided by Coombs [4, p. 471].

There is one final observation which needs to be made concerning the relationship between qualitative and quantitative variables. Quantitative measurements may be used simply to define *ordered* categories to which many of the procedures we shall later discuss for qualitative variables may be applied. For example, we may have age measurements on elements from a population and yet choose to speak only of the two categories "over 40 years of age" and "40 years of age and under"; or we may have measurements of height and choose to speak of the three categories "under 5 feet tall," "between 5 and 6 feet tall," and "over 6 feet tall."

2.3. The Distribution

Once values of a specified variable have been assigned to some or all of the individual elements of a defined population, a statistical investigation is ordinarily no longer concerned with the separate identities of these elements. What is desired is a *population description* and not just a listing of the elements and their values of the variable. For example, the 1953 SCF asked (in early 1953) each of 3,097 consumer spending units the question "How about a year from now, do you think you people will be making more money or less money than you are now, or what do you expect?" The answers to this question, for those spending units among the 3,097 which were *nonfarm*, were presented [1, June, 1953, p. 592] in the following summarized form:

Expected change in income	*Per cent of nonfarm spending units*
Increase	34
None	33
Decrease	10
Uncertain	18
Not ascertained	5
All cases	100

This table gives the percentage distribution of nonfarm spending units among the categories of the qualitative variable "expected change in income." This percentage distribution provides one way of describing the population of nonfarm spending units with respect to the given variable. Another way would be to give the actual number of nonfarm spending units falling in each of the categories. Of course, the foregoing distribution is not based on *all* elements of the population, and this raises special problems which will be outlined in the following sections of this chapter.

Quantitative variables can frequently be dealt with in a manner very similar to the foregoing. For example, the 1953 SCF obtained from each unit in the full sample of 3,097 spending units the amount of money income (before taxes) for the year 1952. Again interest is centered not on the individual spending unit but on the population of spending units. The data obtained were presented in the following form [1, June, 1953, p. 589]:

Money income before taxes in 1952	*Per cent of spending units*
Under $1,000	11
$1,000–1,999	14
$2,000 2,999	16
$3,000–3,999	18
$4,000–4,999	15
$5,000–7,499	17
$7,500–9,999	5
$10,000 and over	4
All cases	100

Here also we have the *percentage distribution* of spending units among categories of a variable, but here the categories are defined in terms of the values of a quantitative variable, and these categories have a definite numerical relationship to one another.

The percentage distribution of spending units among the defined categories of the variable "money income before taxes in 1952" describes the population of spending units with respect to this variable, subject to the limitation that we do not have *all* spending units, but now many new possibilities for description are opened up—in addition, of course, to that of giving the actual number of units in each category. For example:

1. We might present the same data in the following form:

Money income before taxes in 1952	*Per cent of spending units having less than stated amount of money income in* 1952
$1,000	11
$2,000	25
$3,000	41
$4,000	59
$5,000	74
$7,500	91
$10,000	96

This table gives the *cumulative percentage distribution* and is obtained by simple addition from the preceding distribution.

2. We might define our categories of the variable in a different fashion, for example, under $500, $500 to $999, $1,000 to $1,499, $1,500 to $1,999, and so on.

3. We might focus attention on *one particular value* of the variable which measures some characteristic of the distribution in which we are interested. Thus we can determine that value of the variable such that one-half of the spending units have a smaller income than it and one-half have a larger income. This is known as the *median* of the distribution and is useful, among other things, in comparing two distributions.

Any statistical investigation is concerned, in one way or another, with the distribution of one or more variables in a population. Many different types of distributions must be considered, depending upon the variables and upon the populations, and many different types of questions can be raised concerning these distributions. A large number of such considerations will be discussed in detail throughout this book, and they are therefore not raised here. However, it is important to recognize that it is frequently necessary to go beyond the mere numerical description of a distribution and to inquire into the reasons why a special type of distribution is obtained.

A simple illustration of this latter point is given by considering a population of workers (doing similar tasks) in a plant with respect to the variable "number of accidents sustained in a stated period of time." This is a discrete variable which can assume the values 0, 1, 2, and so on. It is a straightforward matter to determine a percentage distribution for such a variable and population—assuming that one can set forth an acceptable definition of an accident and can obtain the proper reports from which to tabulate accidents—but then a more difficult and important statistical problem arises. This is the problem of attempting to determine from the observed distribution something about the manner in which industrial accidents occur. For example, we might try to decide among the following three possibilities: (1) accidents occur "by chance"; (2) some individuals are more liable to accidents than are others; and (3) all individuals are "equally" liable to accidents in the beginning, but once an individual has sustained an accident, he then becomes more or less liable to future accidents. An extended psychological and statistical treatment of this problem has been given by Arbous and Kerrich [5]. In dealing with problems of this general nature, one attempts to derive a *theoretical*, or *model, distribution* on the basis of an analysis of the problem, including the way one thinks the underlying mechanism operates, and then to compare observed distributions with this theoretical distribution.

As a final point in this discussion of distributions, we may note that it is possible to consider the problem of describing a population with respect to *two or more variables simultaneously*, and not just with respect to a single variable as has been done heretofore. It is usually possible to associate values of many variables with a single population element. Thus an individual may be categorized simultaneously with respect to height, weight, age, geographic area of the country in which he lives, answer to an opinion or attitude question, and so on, and we may wish to describe a population with respect to some combination of the variables. For example, we may return to the SCF and say that we wish to specify the per cent of spending units in each of the *combination categories* of the two variables "expected change in income" and "money income before taxes in 1952." Thus we would be interested in the per cent of spending units falling in the category "under $1,000—increase in income," in the category "$10,000 and over—decrease in income," and so on. The table obtained in this way would have 40 entries (five "expected change in income" entries for each of the eight "money income before taxes in 1952" categories). Such a procedure can clearly be extended to any number of variables.

2.4. The Sample

The 1953 Survey of Consumer Finances obtained data from 3,097 spending units for the purpose of describing the population of *all* spending units in the United States, which was estimated to contain approximately 54 million elements at the time of the survey. This group of 3,097 spending units is called a *sample* from the population of all such units, and the use of such a sample, although necessary for reasons of economy (both of time and money) and accuracy, raises many special problems for consideration. In particular, we need to know how well the sample "represents" the population, or alternatively, how closely the sample results approximate the results that would have been obtained if the entire population had been studied. Providing an answer to this question involves the procedure that is used in selecting the sample and the procedures that are used in analyzing and interpreting the survey results. If insufficient attention is paid to these and related considerations, it may be impossible to evaluate properly the conclusions drawn from a study, or even to specify the population to which the conclusions apply. This latter point has been made very well in a review by Dollard [6] of the book *Sexual Behavior in the Human Female* by Kinsey, Pomeroy, Martin, and Gebhard. Dollard says:

> From the scientific standpoint this is a strangely flawed masterpiece. It is magnificent in scope, courage, detail of information and exploratory value; yet, it is impossible to decide how generally applicable its conclusions are. . . . . . . We have 5,940 histories but there is no way of knowing whether they represent American women generally or are just an interesting body of isolated data. This uncertainty is the painful price of abandoning scientific sampling methods. Furthermore this uncertainty cannot just be stated and then dismissed; it must be remembered in regard to each and every bit of data which is claimed as a general fact from the Kinsey survey. It clings to every inference from the study and to every suggestion for action emanating from it.

The use of samples to represent populations is inescapable in almost any type of investigation. There are, however, different degrees of "necessity" which may be distinguished. For example, there are instances in which it is physically impossible to consider every element in the population, even if that course of action were desired. These instances are of three levels, namely:

1. The population has an infinite number of elements and could never be exhausted by considering the elements one by one. Numerous examples of such populations were given in Sec. 2.1.

2. The measurement or testing necessary to determine the characteristics of an element destroys whatever intrinsic value the element possesses. Thus the life of an electric light bulb can be determined only by using the bulb until it burns out. Therefore a manufacturer cannot test all the bulbs which he manufactures in order to determine their quality, and he must be content with the examination of a sample.

3. The sample, or records on it, have been preserved, but otherwise the population is inaccessible, no longer in existence, unobservable, and so on. Thus we study historical documents relating to a particular event or individual, and we are dealing with a sample of such documents—those that have been preserved and are available for study.

Even though we are dealing with a finite population where it would be physically possible to enumerate all elements, there are many circumstances under which it may be desirable to study a sample rather than all elements in the population. Some of the more important of these circumstances are:

1. The uses to which the data are to be put do not require the accuracy of a census (i.e., a complete count). Consequently a census would be wasteful of time and money.

2. Even if the resources are available for a census, it may be advantageous to sample and to use a portion of these resources for improving the methods and agents of observation.

3. There is insufficient time and money for collecting information from the entire population or for processing the material obtained from all elements of the population.

The foregoing points are not mutually exclusive, and more than one of them may apply in any given instance. The Survey of Consumer Finances provides a case in point, as does the work of the U.S. Bureau of the Census that was described in Chap. 1.

As has already been emphasized, the first goal of any statistical investigation must be to secure a sample which represents the population under study. When dealing with *finite* populations, we all have certain intuitive ideas concerning the meaning of the term *representative sample* and concerning procedures of selection that will give rise to representative samples. For example, we frequently see reference to the fact that a sample is a miniature population. Thus a sample of the adult population of the United States might contain the same fraction of men, women, people 40 years of age and over, people under 40 years of age, people living in cities, people living on farms, and so on, as does the entire population, and somehow or other we feel that such a sample should give results "closely" approximating those that would have been obtained from the entire population. Or we feel that a representative sample of workers may be obtained in a plant by checking off every fifth name on a payroll list.

But it is not enough to have only intuitive feelings on these subjects! And what can one do intuitively about the problem of sampling from an *infinite* population where it is not possible to take every fifth, or hundredth, or millionth element? What is really needed is an objective procedure for measuring the degree to which the sample results approximate the corresponding but *unknown* population results—and this in turn will determine the methods which should be used in selecting samples. Much of this book will be devoted to describing and illustrating the general approach which can be applied to problems of this nature, and in setting forth the solutions for various special situations.

In concluding this section on samples, it may be well to point out that statistical investigations are conditioned by practical as well as theoretical considerations, in addition to those of cost and time which have already been mentioned. Thus in the sampling of human populations, it will do little good to have a sample design which cannot be adhered to in the field. The results of the survey will depend not only upon how the sampling procedure looks on paper, but upon the manner in which it is carried out. If the interviewers and other field workers do not carry out their instructions, either because the instructions are too complex or because these human agents are untrustworthy, all the theoretical considerations will be of no avail. In the sampling of human populations, one must also secure the cooperation of those persons chosen for the sample. People who cannot be contacted for interview or who refuse to be interviewed can seriously distort the results if they constitute a large fraction of the sample and are sufficiently different from those who are interviewed. Thus experience has shown that it is generally the younger people who are most difficult to contact for personal interview, and that it is a disinterested group—with respect to the subject of a survey—who are least apt to respond to a mail questionnaire. In survey research, the effects of such groups are frequently referred to as "nonresponse" errors.

2.5. The Measurement

It is not enough that one specify a population, variables, and a sample from which inferences to the population will be drawn. One must also assign a value of the variable to each element in the sample, and we are here referring to an actual numerical value in the case of a quantitative variable or to a category in the case of a qualitative variable. Undetected errors or undetected shortcomings in this measurement process will be carried through to influence the final results, no matter how complex or sophisticated may be the statistical methods that are applied to the original data. Statistical methods may aid in detecting, evaluating, and allowing for errors from *recognized sources*, but they will not automatically make "good" data from "bad" data.

The SCF recognizes that its data are subject to "reporting" errors and refers to these [1, June, 1950, p. 805] as follows:

> . . . The most important source of these errors is thought to be incorrect recall. Both psychological studies and special investigations conducted with repeated financial surveys indicate that frequency and size of memory errors increase with the period of time between an event and its recall. . . . Furthermore, the memory errors are greater with respect to subjectively unimportant events than to important events. . . .
>
> To reduce memory errors, the interviewers frequently call attention to the fact that it is hard to remember some of the information requested. . . .
>
> Reporting errors may also result from intentional falsification of responses. Methodological studies making use of repeated questioning of the same respondents concerning the same problems indicate that the frequency of such action is not great. . . . Nevertheless, intentional understatement of income and assets undoubtedly occurs in some cases.

It has long been recognized that it is difficult to secure accurate personal financial information, for the reasons set forth in the above quotation, and yet similar known or hidden problems of measurement may exist in any type of research investigation. Several "striking" illustrations from social research will now be provided to emphasize this point, and some further references will be found at the end of this chapter.

In 1945 the Monthly Report on the Labor Force of the U.S. Bureau of the Census conducted some experiments to determine the effect of different questions in securing information with which to obtain a complete count of persons at work [7]. Prior to 1945 the following single question was asked, "Was this person at work on a private or government job last week?" A new schedule replaced this single question with two questions. The first simply asked concerning the respondent's own classification of activity status, and the second asked whether or not persons whose major activity was considered to be something other than working did any work during the census week. In July of 1945 both the old and new schedules were used. The new schedule obtained an estimated 1,400,000 additional persons (over those obtained by the single question on the old schedule) at work, and nearly 50 per cent of the additional workers worked 35 or more hours during the census week!

A somewhat different problem was investigated by the University of

Denver Opinion Research Center in 1949 [8]. This study was aimed at discovering the "validity" of responses which individuals give to survey questions of a fairly factual nature. For example, a sample of 920 individuals in the city of Denver were asked the question "Did you yourself happen to contribute or pledge any money to the Community Chest during its campaign last fall?" This question was asked during April and May of 1949. The names and responses of these 920 individuals were checked against the files of the Community Chest with the following results:

VALIDITY OF REPORTS OF COMMUNITY CHEST CONTRIBUTIONS

	Per cent
Reported not giving (statements assumed to be correct, but not checked against records)	31
Reported giving, and did give	25
Reported giving, and might have given through uncheckable source	8
Reported giving, but did not give	34
Don't remember; no answer	2
Total	100

The authors attribute the high fraction of "invalid" responses to social pressures and a belief, on the part of respondents, that the responses would not be checked. There was also possibly some ambiguity in the phrasing of the question. This particular question represents one of the "extremes" in the investigation—much smaller distortions appeared with respect to such items as possession of a library card, driver's license, etc.—but it does make one pause and think about the possible errors that can arise in social measurement.

The existence of difficulties of measurement in research in the social sciences is known. And yet we cannot throw up our hands in dismay and say that nothing can be done. We must allow for these difficulties in our study plans, setting up procedures for detecting, eliminating, controlling, and measuring their effects. Statistical methods can aid in this task. Finally, it may be well to remind the reader that difficulties with measurement are not confined to social research—faulty diagnosis may be made in medicine, instruments may be imperfect in physical research, human errors may be made in reading instruments, and so on.

2.6. The Processing

After a value of the variable or variables has been assigned to every element in the sample, it is next necessary to consider the mechanical procedures that are going to be used to combine these individual bits of information into the desired population description. The complexities of these procedures and the type of equipment that is required to perform them efficiently will depend upon a large number of factors, e.g., the kind and number of variables that are being studied, the size of the sample being used, and the type of description that is desired.

The wide range of possibilities that may occur in this processing operation may be illustrated by a few simple examples. The adjectives "large" and "small" will be used here in a purely relative sense.

Small Sample, Small Number of Qualitative Variables. In this type of situation, one can think of having a single 3 by 5 card for each element in the sample on which is recorded an identification number and the categories of the variable into which the element falls. Hand sorting of these cards will then provide a count of the number of sample elements in each category of each variable, or in categories defined by some combination of the variables.

Larger Sample, Qualitative Variables. At some point, as the size of the sample increases, the hand sorting of cards becomes very laborious and time-consuming. It may then be desirable to transfer the original data on to some form of punched card. For example, McBee Keysort Cards have punching positions around their outer edges, and a long metal needle can be used to select out cards with the desired configuration of punches; also, Remington Rand and International Business Machines Corporation produce machines which will handle their respective cards at an extremely high rate of speed, will count the number of cards possessing stated characteristics, and in some instances will provide a printed record of these counts. This latter type of machine was originally developed to aid the U.S. Bureau of the Census in processing the census returns.

Computational Considerations. Not only may the preparation of population descriptions involve the tabulation of large masses of data, but it may also involve extensive computational programs. At the present time, one has a wide variety of potentially available equipment, ranging from the slide rule and desk calculator through the punched-card calculating machines of Remington Rand and IBM to the extremely high speed electronic calculating machines that have been developed in the past ten or fifteen years.

All the foregoing considerations raise many new practical problems that may have to be dealt with in an extensive statistical investigation. The measurements obtained from the sample elements will have to be put in proper form for machine analysis, and this may introduce one more source of error that will have to be controlled; specially trained individuals may be needed to supervise the use of tabulation and computational equipment; and one must allow in the research budget for these items. It is of little avail to have data and insufficient time and money to analyze them.

2.7. Presentation, Evaluation, and Use

One of the end products of a research investigation is some form of written report, be it an informal account for circulation among individuals working on common problems, a report to be submitted to superiors in a business, industrial, or governmental organization, or a paper to be published in a scientific journal. Thus the SCF presents its findings in the *Federal Reserve Bulletin*, and these findings are available to all who wish to read. The individual or organization preparing such accounts owes certain responsibilities to potential users of the results. In particular, the data should be presented in charts, graphs, and tables in such a manner that they are readily accessible and understandable; the entire setting within which the research was performed should be described—and this includes all the points which have been covered in this chapter; and an objective and quantitative assessment of errors should be given, including those arising from sampling, measurement, and processing.

Provided that an investigation is designed and executed in an appropriate manner, it is possible to give an objective interpretation to the portion of error that arises from sampling—and this is true of infinite populations as well as finite populations. The methods of reasoning and some of the techniques that may be used in specific situations will be treated in the succeeding portions of this book, and we shall here give only one brief example from the SCF. The 1952 Survey of Consumer Finances reported [1, July, 1952, p. 751] that 33 per cent of about 2,800 spending units interviewed in January, February, and early March of 1952 felt that their own financial situation was better than it had been 1 year earlier. This is not necessarily the same value that would have been obtained had *all* spending units in the population been interviewed. However, on the basis of the sampling procedure used and the sample results actually obtained, one can state with some degree of confidence that the population value lies between 30.3 per cent and 35.7 per cent. This degree of confidence, 95 per cent in the present instance, measures the fraction of times that our positive statement would be correct if the same procedure were applied to an indefinitely large number of such surveys, the sample per cent varying from sample to sample and the corresponding limits (30.3 and 35.7 for this one survey) also varying from sample to sample. We can arbitrarily set the degree of confidence which we desire, e.g., 90 per cent, 95 per cent, or 99 per cent, excluding 100 per cent, and we can widen or narrow the length of the interval by decreasing or increasing the size of the sample used.

Actually, it is not only the sampling error that we wish to measure in any given situation. What is really required is a measure of total error, and we shall later attempt to give some attention to this topic.

In concluding this brief survey of the setting within which we apply statistical methods, it seems desirable to reiterate the points that research investigations are rarely if ever conducted just to satisfy idle curiosity and that any investigation requires a certain expenditure of effort, time, and money. Under these circumstances, it seems only natural to attempt to balance against one another the following three factors:

1. The cost of conducting the investigation
2. The errors associated with the final results
3. The consequences that will follow from the basing of decisions on data which are subject to a certain degree of error

If one expends too little effort on conducting the investigation, then the errors will in general be large and one runs the risk of channeling much future research down a blind alley, of investing resources on projects that are doomed to failure, and so on. On the other hand, one may spend too much effort on an investigation in relation to the uses to which the data are to be put. Considerations such as these have led to the study of *experimental design* (in which one attempts to obtain as much information as possible for a fixed expenditure level) and *decision theory* (in which one attempts to obtain a balance between all three of the foregoing factors).

EXERCISES

2.1. A housewife, living in a moderate-sized city, does her weekly shopping for food sometime between 4 and 5 P.M. each Wednesday afternoon. She goes to the nearest supermarket and after selecting the desired items goes to the checkout station having the shortest line. She may, of course, find a station at which no one is waiting. Frequently she finds one or more customers in front of her and must therefore wait her turn.

a. If we observe this housewife's "waiting" experience on a stated Wednesday, of what population is this experience a member? Would we ordinarily regard this population as being finite or infinite in size?

b. There are at least three variables, related to "waiting" experience, that one may use to describe an element from the population specified in *a*, one qualitative, one naturally discrete, and the third conceptually continuous. Name and describe these variables.

c. Give a rough outline of the manner in which you would expect to describe the population in *a* with respect to each of the variables given in *b*.

d. Suppose our housewife decides to change her shopping hour from between 4 and 5 P.M. on Wednesday to between 5 and 6 P.M. on Friday. Assuming no change in the checkout system of the supermarket, how would you expect the distributions given in *c* to change? In answering this, remember that workers paid on a weekly basis usually receive their salary on Friday afternoon and that there is buying for the week end.

e. Suppose the store manager observes the performance of checkout stations on several Wednesday afternoons and on this basis decides how many cashiers he should have. Will he have "satisfied" customers on Friday afternoon? Or, suppose he observes on Friday afternoon and hires accordingly. What will happen to his overhead expenses during the remainder of the week?

2.2. You are given a book by a noted novelist and are asked to provide a quantitative characterization of this author's style. There are numerous and related finite populations contained in a book of this type, depending upon how the fundamental element of the population is defined. For example, there is the finite population of sentences. Name as many of these finite populations as you can. What are some of the qualitative and naturally discrete variables that can be associated with the fundamental elements you have named? Can you think of some easy ways of drawing samples from the populations you have defined?

2.3. The following experimental technique was used by Asch [9] to make a preliminary investigation of the effects of group pressure upon the modification and distortion of judgments: "We employed the procedure of placing an individual in a relation of radical conflict with all other members of a group, of measuring its effect on him in quantitative terms, and of describing its psychological consequences. A group of eight individuals was instructed to judge a series of simple, clearly structured perceptual relations—to match the length of a given line with one of three unequal lines. Each member of the group announced his judgments publicly. In the midst of this monotonous 'test' one individual found himself suddenly contradicted by the entire group, and this contradiction was repeated again and again in the course of the experiment. The group in question had, with the exception of one member, previously met with the experimenter and received instructions to respond at certain points with wrong—and unanimous—judgments. The errors of the majority were large (ranging between ½″ and 1¾″) and of an order not encountered under control conditions. The outstanding person—the critical subject—whom we had placed in the position of a *minority of one* in the midst of a unanimous majority—was the object of investigation. . . . Both the members of the majority and the critical subjects were male college students." Each critical subject, and there were 50 of these, had to make a judgment on the length of 18 lines. The remainder of the group of which he was a member gave correct responses on 6 of the lines and incorrect responses on the remaining 12. The measurement consisted merely of observing whether the critical subject gave a correct or incorrect response on each of the 12 trials on which the group gave an incorrect response.

a. Each critical subject is faced with 12 trials on which the group unanimously gives an incorrect response. What population do these 12 trials "represent"? Is it a finite or infinite population? What is the variable associated with an individual element of this population? What values can it assume? Is it a qualitative or quantitative variable? What variable do we associate with the entire set of 12 trials? What values can it assume? Is it qualitative or quantitative? If quantitative, is it naturally discrete or conceptually continuous?

b. Fifty critical subjects were used in this experiment. What population do these 50 subjects represent? Is it a finite or infinite population? What is the variable which the measurement associates with an individual element of this population? Can you suggest what form of distribution would be used to describe the sample with respect to this variable?

c. Suppose that a school board, consisting of eight members, is debating the problem of new building construction. Seven members are for a certain program; one is opposed. Can any conclusions arrived at in the foregoing study be carried over to this situation? What populations and variables can you suggest for study, as extensions of the preliminary investigation, which would bring the results closer to the day-to-day behavior of groups?

REFERENCES CITED

1. *Federal Reserve Bulletin*, Board of Governors of the Federal Reserve System. Washington. Numerous articles on the Survey of Consumer Finances have appeared since 1946. See especially the article "Methods of the Survey of Consumer Finances" in the July, 1950, issue.
2. Deming, W. Edwards, and Frederick F. Stephan: "On the Interpretation of Censuses as Samples," *Jour. Amer. Stat. Assoc.*, **36**:45–49 (1941).
3. Stouffer, Samuel A., et al.: *The American Soldier*, vol. IV, Princeton University Press, Princeton, N.J., 1950.
4. Festinger, Leon, and Daniel Katz: *Research Methods in the Behavioral Sciences* The Dryden Press, Inc., New York, 1953.
5. Arbous, A. G., and J. E. Kerrich: "Accident Statistics and the Concept of Accident-proneness," *Biometrics*, **7**:340–432 (1951).
6. Dollard, John: "A Review of Sexual Behavior in the Human Female," *New York Herald Tribune Book Review*, Sept. 13, 1953.
7. Bancroft, Gertrude, and Emmett H. Welch: "Recent Experience with Problems of Labor Force Measurement," *Jour. Amer. Stat. Assoc.*, **41**:303–312 (1946).
8. Parry, Hugh J., and Helen M. Crossley: "Validity of Responses to Survey Questions," *Pub. Opin. Quart.*, **14**:61–80 (1950).
9. Asch, S. E.: "Effects of Group Pressure upon the Modification and Distortion of Judgments," pp. 2–11, in Harold Guetzkow (ed.), *Groups, Leadership and Men*, Carnegie Press, Pittsburgh, 1951.

OTHER READING

Bross, Irwin D. J.: *Design for Decision*, The Macmillan Company, New York, 1953. An amusing and informative treatment of the use of statistical methods in aiding one to make wise decisions in the absence of complete information.

Deming, W. Edwards: "On Errors in Surveys," *Amer. Sociol. Rev.*, **9**:359–369 (1944). An extended list of the problems and difficulties that may beset one in conducting sample surveys.

Morgenstern, Oskar: *On the Accuracy of Economic Observations*, Princeton University Press, Princeton, N.J., 1950. An account, with illustrations, of the problems faced in obtaining accurate data relating to economic phenomena.

Yerushalmy, J., et al.: "An Evaluation of the Role of Serial Chest Roentgenograms in Estimating the Progress of Disease in Patients with Pulmonary Tuberculosis," *Amer. Rev. Tub.*, **64**:225–248 (1951). An investigation of the extent of agreement between independent diagnoses of X-ray plates, both between different readers and between the same reader at different times.

CHAPTER 3

THE DISTRIBUTION OF A VARIABLE

A distribution specifies the manner in which the values of a variable are distributed among the elements of a population and thus provides a way of describing the population with respect to the variable. The present chapter will be concerned with the general properties of distributions, including their graphical representation, while the following two chapters will focus attention on certain useful characteristics of distributions. In these chapters we shall treat both populations and samples, although we shall not discuss the selection of samples or the evaluation of sample results. These considerations will be deferred.

3.1. Distributions for Qualitative Variables

3.1.1. *Finite Populations.* A finite population is described with respect to a qualitative variable by counting the number of elements falling in each category of the variable and presenting these results in convenient form. Thus, for the population of males 14 years of age and over, the 1950 Census of Population obtained the distribution given in Table 3.1 for the variable "labor-force status."

TABLE 3.1. LABOR-FORCE STATUS OF MALES 14 YEARS OF AGE AND OVER (April, 1950)

Labor-force status	Number	Per cent
In the labor force, including Armed Forces	43,553,386	78.7
Not in the labor force	11,758,231	21.3
Total	55,311,617	100.0

SOURCE: *1950 Census of Population,* vol. II, pt. 1, table 50.

Although the basic data are given by the "counts," the *percentage distribution* in this table provides a convenient way of showing the division of the total population of 55,311,617 individuals between the two categories of the variable. This manner of presentation is particularly helpful if one wishes to compare two populations of unequal size with respect to labor-force status or with respect to any other qualitative variable. Moreover, it is a necessary form of presentation when dealing with infinite populations.

The size of the civilian labor force is frequently a matter of great concern, and especially so in times of mobilization for war production. However, the manner in which this civilian component is divided into "employed" and "unemployed" workers is of continuing importance since it provides one measure of the economic health of the nation. In April, 1950, the Census of Population obtained the distribution given in Table 3.2 for the variable "employment status," where we are now interested in the population composed of civilian males 14 years of age and over *in the labor force.*

TABLE 3.2. EMPLOYMENT STATUS OF CIVILIAN MALES 14 YEARS OF AGE AND OVER AND IN THE LABOR FORCE
(April, 1950)

Employment status	Number	Per cent
Employed	40,519,462	95.1
Unemployed	2,079,305	4.9
Civilian labor force	42,598,767	100.0

SOURCE: *1950 Census of Population*, vol. II, pt. 1, table 50.

Here again the distribution has been obtained by counting the number of individuals belonging to each of the two categories. Further subdivisions of these categories can be defined if desired, and the interested reader should consult the quoted source.

The foregoing data afford an opportunity to reiterate a point made in Sec. 2.5. The unwary person may think that a complete count of a population provides figures which are free from error. This is not necessarily the case. For example, there are many marginal individuals in the United States to whom it is difficult to apply the defined concepts of labor-force measurement and from whom it is difficult to secure the information on the basis of which the classification into various categories must be made. Thus the quoted source for Tables 3.1 and 3.2 states that "the statistics will understate the size of the labor force and overstate the number of persons not in the labor force"; and "the estimated size of the civilian labor force in the United States based on the Current Population Survey is about 5 per cent above the corresponding figure from the 1950 Census. The discrepancy lies chiefly in the count of incidental workers." The Current Population Survey is a national sample of approximately 25,000 households from which various items of information are secured on a monthly basis. The field staff for the

CPS is much more highly trained and supervised than is the staff of the decennial census, and therefore we see the reason for the above emphasis on the evaluation of population counts on the basis of sample results.

There will be continuing need for general notational conventions, and it is convenient to start their introduction now. The total number of elements in a finite population will always be designated by N. Thus from Table 3.1 we see that $N = 55{,}311{,}617$ = total number of males 14 years of age and over, as of April, 1950; and for Table 3.2, $N = 42{,}598{,}767$ = total number of civilian males 14 years of age and over in the labor force, as of April, 1950. In addition to this convention, we must also be able to specify the manner in which the total population is distributed among the various categories of the variable. Let us arbitrarily assign to each category of the variable one of the integers 1, 2, . . . , k, where k is the *total number of categories*, whatever it may be. Then N_1 will be the total number of population elements in the first category of the variable, N_2 the number in the second category, . . . , N_k the number in the kth and last category.

It is also desirable to have a notation for the fraction of the population in each category. These fractions will be denoted as $p_1, p_2, \ldots, p_k$. For Table 3.2 we then have the following correspondence between symbols and actual numbers: $N = 42{,}598{,}767$ = number of elements in the population; $k = 2$ = number of categories; $N_1 = 40{,}519{,}462$ = number of elements in the "employed" category; $N_2 = 2{,}079{,}305$ = number of elements in the "unemployed" category; $p_1 = N_1/N = .951$ = fraction of elements in the "employed" category; and $p_2 = N_2/N = .049$ = fraction of elements in the "unemployed" category. In preparing tables, it is sometimes convenient to present the fractions as per cents, as was done in Tables 3.1 and 3.2, but the fractions will prove more basic for later work. We may also note the following relationships between these various symbols: $N_1 + N_2 = N$; $p_1 + p_2 = 1.00$; $N_1 = p_1N$; and $N_2 = p_2N$.

3.1.2. *Infinite Populations.* When one must describe an infinite population with respect to a qualitative variable, it is no longer possible to think in terms of the counting operation that was used for a finite population. One cannot specify the size of the population, nor can one specify the number of elements that belong to any particular category of the variable. However, we can think, at least conceptually, in terms of the fraction of the population elements that would be classified into the several categories of the variable. That is, the population description is provided by giving the possible categories of the variable and the fraction of the population belonging to each category. As in the case of a finite population, these fractions will be denoted by the symbols $p_1, p_2, \ldots, p_k$. Since every element in the population must fall in one and only one of these categories, the relationship $p_1 + p_2 + \cdots + p_k = 1$ must hold.

At this juncture one may very naturally raise the question, "But how can we ever get at the values of these p's if there exists no possibility of enumerating the entire population?" Much of this book will be devoted to a discussion of the reasoning and techniques that can be used to answer this and related questions, but an indication can be given here of several of the approaches that are possible. For example:

1. We may attempt to *estimate* the values of the p's on the basis of a sample drawn from the population. As an illustration, consider an indefinitely large group of individuals suffering from pulmonary tuberculosis, a series of X-ray plates being available for each patient. Two technicians independently attempt to evaluate the progress of a patient from his plates. A single element of our population consists of an evaluation of the plates of a single patient by two readers, and the variable is "agreement of diagnosis," with the two qualitative categories "readers agree on diagnosis" and "readers disagree on diagnosis." We wish to *estimate* the fraction of times that the readers would agree, or disagree, if an indefinitely large number of patients were used.

2. We may *arbitrarily fix convenient values* of the p's, attempt to define an infinite population in a manner such that we think these values hold, and then proceed as though these were the true values of the p's. As we shall see later, this is the approach that is customarily used in drawing samples from a finite population. As an illustration, suppose we have a population consisting of the three elements A, B, and C from which we wish to draw samples of size 2. There are three possible samples that may be drawn, AB, AC, and BC. We now consider the infinite population consisting of *all* drawings of a sample of size 2. A single element of this population consists of the drawing of a sample of size 2, and the variable is a qualitative one with three possible categories: sample AB is drawn, sample AC is drawn, or sample BC is drawn. If we now arbitrarily decide that in this infinite population of all drawings sample AB should occur one-third of the time, sample AC should occur one-third of the time, and sample BC one-third of the time, we have what is known as random sampling. We think that such values of the p's will arise if the following mechanical, or "chance," procedure is used for the drawing. Prepare three identical chips with an A marked on one, a B on the second, and a C on the third. Put the chips in a container, mix them up thoroughly, and draw out two without looking at the notations on the chips until after the drawing is completed.

There are numerous other ways in which problems concerning the distribution of a qualitative variable in an infinite population may be phrased, and many of these will be discussed as we proceed.

3.1.3. *Samples.* A sample, whether from a finite or infinite population, always consists of a stated (and therefore finite) number of elements. Consequently we can return to the counting operations of Sec. 3.1.1 and prepare distributions of the possible values of the variable among the *sample elements.* It is of course necessary to recognize that the distributions so obtained are only approximations to the corresponding population distributions and to make use of notational conventions that will emphasize this distinction between sample and population.

In order to make the discussion more concrete, we shall now present one set of data giving a sample distribution of a qualitative variable. In June and August of 1946, the Committee on Social and Economic Aspects of Atomic Energy of the Social Science Research Council sponsored several nationwide surveys concerning American attitudes toward foreign affairs [1]. In one of the August surveys, each individual in a national sample of approximately 600 adults was asked the question, among others, "How do you feel about the general idea of having an organization like the United Nations?" The answers to this question were recorded verbatim and then classified into a set of categories. The results of this classification are given in Table 3.3.

TABLE 3.3. DISTRIBUTION OF ANSWERS TO THE QUESTION "How do you feel about the general idea of having an organization like the United Nations?" (August, 1946—national sample of 600 adults)

Response categories	Number*	Per cent*
Approve	258	43
Approve with qualifications or uncertainty	96	16
Undecided, don't know	12	2
Disapprove with qualifications or uncertainty	6	1
Disapprove	12	2
Opinions not ascertained	12	2
Do not know what the United Nations is	204	34
Total	600	100

* The "per cent" entries were given in the quoted source. The "number" entries were obtained from these per cents and the stated approximate sample size of 600.

SOURCE: *Public Reaction to the Atomic Bomb and World Affairs,* p. 168, Cornell University, Ithaca, N.Y., April, 1947.

There are a number of observations which may be made concerning this table. First of all, the *number* of sample elements falling in any of the response categories is no longer of importance, except that it is used in obtaining the *per cent* or *fraction* of the sample falling in that category. The number of elements in a category depends on the size of the sample and does not have an absolute interpretation, as did the population entries in Tables 3.1 and 3.2. On the other hand, the per cent of the sample falling in a response category is an approximation to the per cent of the population falling in the same category and is therefore of direct interest. It is for this reason that we seldom present the sample number in a category, *although it is essential that one always give the total size of sample on which the per cent distribution is based.* As one might intuitively expect, the error due to sampling of the separate per cents depends on the size of the sample, as well as on the *method* by which the sample was selected. The second point to be kept in mind is that these data, as well as any other set of data, are subject to possible errors other than those of sampling, e.g., errors arising from the response situation (interviewer, form of question, time of day, etc.) or from the *coding* process (classifying verbatim responses into the specified set of categories).

In closing this section, we shall give a brief summary of the notation that will be used to refer to sample distributions of qualitative variables. As before, it is assumed that the qualitative variable has k categories which are numbered from 1 up to k. The total size of the sample will be denoted by n ($n = 600$ for Table 3.3). The number of sample elements falling in the first category of the variable will be n_1, the number in the second n_2, . . . , and the number in the kth and last, n_k. The corresponding sample fractions for the various categories are n_1/n, n_2/n, . . . , n_k/n. Thus for Table 3.3 we have $n_1/n = 258/600 = .43$, $n_2/n = 96/600 = .16$, . . . , $n_7/n = 204/600 = .34$. We also note that

$$n_1 + n_2 + \cdots + n_k = n$$

and

$$\frac{n_1}{n} + \frac{n_2}{n} + \cdots + \frac{n_k}{n} = 1$$

Putting this notation together with that of Secs. 3.1.1 and 3.1.2 we see that n_1/n, the sample fraction in the first category, is an estimate of p_1, the population fraction in the first category, and so on.

One further notational device will also prove helpful. We shall frequently have to write sums such as the above, and we shall abbreviate these as follows:

$$n_1 + n_2 + \cdots + n_k = \sum_{i=1}^{k} n_i$$

and

$$\frac{n_1}{n} + \frac{n_2}{n} + \cdots + \frac{n_k}{n} = \sum_{i=1}^{k} \frac{n_i}{n}$$

The symbol Σ (capital Greek sigma) stands for the word "sum," and i

is a general subscript which can take on any of the specific values 1, 2, . . . , k. Therefore we read $\sum_{i=1}^{k} n_i$ as "Let i take on each value from 1 up to k, take the corresponding values of the n_i's, and add them." If there is no ambiguity concerning the values that the running subscript i can assume, this will be written simply as Σn_i.

3.1.4. *Graphical Representation.* A statistical investigation gathers and analyzes data that are expected to be "interesting" and "useful"

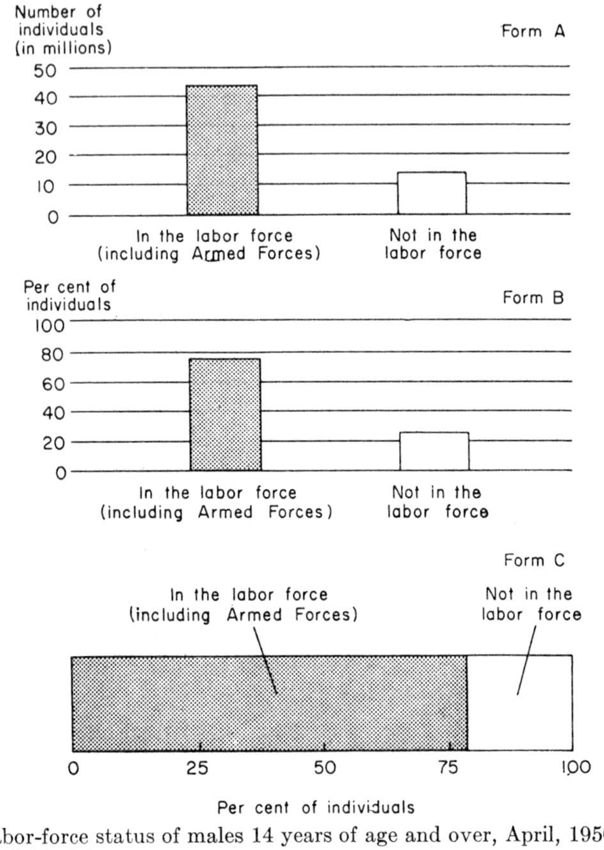

FIG. 3.1. Labor-force status of males 14 years of age and over, April, 1950. SOURCE: See data of Table 3.1.

to an audience that is removed from the individuals or organizations responsible for the investigation. This means that in addition to providing a complete description of the procedures used in collecting and interpreting the data, the results must be presented in such a manner that they are readily accessible and understandable to this wider audience. Any device that can aid in this process is worthy of consideration.

Pictorial or graphical representation is one of the devices which has proved to be helpful. Some of the rudimentary forms of graphical presentation for qualitative variables will be illustrated in this section.

Distributions of qualitative variables are usually presented through some variation of the *bar chart*, or *pie chart*. Thus the distribution of Table 3.1 might be pictured in one of the three forms of bar chart shown in Fig. 3.1. Form A shows pictorially the number of population elements "in the labor force" and the number "not in the labor force," while Form B shows the percentage distribution between these two categories. Form C also shows the percentage distribution, but it emphasizes that the two categories encompass the entire population and that the sum of the separate per cents is therefore equal to 100. All the forms provide the same pictorial impression of the relative sizes of the two categories, and various combinations of the forms are possible. For example, the appropriate percentage values might be entered within the bars of Form A, or the appropriate numbers entered inside the bars of B and C. This form of presentation can clearly be extended to distributions of variables having any number of categories.

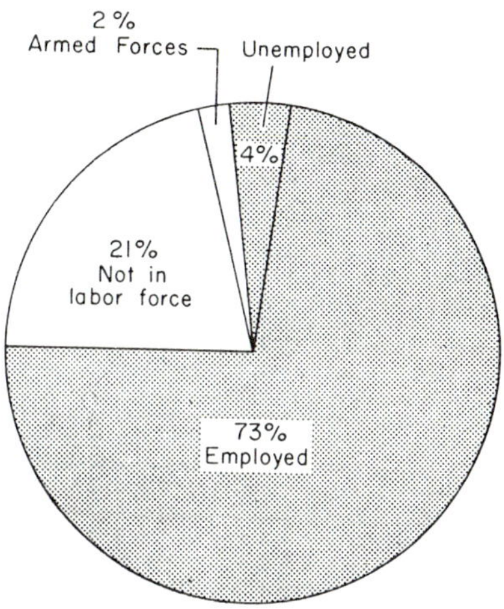

FIG. 3.2. Labor-force and employment status of males 14 years of age and over, April, 1950. (Shaded area represents civilian labor force.) SOURCE: See data of Tables 3.1 and 3.2.

Figure 3.2 uses the pie chart to show the distribution of the population of males 14 years of age and over among the four categories of Tables 3.1 and 3.2. These categories are "not in the labor force," "in the Armed Forces," "unemployed," and "employed." A circle is divided into 100 equal sectors, and these sectors are allocated among the categories of the variable according to their respective per cents. The distribution shown in Fig. 3.2 could also be presented in any of the forms of Fig. 3.1.

When the purpose of an investigation is to compare the distributions of the *same* qualitative variable in *different* populations, it is most convenient to use a form of bar chart. The exact form will depend on a number of factors. Should attention be focused on the percentage distribution or upon the actual number of population elements in the separate categories? Should the distributions be presented in their entirety, or should the comparisons be focused on one or more selected categories of the variable? No general answers to these questions can be given, but Fig. 3.3 does provide a specific illustration of this problem of comparison. These data are related to those that we have been discussing, but their use in the quoted source was in a different context. It is known that the aged population in the United States (e.g., individuals 65 years of age and over) is increasing in both absolute and relative size and that this raises many economic, sociological, and psychological problems. In particular, an increasingly smaller fraction of this aged population belongs to the category "in the labor force." Figure 3.3 emphasizes this fact by showing the per cent of males in each of two age groups—55 to 64 years of age and 65 and over—in the labor

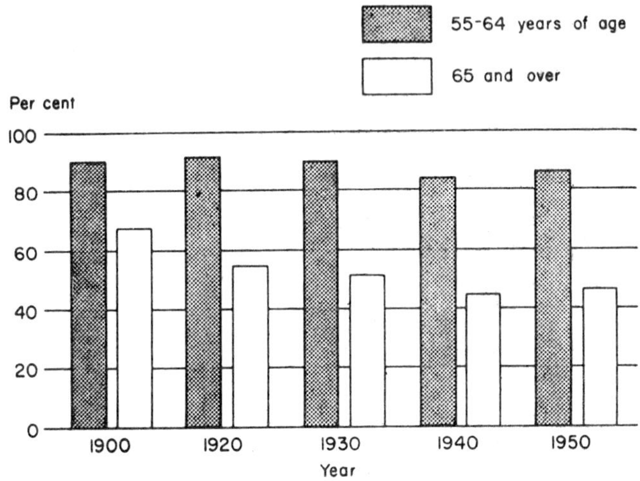

FIG. 3.3. Per cent of males in the labor force for each of two age groups. SOURCE: Philip M. Hauser, "Changes in the Labor Force Participation of the Older Workers," *Amer. Jour. Sociol.*, **59**:312–323 (1954).

force for a number of different years. We immediately see that the labor-force participation of the older males is steadily decreasing while the labor-force participation of males 55 to 64 has remained more or less unchanged over the same period of time. We have thus compared 10 populations (two age groups for each of 5 years) with respect to one category of the variable "labor-force participation," the comparison being made on the basis of the percentage distribution rather than on the basis of the actual number distribution.

If sample distributions are presented graphically, the percentage distribution is usually most appropriate since the sample numbers have no direct population interpretation. However, the same impression of relative size of the separate categories will be given whether one uses sample numbers or sample per cents.

Tables and graphs should always form self-contained units whose contents can be interpreted without extended reference to the text. To this end, inclusive and clearly stated titles and headings must be given, the meanings of the various entries indicated, and a statement made against the ordinate and abscissa concerning the characteristics to which these scales refer. References which consider all these points in some detail will be found at the end of this chapter.

3.2. Distributions of Discrete Quantitative Variables

3.2.1. *Finite Populations and Samples.* Our discussion of qualitative variables has shown that the same techniques can be used for both finite population and sample distributions—just so long as we remember that sample distributions are only approximations to population distributions and employ notation that emphasizes this fact. Consequently we shall now combine the treatment of the two.

Recalling that a "naturally" discrete quantitative variable arises when an integer is assigned to an element corresponding to the *number* of fundamental and nondivisible units which characterize the element, it is easy to extend the treatment of qualitative variables to cover this new case. For example, in the summer of 1941 the Milbank Memorial Fund contacted every white household in the city of Indianapolis in preparation for an extended study of certain selected households. Among other items, data with respect to the variable "number of children ever born" were obtained from each of 6,551 wives who satisfied the following conditions: between 40 and 45 years of age, both husband and wife native-born, and both husband and wife married only once. These conditions, plus that of residence in Indianapolis at the time of the survey, defined a finite population, and every element in this population was enumerated. This population can be described with respect to the variable by counting the number of elements having each possible value of the variable, thereby obtaining the distribution of the variable in the population. The results of performing this operation are given in Table 3.4.

The first column in this table gives an ordered listing, from small to large, of the different values that the variable can assume—0, 1, 2, ... Since we wish to be able to speak about quantitative variables in general, we shall use the symbol x to denote a specific value of such a variable and shall distinguish among different specific values by using subscripts. Therefore, x_1 is the first value that can be assumed and is equal to 0 in this instance; similarly $x_2 = 1$, $x_3 = 2$, ... It should be noted that the last entry in this column is "10 or more." Individual values of the variable contained in the grouping "10 or more" were not listed because there were very few wives in the population (72 out of 6,551) to whom these values were assigned, and the authors undoubtedly felt that the additional detail was not warranted—although a reader might be interested in knowing the largest value the variable assumed. The grouping together of the "tail" values of the variable into a single category is a rather common procedure in the presentation of data of this type.

TABLE 3.4. DISTRIBUTION OF NUMBER OF CHILDREN EVER BORN IN THE POPULATION OF NATIVE-WHITE WIVES
(Wives aged 40–44 with unbroken first marriages)
(Indianapolis, 1941)

Number of children ever born	Number of wives	Per cent of wives	Cumulative per cent
0	1,232	18.8	18.8
1	1,520	23.2	42.0
2	1,547	23.6	65.6
3	963	14.7	80.3
4	531	8.1	88.4
5	295	4.5	92.9
6	170	2.6	95.5
7	105	1.6	97.1
8	66	1.0	98.1
9	52	.8	98.9
10 or more	72	1.1	
Total..............	6,553	100.0	

SOURCE: P. K. Whelpton and Clyde V. Kiser, *Social and Psychological Factors Affecting Fertility*, vol. I, p. 62, Milbank Memorial Fund, New York, 1946.

The second column gives the number of wives having each listed value of the variable and provides the *frequency distribution* of the variable in the population. In general, we shall use the notation of the preceding section, saying that N_1 elements in the population have the value x_1, N_2 have the value x_2, and so on. The sum of all the N_i's is equal to the total number N of elements. The third column presents the per cent of the population having each value of the variable. That is, for the first value of the variable ($x_1 = 0$), we have

$$100\frac{N_1}{N} = 100\frac{1,232}{6,551} = 18.8$$

As before, p_1 is the *fraction* of the population having the first value of the variable, and so on. Thus $p_1 = .188$.

The fourth column in Table 3.4 gives still another form of distribution for the variable "number of children ever born." The cumulative per cent for any given value of the variable is the per cent of the population having values of the variable *equal to or less than* that value. Thus we see that 65.6 per cent of the wives had 2 or fewer children. The cumulative per cent for any value of the variable can be represented either in terms of the N_i's or in terms of the p_i's and for x_4 ($= 3$) it is $100(N_1 + N_2 + N_3 + N_4)/N$, or $100(p_1 + p_2 + p_3 + p_4)$. No meaningful entry can be made for the last category of the variable since we do not know the largest value that it assumed in this population. A cumulative percentage distribution is sometimes practically useful and is extremely useful for theoretical purposes.

One final observation should be made with respect to these data. In the context of this discussion, they refer to a complete enumeration of a defined population and there is no question concerning agreement between sample and population. And yet at the same time the wives studied constitute only a sample of all women in the United States possessing the stated characteristics. Many users of the results of such a study will wish to generalize to this wider population. Then it is necessary to consider how "representative" Indianapolis women are of this wider population. The quoted source recognizes this fact and points out that the 1940 census reported, for example, that 12.4 per cent of "ever-married white women between the ages of 40 and 45" in the United States had borne 6 or more children, as compared with 7.1 per cent for the Indianapolis group. Similar problems of generalization are raised whenever studies whose conclusions are hoped to be widely applicable are carried out in a particular community, in a particular plant, or in some other special setting.

Here, as for qualitative variables, graphical representation is frequently helpful in providing an over-all picture of a distribution. A form of graphical presentation of the percentage distribution of Table 3.4 is given in Fig. 3.4. The values of the variable are marked along the horizontal scale, or abscissa, and the vertical scale, or ordinate, is graduated in per cent. At each value of the variable a vertical line is drawn equal in height to the per cent of the population having that value. If the x axis is regarded as a numerical scale, the per cent for the "10 or more" category cannot be plotted since the individual values of the variable (and their corresponding per cents) that make up this category are not given. However, if the numbers on the x axis are merely regarded as designations of categories, then it is appropriate to plot the per cent for this category. One could also graduate the vertical scale in numbers and plot the entries under "number of wives." Exactly the same visual impression would be given as is obtained from Fig. 3.4 since the vertical lines have the same relative heights, regardless of whether they denote numbers or per cents.

A different form of graphical presentation is customarily used for the cumulative percentage distribution, as is illustrated in Fig. 3.5. At each value of the variable we plot a point representing the corresponding cumulative per cent and then draw a horizontal line to the right from this point, this horizontal line not quite extending to the next value of the variable. If one now takes any point on the x scale, whether it be a possible value of the variable or not, and reads the height of the horizontal line above it, the per cent of the population having a value of the variable less than or equal to this point on the x scale is obtained. Thus reading up from $x = 3$, a value of 80.3 per cent is obtained. Similarly,

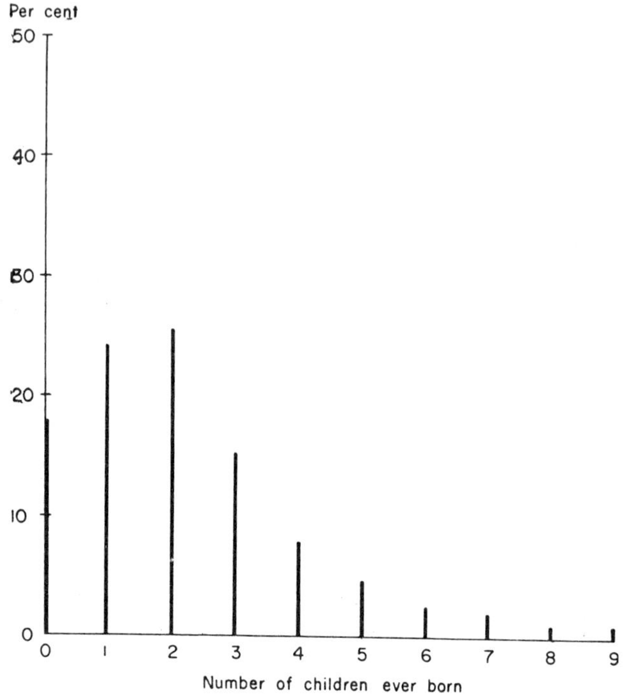

FIG. 3.4. Percentage distribution of number of children ever born in the population of native-white wives. (Wives aged 40–44 with unbroken first marriages in Indianapolis, 1941.) SOURCE: See data of Table 3.4.

if we read up from $x = 3.5$, a value of 80.3 per cent is also obtained since no element of the population can have a value of the variable between 3 and 3.5. At first glance this form of presentation may seem artificial, but it is actually a very powerful one in that it enables one to deal simultaneously with discrete and continuous variables in certain kinds of theoretical work in statistics.

As an example of the practical use of cumulative percentage distributions we may refer to a recent study of advertising media conducted by Alfred Politz Research, Inc. for *Life* magazine. A total of 5,236 respondents were interviewed at six points in time during 1952 and early 1953 and were asked at each point whether they had seen a recent issue of certain magazines, had listened to a recent broadcast of selected radio programs, or had seen a recent showing of a number of television programs. These respondents were selected to be "representative" of the United States noninstitutional population 10 years of age and over. Data on repeat audiences for *Life* are given in Table 3.5, while similar data for

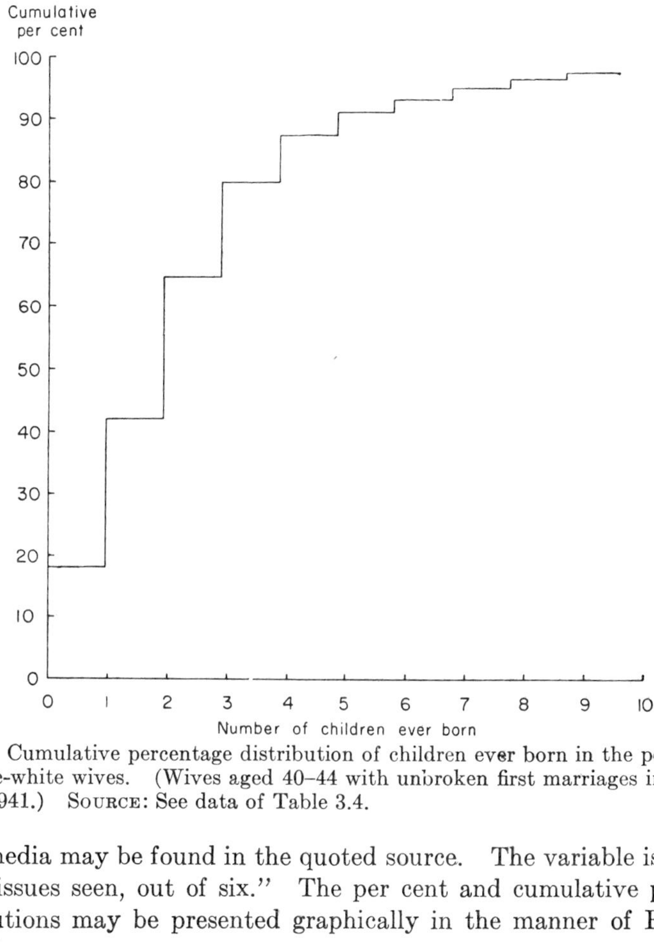

FIG. 3.5. Cumulative percentage distribution of children ever born in the population of native-white wives. (Wives aged 40–44 with unbroken first marriages in Indianapolis, 1941.) SOURCE: See data of Table 3.4.

other media may be found in the quoted source. The variable is "number of issues seen, out of six." The per cent and cumulative per cent distributions may be presented graphically in the manner of Figs. 3.4 and 3.5.

It will be noted that we have cumulated the per cents of Table 3.5 beginning with $x = 1$ rather than with $x = 0$, as was the case in Table 3.4. This has been done because the emphasis of this study was on "seeing." That is, an advertiser's focus will be on the fact that 36.3 per cent of the sample was reached by three, two, or one of the six issues, rather than on the fact that 85.7 per cent of the sample saw three or fewer of the six issues (which includes the 49.4 per cent who saw none of the six issues). This type of situation occurs frequently in studying discrete quantitative variables, and one should always be aware of the possibility that zero values of the variable may require special treatment.

TABLE 3.5. REPEAT AUDIENCES OF SIX SELECTED ISSUES OF *Life* MAGAZINE
(National sample of 5,236 individuals 10 years of age and over)

Number of issues seen	Per cent of sample seeing stated number of issues	Cumulative per cent (starting at 1)
0	49.4	
1	17.6	17.6
2	11.5	29.1
3	7.2	36.3
4	5.5	41.8
5	4.3	46.1
6	4.5	50.6
Total......	100.0	

SOURCE: *A Study of Four Media*, p. 17, conducted for *Life* magazine by Alfred Politz Research, Inc., New York, 1953. The cited source does not provide this complete detail. The research department of *Life* kindly supplied the additional material.

There are two special points which are worth mentioning with respect to this readership example. First, there is the problem of defining a "seer" or a "reader," and these two terms are used interchangeably in the report, and of determining whether any given individual is or is not a reader. This is a question of measurement, as was emphasized in Chap. 2. For purposes of this study a reader was defined as "any person who, after going through an issue with the interviewer, states that he is sure that he has looked into that issue some time previous to the interview." A discussion of this definition and of its application will be found in the cited report. Second, the sample was designed primarily for providing estimates or approximations of the fractions of the population having each possible value of the variable. However, these estimates can be used to obtain approximations for the number of elements in the population having each value of the variable. Provided that the total size N of the population is known or can be estimated, it is only necessary to multiply the sample per cents by N in order to obtain these approximations. The report states that "All audience data are projected on a total of 119,600,000 persons. This is the estimate of the Bureau of the Census of the population age 10 years and over, April, 1952." If desired, it is also possible to design samples that will provide approximations to these population numbers without recourse to external data.

Finally, we shall observe the following conventions regarding sample notation. The total number of elements in the sample will be n, as opposed to N, the total number of elements in the population. Similarly, the number of sample elements having the first value of the variable x_1 will be n_1; the number having the second value x_2 will be n_2; and so on. The corresponding sample fractions will be $n_1/n, n_2/n, \ldots, n_k/n$, and these will ordinarily be taken as estimates of the population fractions $p_1 = N_1/N, p_2 = N_2/N, \ldots, p_k = N_k/N$. The number of possible values of the variable is k, which is equal to 7 in this instance. Thus we see from Table 3.5 that $n_2/n = .176$, and this is the fraction of the sample who had seen exactly one of the six issues of *Life*. The estimated number of individuals in the population who had seen exactly one of the six issues was (.176)(119,600,000) = 21,050,000. One of our main concerns in this book will be to obtain methods for determining how "good" such estimates are.

3.2.2. *Infinite Populations.* The distribution of a discrete quantitative variable for an infinite population can be described only by specifying the fraction of times that each value of the variable occurs. Just as for the qualitative case, it is no longer appropriate to speak of a counting operation since an infinite population exists merely as a concept and not as a practical reality.

As an illustration of the manner in which such a population may be generated and described, consider the following experiment. Five individuals, whom we shall designate as A, B, C, D, and E, are given these instructions: "Each of you take four identical cards; write the name of every other member of the group on a separate card; shuffle the cards thoroughly and without looking at them, draw out one." Thus individual A takes four cards. On one he writes B, on the second C, on the third D, and on the fourth E. Finally, he draws one of the cards in the manner prescribed above, and the selected name is marked down. The other four individuals carry out the same procedure, each using the other (than his own) four names. Choices made in this way are called *random* choices. At the end of this experiment, we note the quantity "number of individuals whose names were not drawn by any of the five." Such individuals may be referred to as isolates. In this situation, there may be either 0, 1, 2, or 3 isolates. It is easy to see that there cannot be more than three isolates by considering one special case. A draws B; B draws A; and either A or B is drawn by each of C, D, and E. Therefore, C, D, and E are isolates (not drawn), and it is impossible to add one more isolate, since if B were also an isolate, A has to choose someone else and that person would be removed from the ranks of isolates.

A single trial of this experiment may be described by the discrete quantitative variable "number of isolates." The possible values of this variable are $x_1 = 0$, $x_2 = 1$, $x_3 = 2$, and $x_4 = 3$. Now let us consider performing this experiment an indefinitely large number of times, thus obtaining an infinite population of trials of the experiment. We can then inquire as to the fraction of times that each value of the variable occurs in this population, namely, p_1, p_2, p_3, and p_4. These fractions can be obtained by advanced methods, and their values are given in Table 3.6.

TABLE 3.6. DISTRIBUTION OF NUMBER OF ISOLATES ON THE BASIS OF RANDOM CHOICE

(Group contains five members; each member makes one random choice.)

Number of isolates	*Fraction* of trials that would show given number of isolates*
0	.04
1	.41
2	.47
3	.08
Total	1.00

* The actual fractions have been rounded to the two decimal places shown.

The distribution shown in Table 3.6 may seem to be of little practical or theoretical interest. Members of any type of formal or informal group, if asked to designate the single individual with whom they would "most like to associate" or whom they feel to be "most competent," and so on, would almost certainly not make their choices on a random basis. However, this random-choice distribution does provide a standard against which to compare actual data and thus becomes extremely important in sociometric investigations that are studying the internal structure of groups. Of course it is necessary to obtain similar distributions for other group sizes and other numbers of choices.

3.3. Distributions for Continuous Quantitative Variables

3.3.1. *Finite Populations and Samples.* As discussed in Sec. 2.2, we speak of a continuous quantitative variable whenever it is possible to think of splitting the fundamental unit of measurement into an indefinitely large or infinite number of subdivisions. Thus such variables as time, weight, length, volume, and the like, are conceptually continuous, even though it is operationally impossible to devise measuring instruments and observational techniques that will distinguish all possible values of these variables. As a matter of fact, we frequently do not even need the most accurate measurement that can be achieved. For example, a person's age in years is sufficient for most practical purposes;

track events and swimming races are usually timed to tenths of seconds because finer distinctions are meaningless; and an individual's weight, as measured to the nearest pound, provides a doctor with all the detail he needs. And yet the physicist, if he so desires, may use instruments that will detect differences in time intervals of a thousandth of a second or less, and the chemist may use balances that will measure weights to within tiny fractions of a gram.

Because of these considerations the actual measurement data used in statistical investigations of one or more continuous variables will be discrete in character. That is, any two elements from a population either will be assigned the same value of the variable, or they will be assigned values which differ by at least some stated amount. This immediately suggests that we can proceed as in Sec. 3.2 where naturally discrete quantitative variables were discussed. Or to put it more concretely, we prepare a distribution by writing down in numerical order the distinguishable values of the variable and then counting the number of elements that have each of these values in a finite population or a sample.

The foregoing procedure can be applied in any situation, but it is usually convenient to modify it in one major respect. This modification consists merely of looking at groups of distinguishable values of the variable rather than at the individual values themselves. This point will now be illustrated with a particular example.

Among the items sometimes recorded and reported by state governments are the lengths of marriages that have ended in annulment or divorce during a given period of time. The usual method of obtaining the length of a terminated marriage is to subtract the day, month, and year of the marriage from the day, month, and year of the divorce or annulment. Thus the variable "length of a terminated marriage" could be reported in days—subject to the realization that some error might be introduced if allowance were not made for the differing numbers of days in a month or in a year. However, there are a number of reasons why one would not follow such a practice: (1) the interest of users of the data will not be centered on the fraction of terminated marriages that lasted exactly 68 days or 5,263 days, or any other specific number, but on the fraction of terminated marriages that lasted less than 1 year, or between 5 and 10 years, and so on; (2) not only would it be a very difficult task to present a table or graph showing all possible values of the variable (if single days were used), but for many of these values there would be few if any terminated marriages having those lengths, even in a very large finite population; and (3) the excessive detail would actually obscure the outstanding characteristics of the distribution of values of the variable among the members of the population.

The manner in which these difficulties may be overcome is illustrated in Table 3.7 which gives the distribution of the variable "duration of marriage" in the population of all marriages terminating through divorce or annulment in the state of Oregon in 1950. We merely count and report the number of elements having values of the variable in each of a number of nonoverlapping intervals. These intervals are customarily referred to as *class intervals.* Thus we see that there were 605 out of a total of 5,927 terminated marriages that had lasted for less than 1 year. Similarly, we see that there were 626 that had lasted for 1 year or more but for less than 2 years, and this number is recorded for the second class interval.

TABLE 3.7. DURATION OF MARRIAGES TERMINATED THROUGH DIVORCE OR ANNULMENT
(Oregon, 1950)

Duration, years	Number of terminated marriages	Per cent of marriages	Cumulative per cent
Under 1	605	10.2	10.2
1 up to 2	626	10.6	20.8
2 up to 3	713	12.0	32.8
3 up to 4	647	10.9	43.7
4 up to 5	559	9.4	53.1
5 up to 6	332	5.6	58.7
6 up to 7	232	3.9	62.6
7 up to 8	237	4.0	66.6
8 up to 9	234	4.0	70.6
9 up to 10	206	3.5	74.1
10 and over	1,536	25.9	
Total.........	5,927	100.0	

SOURCE: U.S. Department of Health, Education and Welfare, *Vital Statistics of the United States,* 1950, vol. II, p. 30, table 9.

It will be observed that the last class interval is labeled "10 and over." This is an open-ended class interval since its upper limit is not indicated. The cited source gives more detail, its open-ended interval being "40 and over," but we have condensed this detail in the interest of saving space. Some of this additional detail will be found in Figs. 3.6 and 3.7. Table 3.7 also shows the percentage and cumulative percentage distributions of the variable "duration of marriage."

The class intervals of Table 3.7 have a number of features which deserve attention. First, the intervals are defined so that any given value of the variable (as measured in years) may be immediately placed without ambiguity in one and only one interval. Thus the interval "under 1 year" receives all observations with a value of the variable

which had lasted for *less than* 3 *years*. Similarly, 74.1 per cent of such marriages lasted for less than 10 years. Thus we see that a cumulative per cent for an interval is associated with the upper boundary of that interval instead of with the interval as a whole. This distribution is therefore graphed in the manner shown in Fig. 3.7. The cumulative per cent for an interval is plotted at the upper boundary of that interval, and these points are then connected by a series of straight lines. The use of

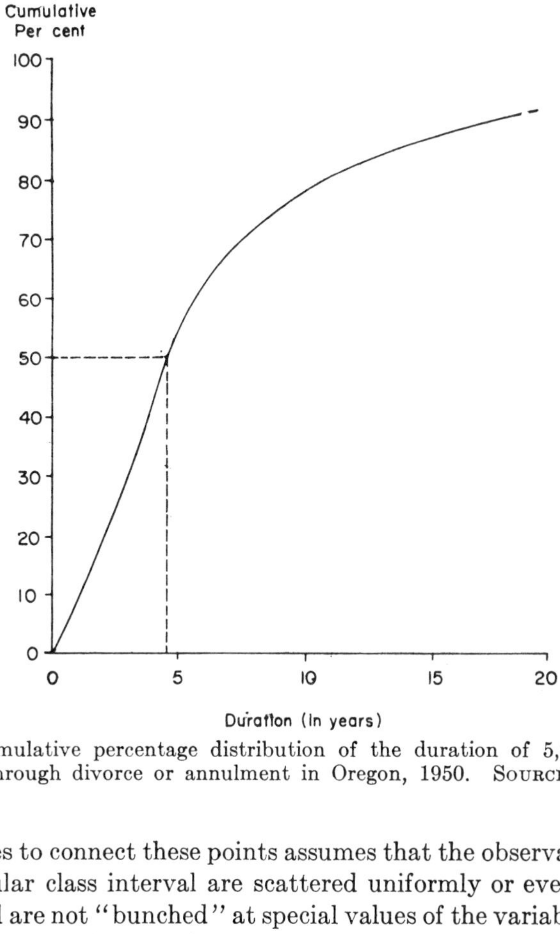

FIG. 3.7. Cumulative percentage distribution of the duration of 5,927 marriages terminated through divorce or annulment in Oregon, 1950. SOURCE: See data of Table 3.7.

straight lines to connect these points assumes that the observations falling in a particular class interval are scattered uniformly or evenly over the interval and are not "bunched" at special values of the variable. Bunching would require the use of graphs with jumps, as in Fig. 3.5. Figure 3.7 is known as a *cumulative percentage polygon*. If so desired, one could graduate the vertical scale in terms of cumulative numbers of elements, but the percentage scale is usually more meaningful and useful.

The cumulative percentage polygon may be read at any value of the variable. If it is read at the boundary of a class interval, then we obtain the per cent of the population having smaller values of the variable. If it is read at any other value of the variable, we obtain an approximation to this percentage, the degree of the approximation depending upon how evenly the observations in that particular interval are spread over the interval. It should be recognized that it can also be read in the opposite direction. For example, we can ask the question, What is the value of the variable such that approximately 50 per cent of the population have smaller values and 50 per cent have larger values? It is only necessary to draw a horizontal line to the polygon from 50 per cent on the vertical scale, drop a vertical line to the scale of the variable, and read off the value of the variable at this point. This procedure is illustrated on Fig. 3.7, and the value obtained is 4.7 years. That is, of all terminated marriages in Oregon in 1950, half had lasted for under 4.7 years and half had lasted for more than 4.7 years.

The preceding example has been discussed as though we were dealing with measurements on all elements of a finite population, and according to our definition of the population this was actually the case. However, all marriages that terminated in the state of Oregon in 1950 constitute only a sample from many other populations of terminated marriages that might be defined. For example, one might be interested in all marriages that terminated in Oregon over an extended period of time and might ask how "representative" the distributions of Table 3.7 are for this larger population. Or one might wish to study the population of all marriages that terminated in the entire United States in 1950. We can do nothing more than make some general observations concerning these points here. Thus we may note the following: (1) one could not expect this distribution to be "representative" of the duration of terminated marriages in states that "specialize" in "quick" divorces; and (2) one could not expect this distribution to be "representative" of the duration of terminated marriages in the entire United States in 1950. As a matter of fact, it is not even possible to obtain reliable and comparable data for all states. Thus the quoted source for Table 3.7 provides such data for only 16 out of the 48 states.

If one has a sample from either a finite or infinite population, then the tabular and graphical procedures which have just been illustrated can be immediately applied to describe the distribution of the sample elements among the various values of the variable. However, it is always necessary to recognize and emphasize that the distributions so obtained are only approximations to the corresponding distributions for the entire population from which the sample was drawn. As in the case of qualitative and discrete variables, it is the sample percentage and cumulative percentage distributions which are of most direct interest, the number of sample elements falling in a class interval being dependent on the size of the sample and not on the size of the population. Our notation will also emphasize the distinction between sample and population, n being the total size of sample (as compared to N, the total size of a population); n_i being the number of sample elements falling in the ith class interval (as compared to N_i, the number of elements in a finite population in the ith class interval); and n_i/n being the fraction of the sample falling in the ith class interval (as compared to $N_i/N = p_i$, the fraction of a finite population in the ith interval or as compared to p_i for an infinite population).

TABLE 3.8. DISTRIBUTION OF ARRIVAL TIMES AT A SUNDAY-MORNING CHURCH SERVICE

Time of arrival, min	Number of arrivals	Per cent	Cumulative per cent
−20 up to −16	2	1.2	1.2
−16 up to −12	14	8.4	9.6
−12 up to − 8	22	13.2	22.8
− 8 up to − 4	16	9.6	32.4
− 4 up to 0	40	23.9	56.3
0 up to 4	34	20.3	76.6
4 up to 8	17	10.2	86.8
8 up to 12	7	4.2	91.0
12 up to 16	8	4.8	95.8
16 up to 20	4	2.4	98.2
20 up to 24	0	0.0	98.2
24 up to 28	1	.6	98.8
28 up to 32	2	1.2	100.0
Total............	167	100.0	

SOURCE: Floyd H. Allport, "The J-curve Hypothesis of Conforming Behavior," *Jour. Soc. Psych.*, **5**:141–183 (1934).

A sample distribution is exhibited in Table 3.8 where the variable is "time of arrival" at a Sunday-morning church service. Time zero represents the start of the church service, negative times represent early arrivals, and positive times represent late arrivals. Observations were reported in terms of 4-minute class intervals extending from 20 minutes before the start of the service to 32 minutes after the start of the service. These data are also given in histogram form in Fig. 3.8. It will be noted that Form A shows the detail for each of the class intervals of Table 3.8, while Form B "lumps" together all early arrivals into the one interval "−20 up to 0." Form B illustrates the manner in which histograms are prepared when the class intervals are of unequal length. The vertical scale is labeled "per cent for 4-minute interval." Since there are *five* 4-minute intervals in the range "−20 up to 0," the plotted height must be the per cent of the sample falling in this interval (56.3) divided by the number of 4-minute intervals (5), or 11.3. By this procedure, we have satisfied our previously stated requirement that the area of the rectangle placed over a class interval should be equal to the per cent (or number) of the sample or population falling in that interval.

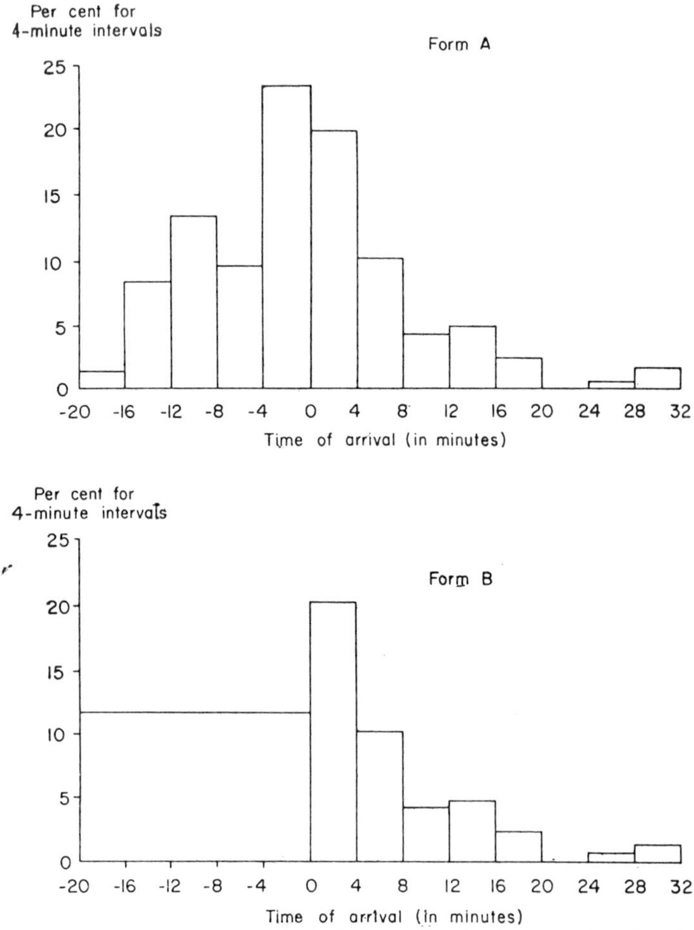

FIG. 3.8. Percentage distribution of arrival times at a Sunday-morning church service. (167 arrivals.) SOURCE: See data of Table 3.8.

Since each of the other class intervals is of length 4 minutes, the per cents for these intervals can be plotted directly. Although the total area of the histogram over the interval "-20 up to 0" is the same for both Form A and Form B, the detail of Form A emphasizes that most early people were only "moderately early," while Form B focuses entirely on the per cent of the sample that was early, without distinguishing degrees.

The foregoing discussion has indicated that there is one more or less arbitrary feature associated with showing the distribution of a continuous quantitative variable among the elements of a finite population or sample, namely, the specification of the class intervals. Although there are no fixed rules that one can or must follow in these situations, there are a number of principles of good practice that may be used as guides. Among these are the following:

1. The number of class intervals should be such that the outstanding and regular features of a distribution are brought out. Too few intervals will obscure important points, while too many will introduce irregularities which are a result of the small number of elements that can be expected to fall in each interval rather than of the inherent nature of the distribution. Thus Form B of Fig. 3.8, by using an interval of "−20 up to 0," hides the fact that early arrivals are in general building up gradually, with the greatest fraction coming in the 4-minute interval immediately preceding the start of the service. Similarly, the 4-minute interval with no arrivals, "20 up to 24," is not an interval which will never show arrivals but only one which did not happen to have any in the observed sample. Intervals of length 4 minutes are getting rather short for the "right tail" of this distribution, considering that there were only 167 elements in the sample. The number of intervals is usually taken as somewhere between 10 and 25, but this depends upon the data with which one is working and not upon any well-defined rule.

2. The class intervals must be defined in an unambiguous fashion so that each observation may be placed in one and only one interval. This was accomplished in Tables 3.7 and 3.8 by connecting the two boundaries of an interval with the phrase "up to," indicating that values of the variable falling on the lower boundary or in the interval (up to but not including the upper boundary) are to be counted as belonging to that interval. However, there are other ways of accomplishing the same end. For example, income classes (where income is measured in dollars) are frequently defined as $0–499, $500–999, and so on. Again there is no question concerning where a particular measurement belongs. Or, class boundaries may be written to one more decimal place than is contained in the original measurements. Thus with measurements such as 1.28, 1.29, 1.31, 1.34, and 1.35, class intervals might be defined as 1.275–1.325, 1.325–1.375, and so on.

3. It is convenient to define intervals having simple boundaries and mid-points. Not only will this make it easier for others to grasp the significance of a distribution, but it will also make it easier to carry out the computations that will be discussed in the following chapters.

4. If possible, the intervals should be of the same length, and open-ended intervals should be avoided. Unequal intervals may not be noticed by the unwary reader and complicate to some extent the prepara-

tion of a histogram. Data referring to an open-ended interval cannot be included in a histogram or a cumulative-percentage polygon, and certain types of computations cannot be performed for distributions with open-ended intervals.

5. Since the histogram and cumulative-percentage polygon assume that observations are more or less uniformly scattered over an interval, we may sometimes have to watch for the appearance of special bunchings or groupings at certain values of the variable. For example, reported ages frequently tend to concentrate on particular values. This is illustrated in Table 3.9. We expect these per cents to decrease gradually in

TABLE 3.9. PERCENTAGE DISTRIBUTION OF REPORTED AGES IN COMPLETED YEARS
(All persons reporting ages from 40 through 50)

Reported age	*Per cent*
40	11.1
41	8.8
42	10.3
43	9.2
44	8.5
45	9.3
46	8.2
47	8.1
48	8.2
49	8.9
50	9.4
Total	100.0

SOURCE: *1950 Census of Population*, vol. II, pt. 1, table 94. Data based on a 20 per cent sample.

going from 40 to 50, because of deaths in the population, but we do not expect them to jump around as in this table. The jumps are assumed to arise from inaccuracies in age reporting and occur primarily on the 0's, on the 2, on the 5, and on the 8. If such bunching occurs because some values of the variable are essentially borrowing from their neighbors, then class intervals should be defined so that these special values are as near as possible to the centers of the class intervals.

6. In defining class intervals, it is better to err on the side of making the subdivision too fine rather than too coarse. Adjacent intervals can always be combined, but the number of elements falling in an interval can never be partitioned accurately among various parts of the interval without going back to the original measurements.

3.3.2. *Infinite Populations.* The distribution of the elements of an infinite population among the distinguishable values of a continuous quantitative variable can only be specified by giving the fraction of the population that possesses each value, or the fraction that falls in a defined class interval. This type of specification of the distribution is frequently all that is required by practical and theoretical considerations. However, if we now imagine that measuring instruments can be devised to distinguish between all possible values of the variable, new avenues for description are opened.

The general form of approach used in this situation is illustrated with hypothetical data in Fig. 3.9. We assume that we are dealing with an indefinitely large population of births and that associated with each birth is a value of the variable "number of minutes after the hour (whatever it may be) at which the birth occurred." Form A shows a distribution such as one might expect to obtain for a sample of 75 births, where the observations are classified into intervals of 20 minutes. The vertical scale is graduated so that the area of the cell over any interval is equal to the per cent of the sample falling in that interval. Thus the area of the first cell is equal to 40, (20)(2.0), the per cent of the sample falling in the interval 0–20. The total area of the cells for Form A is 100 per cent, representing the total sample.

Form B illustrates what we might expect if a sample of 600 births was observed and the birth times classified into 5-minute intervals. Here again the area of a cell over an interval represents the per cent of the sample falling in that interval, and the total area of all cells is equal to 100 per cent. We note that the differences in height of adjacent cells are not as great as in Form A and also that the difference between the maximum height and minimum height of cells is not quite as great in Form B as in Form A.

As the final step in our argument, we can think of drawing larger and larger samples from this population. With successively larger samples and the assumed perfect measurement, the class intervals can be made smaller and smaller. If we now think of carrying on this process indefinitely (larger and larger samples coupled with smaller and smaller class intervals), it seems reasonable to imagine that the differences in height between adjacent cells will disappear and that we shall be left with the straight, horizontal line shown in Form C for the tops of the cells.

Form C is known as a *uniform*, or *rectangular*, *distribution*, and the derivation of these names is clear from the preceding remarks. We can never realize such a distribution in a practical situation because we can never devise perfect measuring instruments, nor can we ever observe the totality of an infinite population. Nevertheless, it is an abstraction which may be used as follows. First of all, we note that the total area under the straight line in Form C is 100 per cent—(60)(1.667). Second, the per cent of the population having values of the variable between any two fixed values x_1 and x_2 is equal to the area over the interval from x_1

to x_2. This is illustrated by the shaded area in Form C, and the amount of this area will always be equal to $(x_2 - x_1)1.667$.

We also have the corresponding cumulative percentage distributions for the three forms of Fig. 3.9. These are shown in Fig. 3.10. The

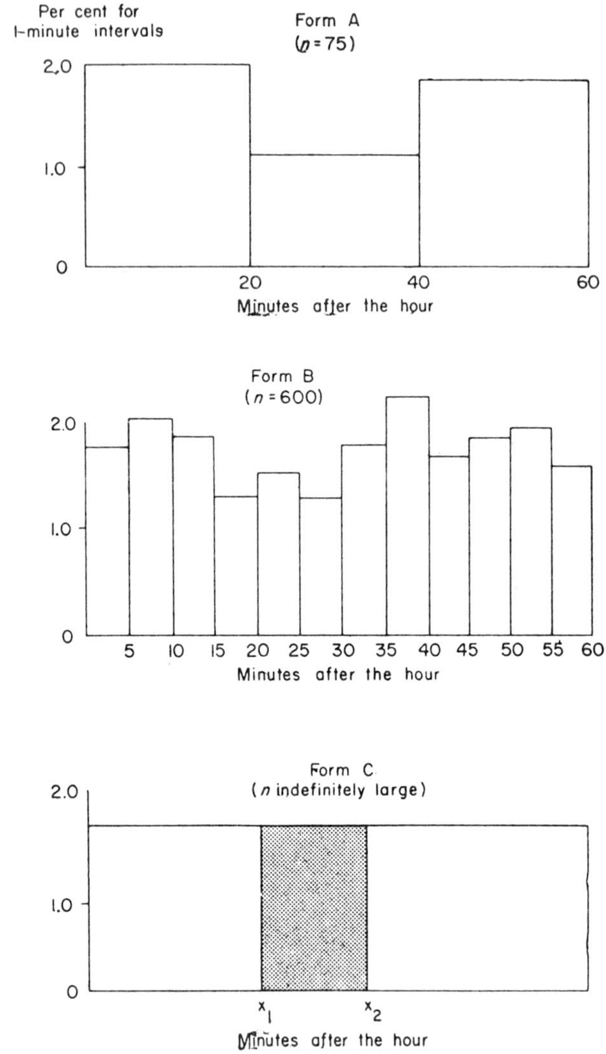

FIG. 3.9. Percentage distributions of occurrence times of births. (Hypothetical data.)

cumulative distribution for Form C refers to the infinite population, and in this instance is a straight line. If we take any value of the variable x and read up to this line, then we obtain the per cent of the population having values of the variable less than or equal to x. Notice particularly that there is no area interpretation associated with the forms of Fig. 3.10. We simply read to the line and obtain a value.

The Form C distributions of Figs. 3.9 and 3.10 were obtained by an intuitive interpretation of what we think would happen if we had an indefinitely large population of births and their associated occurrence times. Moreover, such rectangular distributions are very easy to deal with since we can easily compute areas and heights (for the cumulative distribution). There are many other situations in which we can think of the same limiting process (i.e., larger and larger samples and smaller and smaller intervals), but where the exact specification and use of the ultimate distributions are not as simple as this.

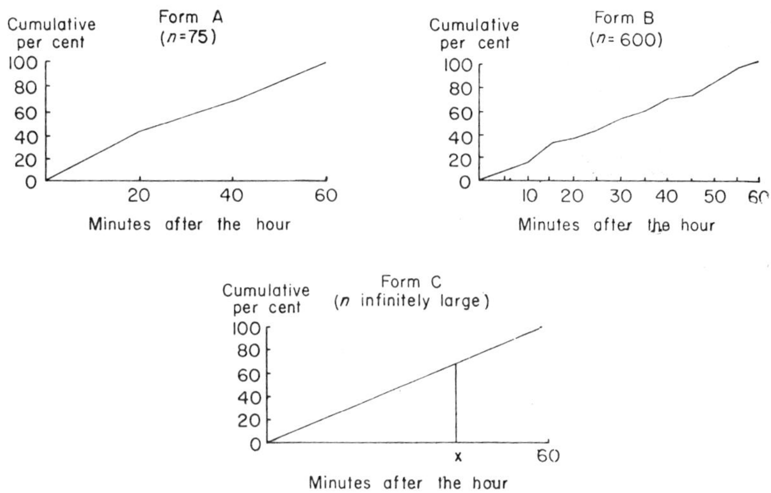

FIG. 3.10. Cumulative percentage distributions of occurrence times of births. (Hypothetical data.)

As an example of this point, consider the distributions of duration of terminated marriages given in Table 3.7. It is possible to think of an infinite population of terminated marriages, similar in character to the finite population represented in that table, for which this limiting process would be appropriate. One might then obtain a "smooth" distribution and cumulative distribution like those exhibited in Fig. 3.11. The percentage distribution starts at zero years, rises imperceptibly at first—since it takes some time to complete the formalities for a divorce or annulment—and then smooths out the irregularities of the histogram of Fig. 3.6. Similarly, the cumulative distribution of Fig. 3.11 smooths out the corresponding distribution of Fig. 3.7. If we now wish to obtain the per cent of terminated marriages which lasted between x_1 years and x_2 years, where x_1 and x_2 may assume any values, it is only necessary to find the area lying under the curve and over the interval from x_1 to x_2. This is illustrated on Fig. 3.11. However, this area can no longer be found by simple means as in the case of the rectangular distribution unless one can specify the curve in sufficient detail and can then operate

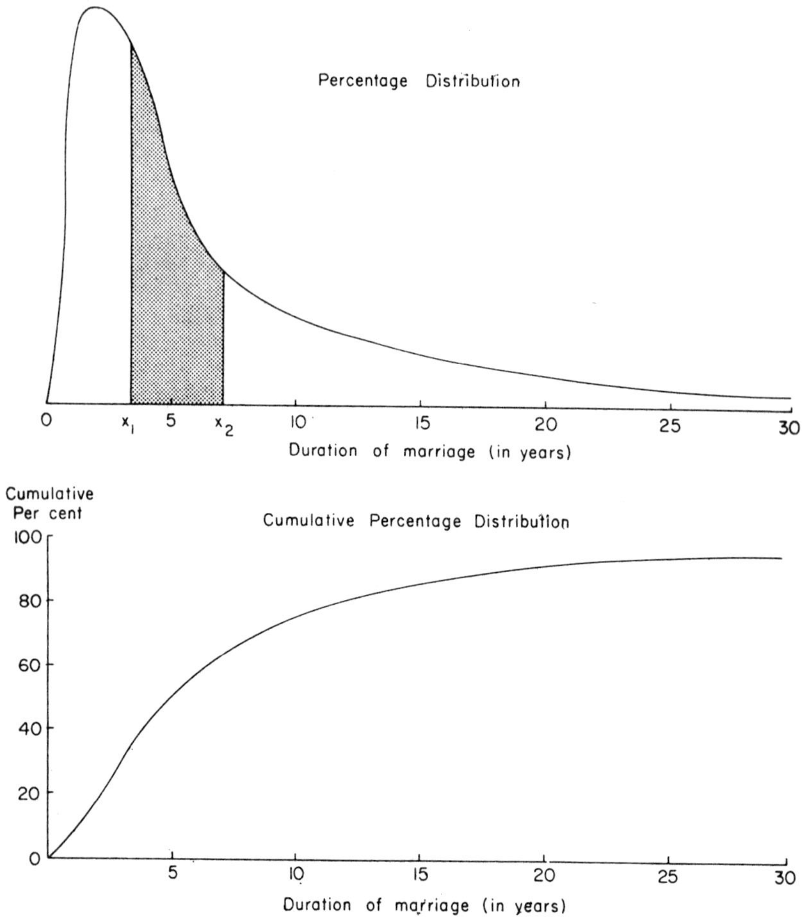

FIG. 3.11. Hypothetical distributions for duration of marriages terminated through divorce or annulment. (See Table 3.7 and Figs. 3.6 and 3.7.)

with this specification. We shall later discuss special situations in which this can be done.

Figure 3.12 illustrates a similar process carried out with the data of Table 3.8 and Fig. 3.8.

"Smooth" percentage distributions similar to those presented in Form C of Fig. 3.9 and in Figs. 3.11 and 3.12 can never actually be observed since they refer to infinite populations and to perfect measurement of a continuous quantitative variable. However, there are many ways in which such distributions may be studied and used. For example:

1. We may arrive, by a deductive argument and without observing any data, at what we think the distribution is for a particular population and

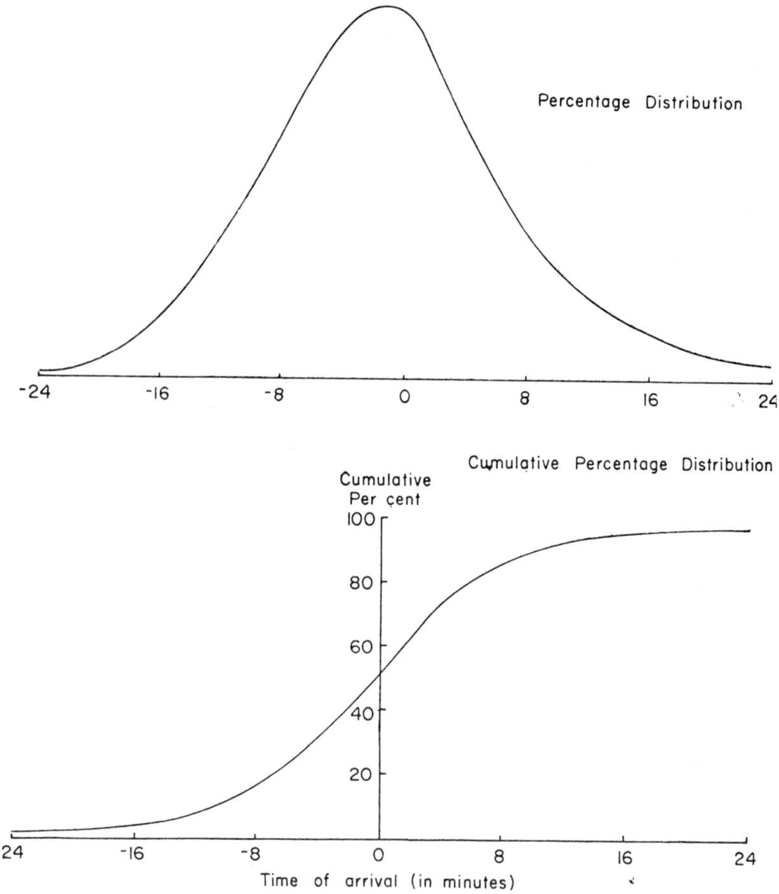

FIG. 3.12. Hypothetical distributions of arrival times at a Sunday-morning church service. (See Table 3.8 and Fig. 3.8.)

variable. Thus the rectangular distribution of birth times can be obtained by arguing that there is no reason to expect births to occur more frequently in one part of the interval from 0 to 60 minutes than in any other part of the interval, assuming the two parts being compared are of equal length. We can then check this argument by seeing whether a sample distribution of occurrence times is compatible with what would be expected if the parent population actually did have this rectangular distribution.

2. We may wish to compare on the basis of samples the "shapes" of distributions in two different populations—the same variable being observed in each population. Thus Fig. 3.12 shows the distribution of "arrival times" at a Sunday-morning church service. Suppose we also considered the distribution of "arrival times" at work for some defined population of workers. Would these two distributions have the same shape?

These problems are only two out of many that may be raised in connection with smooth percentage distributions. Some of these problems will be stated and discussed in detail in later portions of this book.

3.4. Some General Terminology Concerning Distribution Shapes for Quantitative Variables

The distributions with which one must deal in statistical investigations assume an almost endless variety of shapes, each differing in some respect from all others. It is customary and desirable to single out with special terminology a number of broad classes of shapes. The rectangular distribution of the preceding section provides a case in point. The terms which will be introduced in this section may, with the exceptions to be noted, be applied to any distribution, whether the population be finite or infinite, whether the distribution refers to the entire population or only to a sample, and whether the quantitative variable be discrete or continuous.

One of the basic standard forms of distribution, against which other forms are frequently contrasted, is the *symmetric* distribution with a *central maximum.* Two such distributions are pictured in Fig. 3.13, Form A referring to an infinite population (continuous variable) and Form B to an infinite population (discrete variable). The total area under the curve in Form A is 100 per cent, this curve being interpreted as in the preceding section, and the sum of the individual per cents in Form B is equal to 100. Both Form A and Form B are symmetric in that we can think of folding over the right-hand side of each distribution (about the center vertical line in Form A and about $x = 2.5$ in Form B) and thus obtain exact correspondence between the right- and left-hand sides of the two distributions.

Although there are many combinations of quantitative variables and infinite populations for which one expects to find perfectly symmetric distributions, one can never realize such distributions in actual practice. The distributions for any finite population or for any sample will always show deviations from perfect symmetry, simply because it is not possible to obtain a large enough number of measurements, or precise enough measurements, to "smooth" out irregularities entirely. Among the variables we expect to have distributions like those in Fig. 3.13 are the following:

1. *Physical characteristics* of adult American males like height, weight, and head circumference. Thus most heights will tend to cluster around some central value. As one moves further and further from this central value, smaller and smaller fractions of males will fall in intervals of equal length, and this falling off will occur somewhat symmetrically on both sides of the central value.

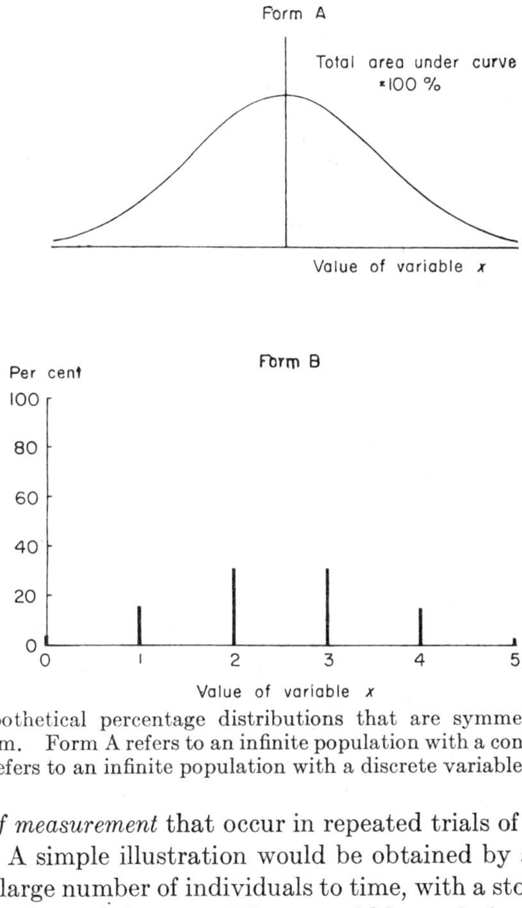

FIG. 3.13. Hypothetical percentage distributions that are symmetric and have a central maximum. Form A refers to an infinite population with a continuous variable, while Form B refers to an infinite population with a discrete variable.

2. *Errors of measurement* that occur in repeated trials of some physical experiment. A simple illustration would be obtained by asking each of an extremely large number of individuals to time, with a stop watch, some event like an eclipse of the sun. There would be variations in the points at which the watches were started and variations in the points at which they were stopped, thus leading to a distribution of "length of eclipse." Empirical evidence has shown that this distribution would be approximately symmetric in shape, more or less like that shown in Form A of Fig. 3.13.

3. *The results of simple two-choice guessing experiments.* The distribution given in Form B could arise under the following circumstances. An individual is given five statements concerning some field of knowledge (about which he is completely uninformed). Each statement is to be labeled as "true" or "false," and the individual, knowing nothing about the field, can only "guess" on each of the five statements. The individual is assigned a value of the variable "number of correct guesses," and this variable takes on the values 0, 1, 2, . . . , 5. If an indefinitely large number of individuals are now asked to do this, Form B gives the per cent of this population who would be expected to have the various values of the variable. It is symmetric in that $x = 2$ and $x = 3$ occur equally frequently, $x = 1$ and $x = 4$ occur equally frequently, and $x = 0$ and $x = 5$ occur equally frequently. We shall see later how to compute these per cents, using only the assumption that each person guesses on each of his five statements.

If the per cents of a population or sample falling in intervals of equal length, or the per cent having the separate values of a discrete variable, decrease with greater rapidity on one side of the point of maximum concentration than on the other, then we speak of an *asymmetric*, or *skewed*, distribution. We can think of obtaining such a distribution by taking the symmetric distribution of Form A of Fig. 3.13 and "pulling out" either the right or left tail of the distribution. If we pull out the right tail, then we speak of a distribution that is skewed to the right, and similarly for the left tail—the direction of the skewness being that of the long tail of the distribution.

Several illustrations of right-skewed distributions have already been given in this chapter, namely, for the variable "number of children ever born" in Fig. 3.4 and "duration of marriages terminated through divorce or annulment" in Figs. 3.6 and 3.11. Reference to these figures will illustrate the essential characteristics of such distributions. Right-skewed distributions arise very frequently in statistical investigations of social phenomena, and among the variables that give rise to this type of distribution are the following: age at marriage, incomes for all families in the United States, number of rooms per house, number of houses per block, number of acres per farm, and so on.

One further type of distribution shape which we shall distinguish is the J-shaped distribution. This distribution shows the maximum concentration at one end of the range of the variable and then decreases gradually to the other end of the range. As an illustration, we may consider some data from a study made by the Bureau of Labor Statistics in the spring of 1951. A national sample of approximately 1,700 tool-and-die makers was interviewed and each individual was asked questions concerning his work history during the 11-year period 1940 to 1950, inclusive. Among

TABLE 3.10. DISTRIBUTION OF NUMBER OF JOB CHANGES MADE BY TOOL-AND-DIE MAKERS IN THE PERIOD 1940 THROUGH 1950
(National sample of tool-and-die makers)

Number of job changes	Number of tool-and-die makers	Per cent
0	979	57.2
1	216	12.6
2	193	11.3
3	95	5.5
4	83	4.8
5	62	3.6
6 or more	84	5.0
Total.......	1,712	100.0

SOURCE: U.S. Department of Labor, *The Mobility of Tool and Die Makers, 1940–1951*, Bulletin 1120, table 4, p. 32.

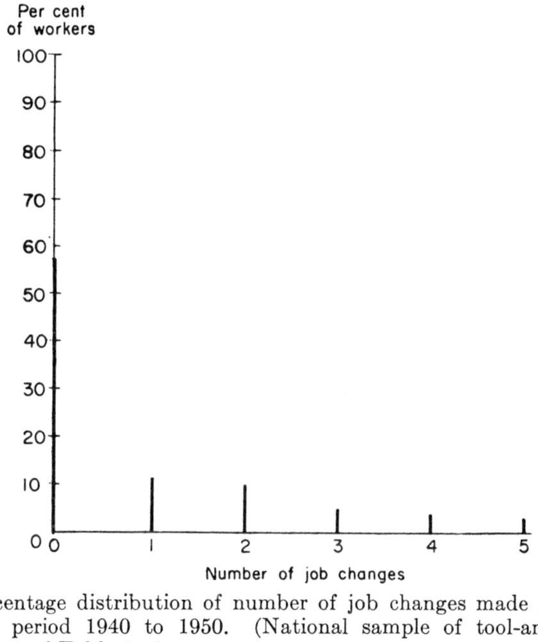

FIG. 3.14. Percentage distribution of number of job changes made by tool-and-die makers in the period 1940 to 1950. (National sample of tool-and-die makers.) SOURCE: See data of Table 3.10.

other items, each worker was asked to state the number of job changes that he had made in the period 1940 through 1950. The results are given in Table 3.10 and Fig. 3.14. It will be observed from this table and figure that the majority of workers had shown no job changes in the studied period and that smaller and smaller percentages of the workers showed more and more job changes.

Although various other general shapes of distributions do occasionally arise, e.g., U-shaped distributions and distributions with two points of maximum concentration, we shall not attempt to discuss and illustrate these here. This will be done as the occasion requires.

EXERCISES

3.1. In the summer of 1951 a sample of 1,500 New York State farmers (men who received one-half or more of their income from the operation of a farm) was interviewed concerning views on selected agricultural policies and programs.

a. The sample distribution of the variable education was:

Education	*Number of farmers*
Some grade school	236
Completed grade school	522
Some high school	368
Completed high school	246
College	116
Not ascertained	12
Total	1,500

SOURCE: Edward O. Moe, *New York Farmers' Opinions on Agricultural Programs,* Cornell Extension Bulletin 864, p. 12, 1952.

Present the sample percentage distribution of the variable education in bar-chart form.

b. What is the correspondence between the numbers given in *a* and the symbolism introduced in this chapter?

c. Each farmer in the sample was asked his opinion concerning government support of farm prices. The answers were studied separately for each category of the variable education with the following results:

Education	Number of farmers	Percentage of farmers saying "prices should be supported"
Some grade school	236	55.1
Completed grade school	522	46.3
Some high school	368	44.3
Completed high school	246	36.2
College	116	32.8

Present these results in a bar chart that emphasizes the difference between education categories with respect to the variable "opinion on price-support programs."

d. Define the population which the foregoing sample is supposed to represent. Do you feel that "time" should be an essential element of this definition? Would you want to use this sample as a basis for drawing inferences concerning the opinions of Georgia farmers on price-support programs in the summer of 1951?

3.2. The assets of United States life insurance companies were distributed in the following manner in 1942 and 1952:

Form of asset	Assets in millions of dollars	
	1942	1952
United States government securities......	11,851	12,774
Securities of business and industry........	10,315	31,646
Mortgages and real estate................	8,389	23,154
All other assets.........................	4,376	5,801
Total................................	34,931	73,375

SOURCE: *Life Insurance Fact Book 1953*, p. 56, Institute of Life Insurance, New York. Sketch these data (using pie charts and bar charts) in manners that will emphasize the differences between 1942 and 1952 in distribution of assets.

3.3. The 1950 U.S. Census of Housing found the following percentage distribution of the variable "number of persons" living in occupied dwelling units (both urban and rural):

Number of persons	*Per cent of occupied dwelling units*
1	9.3
2	28.0
3	22.8
4	18.4
5	10.4
6	5.3
7	2.7
8	1.4
9	.8
10 or more	.9
Total	100.0

SOURCE: *1950 Census of Housing*, vol. 1, pt. 1, table 10.
Prepare graphs of this percentage distribution and of the corresponding cumulative percentage distribution.

3.4. The British Transport Commission has made a study, among others, of the use of public transportation facilities in the city of Bristol. A sample of 3,179 people living in 955 households was interviewed concerning travel during the 7 days prior to the interview. A "journey" was defined as going from one place to another for some definite purpose, while a "journey stage" is a unit representing the part of one person's journey spent in one vehicle. The following distribution of "length of journey stage" was obtained.

DATA FOR EXERCISE 3.4

Length of journey stage, min	Number of journey stages	Per cent of journey stages
Less than 5	3,579	23.0
5 up to 10	4,824	31.0
10 up to 15	2,941	18.9
15 up to 20	2,490	16.0
20 up to 25	887	5.7
25 up to 30	467	3.0
30 up to 45	296	1.9
Over 45	16	.1
Not stated	61	.4
Total.........	15,561	100.0

SOURCE: British Transport Commission, *Bristol on the Move*, London, 1953.

a. Draw a histogram showing the percentage distribution of "length of journey stage." Notice that there is an open-ended interval and that the interval 30 up to 45 is not of the same length as the others.

b. Draw the cumulative-percentage polygon for this distribution.

c. Sketch on the figures drawn for *a* and *b* the "smooth" and hypothetical distributions that might be expected if an indefinitely large number of journey stages were available for study.

REFERENCE CITED

1. Cottrell, Leonard S., Jr., and Sylvia Eberhardt: *American Opinion on World Affairs in the Atomic Age*, Princeton University Press, Princeton, N.J., 1948.

OTHER READING

Brinton, Willard C.: *Graphic Presentation*, Brinton Associates, New York, 1939. Many illustrations of dramatic ways to present data to the general public.

Croxton, Frederick E., and Dudley J. Cowden: *Practical Business Statistics*, 2d ed., pp. 74–135, Prentice-Hall, Inc., Englewood Cliffs, N.J., 1948. A discussion of the standard types of charts.

Hauser, Philip M., and William R. Leonard: *Government Statistics for Business Use*, 2d ed., John Wiley & Sons, Inc., New York, 1956. A useful guide to government sources.

Jenkinson, Bruce L.: *Bureau of the Census Manual of Tabular Presentation*, U.S. Government Printing Office, Washington, 1949. Useful for those who must make up large numbers of tables of basic social and economic data.

Rosander, A. C.: *Elementary Principles of Statistics*, chap. 3, D. Van Nostrand Company, Inc., Princeton, N.J., 1951. A wealth of material illustrating various types of distributions.

Wilks, S. S.: *Elementary Statistical Analysis*, pp. 1–33, Princeton University Press, Princeton, N.J., 1948. A precise elementary discussion of the material covered to this point. Has many good problems.

CHAPTER 4

THE LOCATION OF A DISTRIBUTION

4.1. Special Characteristics of a Distribution

A population is described with respect to a quantitative variable by specifying the distribution of the values of the variable among the population elements. In the preceding chapter we were concerned with the full detail of each distribution. That is, we considered tables and graphs which showed the per cent of the population having any given value of a discrete variable or the per cents of the population having values of a continuous variable in stated intervals. Actually it is not always necessary to consider this complete detail. A distribution may be summarized or described in "capsule" form by focusing attention upon a few of its major characteristics. Measures of these major characteristics are especially helpful in comparing distributions of the same variable in different populations. This chapter and the one that follows will be devoted to the measurement of some of these characteristics.

The characteristics of a distribution that will be considered may be easily introduced graphically. Figure 4.1 shows three percentage distributions. Each distribution refers to the same variable, namely, total money income received by individuals 14 years of age and over. However, three different populations are represented by the distributions. Form A is for individuals who were classified occupationally as "farm laborers and foremen"; Form B is for "service workers, except private household"; and Form C is for "professional, technical, and kindred workers." In all three instances the general distribution shape is more or less the same—skewed to the right—but the similarity among the distributions ends at this point. We shall now examine some of the ways in which they differ.

First, the distributions appear to have different *locations* on the scale of the variable, this scale being identical for all three forms. Form B seems to be located to the right of Form A and Form C to the right of Form B. We would perhaps summarize these differences in a crude fashion by saying that "professional, technical, and kindred workers" in general have a higher income level than "service workers," and that these in turn have, in general, a higher income level than "farm laborers and foremen." These statements are clearly of a subjective nature, and it therefore seems reasonable to attempt to make them more precise by assigning *quantitative measures of location* to each distribution. Provided that such measures can be found, we may then say that Form B is located so many units to the right to Form A and that Form C is located so many

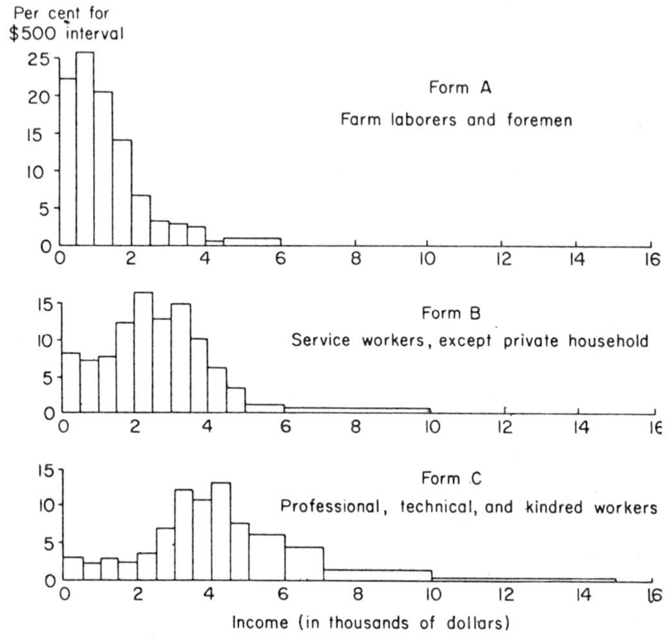

FIG. 4.1. Percentage distributions of total money income received in 1951 by males 14 years of age and over for each of three major occupation groups. (Data based on a sample.) SOURCE: *Statistical Abstract of the United States*, 1953, pp. 294–295, table 326, Bureau of the Census, Washington. (*Note:* Form A does not show .7 per cent having incomes of $15,000 and over; Form B does not show .3 per cent having a loss for the year 1951 or .2 per cent having incomes of $15,000 and over; and Form C does not show .6 per cent having a loss for the year 1951 or 2.3 per cent having incomes of $15,000 and over.)

units to the right of Form B. This chapter will introduce and discuss a number of different measures of location that may be used for this purpose.

Second, differences between Forms A, B, and C do not end with location. For example, the incomes of "farm laborers and foremen" are heavily concentrated in the range $0–2,000, while those of "professional, technical, and kindred workers" are somewhat evenly spread over the range from $0–10,000 with an appreciable per cent being over $10,000.

We might express this difference by saying that incomes of "farm laborers and foremen" tend to be more similar or homogeneous than do incomes of "professional, technical, and kindred workers." If we now wish to make this statement quantitatively precise, we shall have to define quantities which will measure the extent to which the elements making up a distribution have similar values of the variable. This can also be put in the converse form by asking for quantities which will measure the extent to which elements have different values of the variable. In the present instance, we might expect such quantities to be largest for Form C and smallest for Form A, with those for Form B falling between these two. These quantities are frequently referred to as measures of *spread* and will be considered in Chap. 5.

Characteristics of a distribution other than location and spread can also be studied. One may be interested in the amount by which a distribution deviates from a symmetric shape and attempt to obtain a *measure of skewness;* or it may be desirable to have a *measure of peakedness* to distinguish between distributions having a "flat" top and those having a "pointed" top. Although these and other characteristics are occasionally useful, they do not have the general importance of location and spread and will not be considered in this book.

4.2. The Arithmetic Mean as a Measure of Location

4.2.1. *Samples and Finite Populations.* The data presented in Table 4.1 will serve to illustrate the computation and interpretation of the arithmetic mean. These data were obtained through monthly interviews with white families of moderate economic circumstances who lived on 34 city blocks in Baltimore at any time during the period 1938 to 1943. For purposes of the quoted study, a disabling acute (as opposed to chronic) illness was taken as "any affection or disturbance of health persisting for a considerable part of one or more days . . . [which] caused inability to pursue usual activities, such as working, attending school, or performing housework." Table 4.1 gives, for employed males between the ages of 16 and 64, the distribution of number of days of disability associated with respiratory illnesses and nonoccupational accidents. Such distributions are very useful for estimating the effect of different waiting and benefit periods on the rates used for disability insurance and other sickness or payment plans.

For present purposes, we shall consider that the distributions of Table 4.1 are for two finite populations and that the variable "number of days of disability" is a discrete variable. Although these distributions do refer to all acute disabling illnesses arising in the defined group in the stated time period, they may also be regarded as distributions for two

TABLE 4.1. DISTRIBUTION OF NUMBER OF DAYS OF DISABILITY FOR ACUTE RESPIRATORY ILLNESSES AND NONOCCUPATIONAL ACCIDENTS AMONG EMPLOYED MALES
(Eastern Health District of Baltimore, 1938–1943)

Days of disability	Number of respiratory illnesses	Number of nonoccupational accidents
1	109	16
2	114	9
3	132	12
4	103	7
5	82	5
6	60	6
7	84	4
8	57	9
9	28	4
10	20	0
11	26	3
12	10	1
13	23	1
14	17	3
15	17	5
16	10	0
17	5	0
18	4	1
19	3	2
20	5	0
21	8	1
22	4	3
23	3	1
24	3	0
25	2	3
26	2	0
27	3	0
28	19	4
35	4	4
42	1	3
49	5	3
56	0	2
63	0	2
70	0	2
77	0	1
91	1	1
105	3	2
119	0	2
147	1	0
Total.....	968	122

SOURCE: Elizabeth H. Jackson, "Duration of Disabling Acute Illness among Employed Males and Females—Eastern Health District of Baltimore, 1938–1943," *Milbank Mem. Fund Quart.*, **29**:294–330, Appendix, table 2 (1951).

samples chosen from much larger populations. Thus the cited paper* states:

> Data . . . cannot be taken to indicate specifically the disability rates to be expected for other population groups, especially with the adoption of disability insurance. Disability rates differ with . . . nature of employment, the passage of time, and the existence of disability insurance. . . . [they] can be used, however, to give a general indication of disabling illness in an employed urban population.

One may also note from the distribution for respiratory illnesses that there was evidently some tendency for respondents to round to whole weeks. This is particularly evident for "7 days" and "28 days."

In order to keep our preliminary discussion of the arithmetic mean simple, it is convenient to draw a sample from each of these two distributions and to work with these samples. One can think of drawing a *random sample* from the distribution for respiratory illnesses in the following manner. A total of 968 identical cards are prepared, corresponding to the 968 illnesses, and on a single card is recorded the number of days of disability for one illness. Thus 109 cards have the number 1; 114 have the number 2; 132 have the number 3; etc. These cards are then thoroughly shuffled, and a number (equal to the required sample size) of cards are drawn. In this instance we chose for convenience to draw a sample of size 10. The resulting number of days of disability were:

Acute respiratory illnesses: 2, 3, 13, 7, 1, 17, 5, 4, 12, 4

A similar process was carried out for nonoccupational accidents with the following results:

Nonoccupational accidents: 8, 12, 7, 56, 1, 5, 119, 63, 3, 5

The preparation and shuffling of cards is not the efficient way of drawing a random sample from a finite population, but we shall defer our discussion of the preferred procedure until Chap. 6.

The *arithmetic mean*, or *average*, of a sample (or finite-population) distribution is defined to be the sum of the values of the variable associated with all elements in the sample (population) divided by the number of elements in the sample (population). Thus for the sample of 10 respira-

* Elizabeth H. Jackson, "Duration of Disabling Acute Illness among Employed Males and Females—Eastern Health District of Baltimore, 1938–1943," *Milbank Mem. Fund Quart.*, **29**:294–330 (1951).

tory illnesses, we have

$$\bar{x}_R = \frac{2 + 3 + 13 + 7 + 1 + 17 + 5 + 4 + 12 + 4}{10}$$

$$= \frac{68}{10} = 6.8 \text{ days}$$

Similarly, for the sample of 10 nonoccupational accidents,

$$\bar{x}_A = \frac{8 + 12 + 7 + 56 + 1 + 5 + 119 + 63 + 3 + 5}{10}$$

$$= \frac{279}{10} = 27.9 \text{ days}$$

The symbol $\bar{x}$ is commonly used for the sample mean, and the subscripts R and A have been added here to $\bar{x}$ to distinguish between respiratory illnesses and nonoccupational accidents. Two immediate observations may be made with respect to these means: (1) they have the same *units* as the variable, days in this example; and (2) the value of the sample mean is not necessarily a value of the variable which is assumed by an element in the sample, or even a value of the variable that can be assumed.

The arithmetic mean provides a quantitative measure of the location of a distribution. This can be seen most easily from the following simple plot of the observations contained in these two samples:

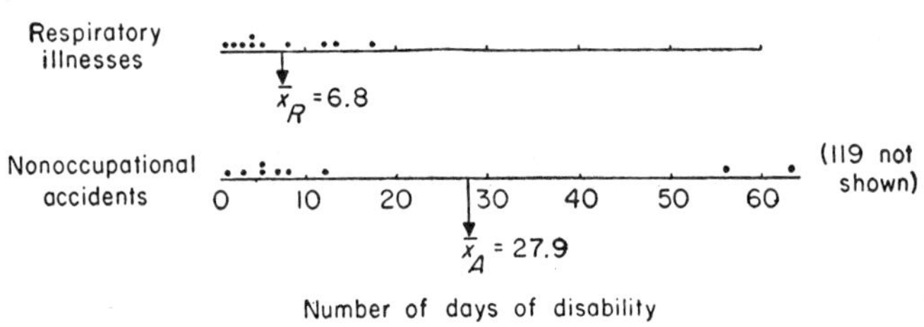

Larger values of the variable appear in the sample of nonoccupational accidents than in the sample of respiratory illnesses, and since the arithmetic mean depends directly on each value of the variable, $\bar{x}_A$ is larger than $\bar{x}_R$. The difference, $\bar{x}_A - \bar{x}_R = 27.9 - 6.8 = 21.1$ days, can be used as a quantitative measure of the difference in location of the two sample distributions. Actually, $\bar{x}_A$ and $\bar{x}_R$ represent only approximations to the true means of the two distributions shown in Table 4.1, and we shall be better able to appreciate the full significance of the mean as a measure of location after we have examined some population distributions.

Since it will be necessary to deal with many different variables and with many different sample sizes, it is convenient to make use of some of the general notational conventions introduced in the last chapter. For a sample of size n, where x_1 is the value of the variable assigned to the first sample element, x_2 is the value for the second element, . . . , and x_n is the value for the last element, we have

$$\bar{x} = \frac{x_1 + x_2 + \cdots + x_n}{n}$$

$$= \frac{\sum_{i=1}^{n} x_i}{n} \qquad (4.1)$$

As previously explained, we read $\sum_{i=1}^{n} x_i$ as "Let i assume successively the values from 1 up to n, take the corresponding values of the x's, and add them." For the sample of respiratory illnesses, $n = 10$, $x_1 = 2$, $x_2 = 3$, $x_3 = 13$, . . . , $x_{10} = 4$. Notice particularly that the subscript in this instance identifies the sample element instead of a particular value of the variable and that one or more of the x's can be equal. Thus $x_8 = 4$ and $x_{10} = 4$ for the respiratory-illness sample.

The definition of the sample arithmetic mean given by formula (4.1) can be immediately extended to a finite population containing N elements. Thus the arithmetic mean of a finite population is

$$\mu = \frac{\sum_{i=1}^{N} x_i}{N} \qquad (4.2)$$

where μ is the Greek letter mu, x_i is the value of the variable for the ith population element, and i takes on the successive values 1, 2, . . . , N.

When one has obtained the actual distribution of a variable in a population, as is the case for Table 4.1, formula (4.2) can be rewritten in a form that is simpler for computational purposes and that is easier to generalize to infinite populations. Consider the distribution for respiratory illnesses. We wish to find the sum of the values of the variable for all 968 elements. We note that there were 109 illnesses, each having 1 day of disability. These illnesses will therefore contribute an amount (109)(1) = 109 days to the sum. Similarly, there were 114 illnesses with 2 days of disability, and these will contribute (114)(2) = 228 days to the sum. This process can clearly be extended throughout the entire distribution, and we obtain

$$\sum_{i=1}^{968} x_i = (109)(1) + (114)(2) + (132)(3) + \cdots + (3)(105) + (1)(147)$$

$$= 7{,}175$$

Therefore,

$$\mu_R = \frac{\sum_{i=1}^{968} x_i}{968} = \frac{7{,}175}{968} = 7.4 \text{ days}$$

We find in like manner that the mean of the distribution for nonoccupational accidents is

$$\mu_A = \frac{(16)(1) + (9)(2) + (12)(3) + \cdot \cdot \cdot + (2)(105) + (2)(119)}{122}$$
$$= \frac{2{,}230}{122} = 18.3 \text{ days}$$

The numerators for μ_R and μ_A can be obtained very efficiently with any desk calculating machine that will multiply and cumulate the products.

The foregoing definitions and computations have given, for each of two finite populations, the mean of a random sample of size 10 and the mean of the entire population. That is, $\bar{x}_R = 6.8$ and $\mu_R = 7.4$, while $\bar{x}_A = 27.9$ and $\mu_A = 18.3$. These sample means are only approximations to their respective population means, and we note that there can be a marked difference between the two. In practice we are usually dealing only with sample means, and it is necessary that procedures be devised for telling how close we expect the sample mean to be to the unknown population mean. It is customary to refer to a quantity which measures some characteristic of a population distribution as a *parameter*, while the corresponding sample quantity is called a *statistic*. Thus μ_R is a parameter, while $\bar{x}_R$ is a statistic that estimates the parameter μ_R.

The computational procedure used for obtaining μ_R and μ_A can be written in general terms if we change the meaning of the subscripts back to that employed in the preceding chapter. Let x_1 be the first value the variable can assume, x_2 the second value, . . . , and x_k the last value. Then the distribution of x in a finite population containing N elements is specified by giving the number of elements N_1 that have the first value, the number N_2 that have the second value, . . . , and the number N_k that have the last value. With this notation, the population mean is

$$\mu = \frac{N_1x_1 + N_2x_2 + \cdot \cdot \cdot + N_{k-1}x_{k-1} + N_kx_k}{N}$$
$$= \frac{\sum_{i=1}^{k} N_ix_i}{N} \tag{4.3}$$

Finally we note that this relation can be rewritten as

$$\begin{aligned}\mu &= \frac{N_1}{N}x_1 + \frac{N_2}{N}x_2 + \cdots + \frac{N_k}{N}x_k \\ &= p_1x_1 + p_2x_2 + \cdots + p_kx_k \\ &= \sum_{i=1}^{k} p_ix_i \qquad (4.4)\end{aligned}$$

where p_i is the fraction of the population having the ith value of the variable (see Sec. 3.2.1). Returning to our computations for μ_R we have as an application of formulas (4.3) and (4.4):

$$\begin{aligned}\mu_R &= \frac{(109)(1) + (114)(2) + (132)(3) + \cdots + (3)(105) + (1)(147)}{968} \\ &= (109/968)(1) + (114/968)(2) + (132/968)(3) \\ &\qquad + \cdots + (3/968)(105) + (1/968)(147) \\ &= (.113)(1) + (.118)(2) + (.136)(3) + \cdots + (.003)(105) \\ &\qquad + (.001)(147) \\ &= 7.4 \text{ days}\end{aligned}$$

Formulas (4.3) and (4.4) can be immediately carried back to the case of the sample mean, and we obtain, as variations of (4.1),

$$\begin{aligned}\bar{x} &= \frac{\sum_{i=1}^{k} n_ix_i}{n} \\ &= \sum_{i=1}^{k} \frac{n_i}{n} x_i \qquad (4.5)\end{aligned}$$

where n is the total size of sample, x_i is the ith value of the variable [and not the value of the variable associated with the ith sample element as in (4.1)], n_i is the number of elements in the sample having the ith value of the variable, and n_i/n is the fraction of the sample having the ith value of the variable. The complete details of an application of both forms of (4.5) are given in Table 4.2. The variable in this instance is "number of issues seen, out of six," and a description of the study from which these data were taken has already been given in conjunction with Table 3.5. The variable can assume seven possible values ($k = 7$) with $x_1 = 0$, $x_2 = 1, \ldots, x_7 = 6$. The arithmetic mean $\bar{x}$ for this sample distribution is found to be 1.3 issues.

In the event that a sample or population distribution is presented in terms of class intervals (see Sec. 3.3.1) rather than in terms of the individual values of the variable, it is still possible to make use of formulas (4.3), (4.4), and (4.5). It is necessary only to define appropriately the values x_i of the variable appearing in these expressions. The entire procedure is illustrated in Table 4.3 with the distribution of arrival times at a Sunday-morning church service (see Table 3.8). This distribution is specified in relation to 4-minute class intervals extending from −20 (arrival 20 minutes before the beginning of the service) up to +32 (arrival 32 minutes after the start of the service). Thus we see, for

TABLE 4.2. REPEAT AUDIENCES OF SIX SELECTED ISSUES OF *Life* MAGAZINE
(National sample of 5,236 individuals 10 years of age and over)

Number of issues seen	Number of sample elements seeing stated number of issues	Fraction of sample seeing stated number of issues	Computations for determining the arithmetic mean	
x_i	n_i^*	n_i/n	n_ix_i	$(n_i/n)x_i$
0	2,586	.494	0	0
1	922	.176	922	.176
2	602	.115	1,204	.230
3	377	.072	1,131	.216
4	288	.055	1,152	.220
5	225	.043	1,125	.215
6	236	.045	1,416	.270
Total....	5,236	1.000	6,950	1.327

$$\bar{x} = \frac{\sum_{i=1}^{7} n_ix_i}{n} = \frac{6{,}950}{5{,}236} = 1.3 \text{ issues}$$

$$= \sum_{i=1}^{7} \frac{n_i}{n} x_i = 1.3 \text{ issues}$$

* The quoted source does not give the individual n_i's. They were obtained from n and from the n_i/n.

SOURCE: See data of Table 3.5.

example, that there were 40 arrivals—representing a fraction .239 of all arrivals—in the interval "−4 up to 0". We are not told the exact value of the variable for each of these 40 arrivals, and it may well be that the investigator did not even record them. However, it is required that a value of the variable be assigned to each of them if the arithmetic mean of the distribution is to be computed.

One way of accomplishing this assignment is to say that arrivals occur at the mid-points of their respective class intervals. That is, the 40 arrivals recorded in the interval "−4 up to 0" are each assumed to have entered church at −2 minutes. The mid-points of the 13 class intervals are given in Table 4.3, and these now become the values x_i of formula

(4.5), k being the number of class intervals. We therefore have $x_1 = -18$ and $n_1 = 2$, $x_2 = -14$, and $n_2 = 14$, . . . , $x_{13} = 30$, and $n_{13} = 2$. The computations then proceed exactly as in Table 4.2, and the result is obtained that the mean arrival time $\bar{x}$ for this distribution is $-.7$ minute. Notice particularly that both positive and negative values of the variable occur in this example and that it is necessary to distinguish between them throughout all computations.

TABLE 4.3. DISTRIBUTION OF ARRIVAL TIMES AT A SUNDAY-MORNING CHURCH SERVICE

Time of arrival, min	Mid-points of class intervals	Number of arrivals	Fraction of arrivals	Computations for determining the arithmetic mean	
	x_i	n_i	n_i/n	$n_i x_i$	$(n_i/n)x_i$
−20 up to −16	−18	2	.012	− 36	− .216
−16 up to −12	−14	14	.084	−196	−1.176
−12 up to − 8	−10	22	.132	−220	−1.320
− 8 up to − 4	− 6	16	.096	− 96	− .576
− 4 up to 0	− 2	40	.239	− 80	− .478
0 up to 4	+ 2	34	.203	+ 68	+ .406
4 up to 8	+ 6	17	.102	+102	+ .612
8 up to 12	+10	7	.042	+ 70	+ .420
12 up to 16	+14	8	.048	+112	+ .672
16 up to 20	+18	4	.024	+ 72	+ .432
20 up to 24	+22	0	0	0	0
24 up to 28	+26	1	.006	+ 26	+ .156
28 up to 32	+30	2	.012	+ 60	+ .360
Total.		167	1.000	−118	− .708

$$\bar{x} = \frac{\sum_{i=1}^{13} n_i x_i}{n} = \frac{-118}{167} = -.7 \text{ min}$$

$$= \sum_{i=1}^{13} \frac{n_i}{n} x_i = -.7 \text{ min}$$

SOURCE: See data of Table 3.8.

There are several observations which can be made concerning this procedure for computing the arithmetic mean of a distribution which is specified in terms of class intervals. These are:

1. The class intervals do not have to be equal in length. It is only necessary to take the mid-point of each interval and to use it in the computations.

2. The arithmetic mean cannot be computed if a distribution is pre-

sented with one or two intervals having open ends. Since the length of such an interval is not specified, the mid-point of the interval is also unknown.

3. It can be shown that an arithmetic mean computed from a class-interval distribution will be equal to the mean that would be obtained by using the individual values of the variable assigned to the sample or population elements if, for each class interval, either (*a*) all elements falling in the interval have the mid-point as their associated value of the variable, or (*b*) the arithmetic mean of the observations classified into the interval is equal to the mid-point of the interval. Case *b* will be satisfied in particular if the observations falling in the interval are uniformly or evenly spread over the entire length of the interval. Case *a* means that in defining class intervals one should watch for tendencies to "pile up" on values of the variable that are not close to the center of an interval. If the lengths of the class intervals are "small," then the two values of the arithmetic mean will be "close" to one another no matter what happens within the class intervals.

4.2.2. *Infinite Populations and Discrete Variables.* When dealing with the distribution of a discrete variable in an infinite population it is no longer possible to define an arithmetic mean in terms of adding up the values of the variable for all elements in the population and then dividing by the number of elements. Formula (4.3) for the mean of a finite population cannot be applied under these circumstances since both its numerator and denominator would be indefinitely large. However, we can still speak of the fraction of the population having each of the possible values of the variable. Therefore the arithmetic mean of such a population distribution can be defined according to the final form of formula (4.4).

That is, we have a discrete variable which can assume one of k possible values, $x_1, x_2, \ldots, x_k$, and a population containing an infinite number of elements. The distribution of the values of the variable among the elements of this population has been specified by giving the fraction of the population p_1 having the value x_1, the fraction p_2 having the value x_2, $\ldots$, and the fraction p_k having the value x_k. Then the arithmetic mean of this distribution is defined to be

$$\mu = p_1x_1 + p_2x_2 + \cdots + p_kx_k$$
$$= \sum_{i=1}^{k} p_ix_i \tag{4.6}$$

As an illustration of the use of (4.6), consider the distribution given in Table 3.6 (see discussion accompanying this table for a description of the experiment leading to this distribution). The computations required for obtaining the mean of this distribution are summarized in Table 4.4. The value of μ is found to be 1.6 isolates.

At this stage in our development of statistical methods the mean of a hypothetical distribution like that of Table 4.4 may seem to have little practical value. However, as we proceed we shall have to make continual use of such distributions and their characteristics, including location. For example, we might have 25 production teams (each having five members) working in a plant. After the teams had been functioning for a period of time, we might ask each individual to name the one other member of his team who had contributed most to its efficient operation.

TABLE 4.4. COMPUTATION OF THE ARITHMETIC MEAN OF THE DISTRIBUTION OF NUMBER OF ISOLATES ON THE BASIS OF RANDOM CHOICE
(Group contains five members; each member makes one random choice.)

Number of isolates x_i	Fraction of experiments that would show given number of isolates p_i	Computation of mean $p_i x_i$
0	.04	0
1	.41	.41
2	.47	.94
3	.08	.24
Total.....	1.00	1.59

$$\mu = \sum_{i=1}^{4} p_i x_i = 1.6 \text{ isolates}$$

SOURCE: See Table 3.6 and accompanying descriptive material.

The number of individuals not mentioned could be recorded for each of the 25 teams and an arithmetic mean $\bar{x}$ computed of the 25 values thus obtained. Comparison of $\bar{x}$ with the theoretical value of 1.6 isolates would then provide a measure of how far from random the actual choice process was.

In closing this section we may note that the arithmetic mean of a distribution for an infinite population is frequently referred to as the *expected value*, or *mathematical expectation*, of the variable x, abbreviated $E(x)$. Thus the symbols $E(x)$ and μ are used interchangeably in this situation.

4.2.3. *Infinite Populations and Continuous Variables.* If interest is centered on the arithmetic mean as a measure of location for the distribution of a continuous variable in an infinite population (see Sec. 3.3.2), then it is necessary to extend still further the considerations of the previous section. One can no longer speak of the fraction of the population having any single value of the variable and must focus attention on the fractions of the population having values of the variable in designated intervals.

The procedure that is used to define the arithmetic mean of this type of distribution may most easily be presented graphically. A distribution

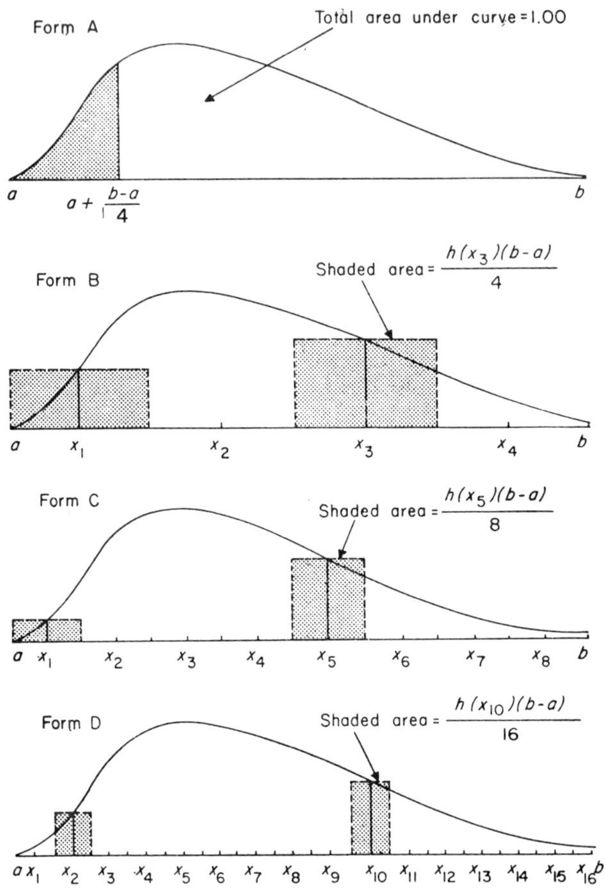

FIG. 4.2. Preliminary steps involved in defining the mean of a hypothetical distribution of a continuous variable.

whose mean we would like to define is pictured in Form A of Fig. 4.2. We assume that the variable x can take on values between a and b only, that the total area under this curve is equal to 1, and that the area under the curve between any two values on the x axis is equal to the fraction of the population having values of the variable in this interval. Thus the shaded area in Form A is the fraction of the population having values of the variable between a and $a + (b - a)/4$.

Let us first attempt to approximate the mean of this distribution by dividing the interval from a to b into four equal parts as pictured in Form B. The mid-points of these intervals are x_1, x_2, x_3, and x_4. The fraction of the population falling in one of these intervals may be approximated by taking a rectangular area over each interval, the base of the rectangle being the length of the interval $(b - a)/4$ and the height of the rectangle being the height of the curve at the center of the interval. Let the height at a particular value x_i be represented by $h(x_i)$. Thus in Form B, the area associated with x_3 is $h(x_3)(b - a)/4$. Similarly the area associated with x_1 is $h(x_1)(b - a)/4$. If we now apply formula (4.6), we have as an approximation to μ,

$$\mu_{\text{approx}} = \sum_{i=1}^{4} x_i \frac{h(x_i)(b - a)}{4}$$

This approximation is clearly not very good because the rectangles do not "fit" the curve at all well. However, suppose we increase the number of intervals into which the total interval is divided. Form C uses 8 intervals, while Form D uses 16 intervals. For Form D we have

$$\mu_{\text{approx}} = \sum_{i=1}^{16} x_i \frac{h(x_i)(b - a)}{16}$$

and this will be a better approximation than was obtained with four intervals since the rectangles now fit the curve more closely.

We can think of continuing the foregoing process indefinitely, next using 32 intervals, then 64, then 128, and so on. The arithmetic mean μ of the distribution of Form A is defined to be the value obtained when this process is continued without end. The actual determination of such a mean is beyond the scope of this book except for one simple case which will now be discussed.

One theoretical distribution whose mean we can actually determine by this approach has already been discussed in Sec. 3.3.2. There we argued that in an indefinitely large population of births the variable "number of minutes after the hour at which the birth occurred" would have a rectangular distribution as pictured in Form C of Fig. 3.9. This distribution is again shown in Fig. 4.3. The interval from 0–60, in which any value of the variable must fall, has been divided into four intervals of equal length (compare this with Form B of Fig. 4.2). The mid-points of these intervals are $x_1 = 7.5$, $x_2 = 22.5$, $x_3 = 37.5$, and $x_4 = 52.5$. The approximation formula for μ, where the value of $h(x_i) = 1/60$ for any x_i, is

$$\mu_{\text{approx}} = \sum_{i=1}^{4} x_i \frac{h(x_i)(b-a)}{4}$$

$$= 7.5\frac{(1/60)(60-0)}{4} + 22.5\frac{(1/60)(60-0)}{4} + 37.5\frac{(1/60)(60-0)}{4} + 52.5\frac{(1/60)(60-0)}{4}$$

$$= \frac{7.5 + 22.5 + 37.5 + 52.5}{4} = \frac{120}{4}$$

$$= 30 \text{ min}$$

The characteristic that makes this case simple is that $\mu_{\text{approx}} = 30$ min, a result which does not depend on the number of intervals (four) that were used in the computation. The same answer would be obtained for any number of intervals. Therefore the true mean μ of this distribution is also equal to 30 minutes, and our argument need be carried no further.

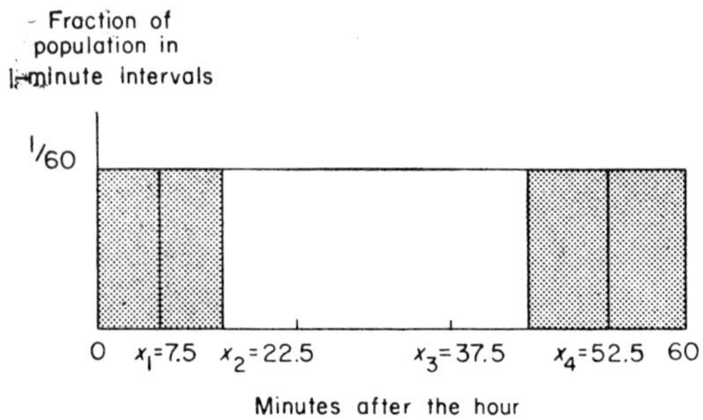

FIG. 4.3. Determination of the arithmetic mean for hypothetical distribution of births according to occurrence time.

4.2.4. *Some General Observations Concerning the Arithmetic Mean.* Now that the arithmetic mean has been defined and computed for a number of different situations we are in a position to gain added insight concerning its use as a measure of location. Some of the more important properties of the mean will be discussed in this section.

An important property of the arithmetic mean is obtained if we consider its behavior when a constant amount is added to or subtracted from each observation in a distribution. This behavior can be easily demonstrated by a numerical example. Let us consider the distribution of arrival times at a Sunday-morning church service given in Table 4.3. Suppose that each of these 167 individuals comes to church on the following Sunday exactly 5 minutes earlier than on the Sunday to which the distribution applies and that no new people come to the service. We may then ask for the mean arrival time in this case. Everyone who arrived in the interval "-20 up to -16" now falls in the interval "-25 up to -21"; those who arrived in the interval "-4 up to 0" now come in the interval "-9 up to -5"; . . . ; and those arriving in the interval "28 up to 32" now arrive in the interval "23 up to 27." The mid-points of these new intervals are therefore -23, -19, . . . , $+25$, and these mid-points can be written as $(-18 - 5), (-14 - 5), (-10 - 5), \ldots, (30 - 5)$. That is, the new mid-points are equal to the old minus 5. The fractions of the sample falling in each interval are the same as in the original distribution.

The arithmetic mean of this new distribution, say $\bar{x}'$, is by definition [see formula (4.5)]

$$\begin{aligned}\bar{x}' &= (.012)(-18 - 5) + (.084)(-14 - 5) + (.132)(-10 - 5) \\ &\qquad + \cdots + (.006)(26 - 5) + (.012)(30 - 5) \\ &= [(.012)(-18) + (.084)(-14) + (.132)(-10) + \cdots + (.006)(26) \\ &\quad + (.012)(30)] + [(-5)(.012 + .084 + .132 + \cdots + .006 + .012)]\end{aligned}$$

The second form of $\bar{x}'$ is obtained from the first by performing the indicated multiplications and then rearranging terms. It will be noted that the first term in square brackets is nothing more than the mean of the original distribution. From the computations of Table 4.3 this is equal to $-.7$ minute. The second term in square brackets has the value -5 since $.012 + .084 + \cdots + .012 = 1$. We therefore obtain the result that $\bar{x}' = \bar{x} - 5 = -.7 - 5 = -5.7$ minutes.

The general property of the arithmetic mean of which the foregoing is a particular illustration is: *If the same amount is added to or subtracted from each observation in a distribution, the mean of the new distribution is equal to the mean of the original distribution plus or minus this constant amount.* This holds for all distributions and is especially easy to demonstrate formally for a finite population. Let μ_x be the mean of a finite population and μ_{x+c} the mean for the same population after a constant c (which may be either plus or minus) has been added to each value of the variable. Using formula (4.4) we have

$$\begin{aligned}\mu_{x+c} &= \sum_{i=1}^{k} p_i(x_i + c) \\ &= \sum_{i=1}^{k} (p_i x_i + p_i c) \\ &= \sum_{i=1}^{k} p_i x_i + c \sum_{i=1}^{k} p_i \\ &= \mu_x + c \qquad (4.7)\end{aligned}$$

The procedure used in going from step 2 to 3 follows the arithmetic of our numerical example and follows readily when we write out the summations. That is,

$$\begin{aligned}\sum_{i=1}^{k} p_i(x_i + c) &= p_1(x_1 + c) + p_2(x_2 + c) + \cdots + p_k(x_k + c) \\ &= (p_1x_1 + p_1c) + (p_2x_2 + p_2c) + \cdots + (p_kx_k + p_kc) \\ &= (p_1x_1 + p_2x_2 + \cdots + p_kx_k) \\ &\qquad\qquad + (p_1c + p_2c + \cdots + p_kc) \\ &= \Sigma p_ix_i + c(\Sigma p_i) \\ &= \mu_x + c\end{aligned}$$

The problem of response errors provides a simple application of this behavior of the mean. For example, suppose a sample of people is to be asked to give their ages on their last birthdays. The responses are recorded, and the sample arithmetic mean $\bar{x}$ is found to be 42.5 years. Suddenly it is discovered that the question was misinterpreted and that individuals gave their ages on their next birthdays. In this situation we wish to know what the mean age for this sample would have been if the responses had been in terms of last birthday. That is, each individual observation is to be decreased by 1 year. From our previous work we know that the mean must be changed by this same amount. Therefore the corrected mean is $\bar{x} - 1 = 42.5 - 1 = 41.5$ years.

Actually, we would never expect that each individual in a sample would make the same response error. However, it can be shown that the shift in the mean of a distribution, due to errors in individual observations, is equal to the *arithmetic mean of the individual errors* (see Exercise 4.7 for a numerical demonstration). Again considering a finite population, let x_i be the true value associated with the ith element and let e_i be the error involved in observing this value. That is, we observe, or this element reports, $(x_i + e_i)$ instead of x_i. Then the observed mean is, using formula (4.2),

$$\frac{\sum_{i=1}^{N} (x_i + e_i)}{N} = \frac{\Sigma x_i + \Sigma e_i}{N} = \frac{\Sigma x_i}{N} + \frac{\Sigma e_i}{N}$$
$$= \mu + \bar{e}$$

Thus we have the result that the observed mean is equal to the true mean μ plus the average error, here designated as $\bar{e}$.

The foregoing results are particularly meaningful if we examine their graphical implications. They say essentially that if a distribution is shifted bodily by an amount c along the scale of the variable (to the right if c is positive and to the left if c is negative), then the arithmetic mean of the distribution is shifted to the right or to the left by the same amount. Thus the mean mirrors or measures the change in location of the distribution.

As an illustration of the shift in location of distributions, consider the frequency distributions pictured in Fig. 4.4. These distributions refer

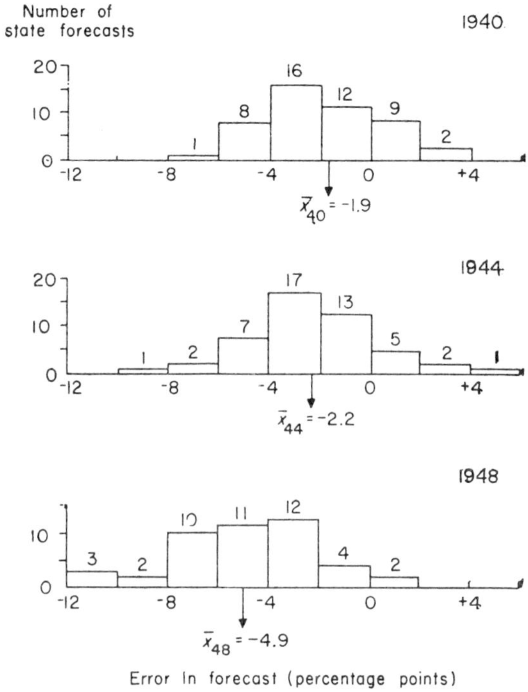

FIG. 4.4. Frequency distributions of the errors in state forecasts of the Presidential vote made by Gallup Poll in 1940, 1944, and 1948. (Alabama, Louisiana, Mississippi, and South Carolina are omitted from the 1948 distribution.) SOURCE: Frederick Mosteller et al., *The Pre-Election Polls of 1948*, Bulletin 60, table V-3, p. 64, Social Science Research Council, New York, 1949. [*Note:* Errors are given in percentage points. They are: percentage of two-party vote (1940 and 1944) or four-party vote (1948) forecast for the Democratic candidate minus the percentage he actually received. Thus positive errors represent overestimates of the Democratic vote, while negative errors represent underestimates.]

to the errors (in percentage points) made by the Gallup Poll in forecasting the Democratic per cent of the Presidential vote for states in the 1940, 1944, and 1948 elections. For example, in the 1948 Presidential election, Gallup forecast that Truman would receive 39 per cent of the four-party vote in the state of New York. Truman actually received 45.3 per cent of the vote. The error in this forecast was therefore (forecast − actual vote) = 39 − 45.3 = −6.3 percentage points. A negative error indicates as underestimate of the Democratic vote, while a positive error indicates an overestimate.

It will be observed that the distributions of Fig. 4.4 are quite similar in shape but that they have somewhat different locations. These differences in location may be measured by the differences in means, where $\bar{x}_{40} = -1.9$, $\bar{x}_{44} = -2.2$, and $\bar{x}_{48} = -4.9$. Thus the distributions tend to shift to the left as we go from 1940 to 1948.

When dealing with any distribution of errors, one would ordinarily hope that the mean error would be zero, or very nearly so. In other words, we hope that positive and negative errors will tend to compensate or cancel one another. We see that this is not the case for the three distributions of Fig. 4.4 and that the Gallup Poll forecasts tended, on the average, to underestimate the Democratic percentages of the state Presidential votes. When such a situation as this arises, we refer to the value of the mean error as the *bias*, or *systematic error*.

One final observation should be made concerning these distributions. The fact that $\bar{x}_{40} = -1.9$ and $\bar{x}_{44} = -2.2$ does not mean that the forecast was $-.3$ percentage points more in error in 1944 than in 1940 for each state—although this might conceivably be the case. Rather it means that the average shift in the errors of state forecasts from 1940 to 1944 was $-.3$.

Although the point has not previously been emphasized, the arithmetic mean has been used in this chapter to compare the locations of distributions only when they have had approximately the same general shape. The importance of this consideration will be apparent if we recognize that distributions may have the same arithmetic mean even though they have widely dissimilar shapes. An illustration of this fact is given in Fig. 4.5. Distribution A is the actual distribution of errors in state forecasts of the Presidential vote made by the Gallup Poll in 1940, while distributions B, C, D, and E are fictitious, each having exactly the same mean as A, namely, -1.9 percentage points. A and B are both roughly symmetric in shape about a central maximum, but B is more "spread out" than A. Distribution C approximates a rectangular distribution, D is U-shaped, and E is a right-skewed distribution. One can, of course, use the mean to compare formally the locations of two dissimilar distributions (A and D, for example). However, there is a risk that the difference in means may be misinterpreted since this is not a simple case of shifting a distribution from one location to another.

There is another property of the arithmetic mean which has a simple geometric interpretation. Consider the random sample of 10 nonoccupational accidents which was used for illustrative purposes in Sec. 4.2.1 and their associated number of days of disability: 8, 12, 7, 56, 1, 5, 119, 63, 3, 5. The mean of the sample is 27.9 days. Let us now compute

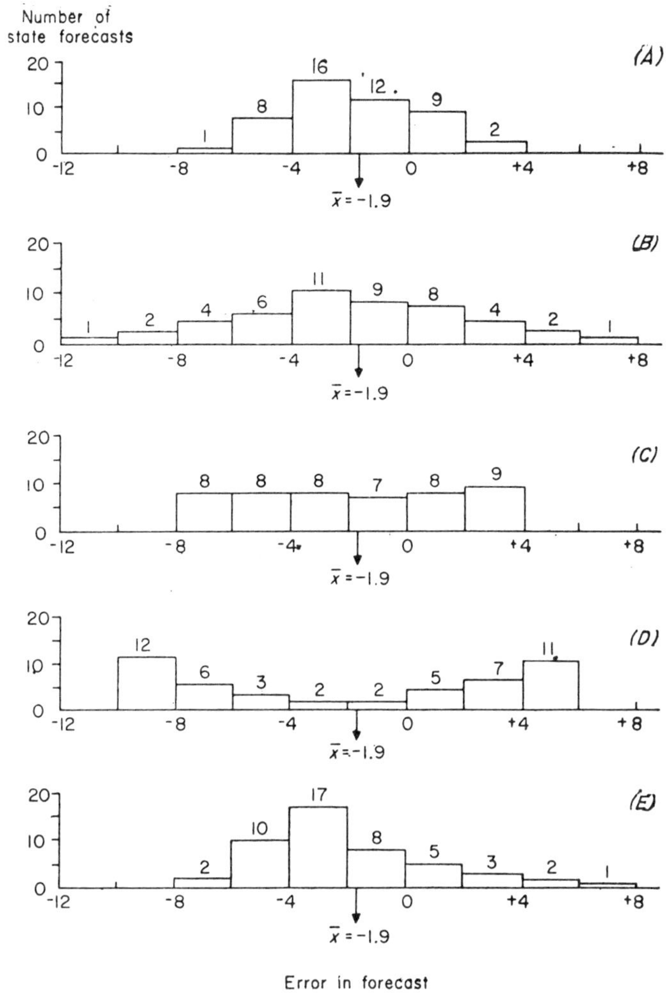

FIG. 4.5. Some frequency distributions of errors in state forecasts which have identical means but different shapes. (Distribution *A* is the actual distribution of errors in state forecasts of the Presidential vote made by the Gallup Poll in 1940, while *B*, *C*, *D*, and *E* are fictitious.)

the difference between each observation and the mean: (8 − 27.9), (12 − 27.9), . . . , (3 − 27.9), and (5 − 27.9). The sum of these algebraic (i.e., plus and minus signs are taken into account) deviations is

$$-19.9 - 15.9 - 20.9 + 28.1 - 26.9 - 22.9 + 91.1 + 35.1 - 24.9 - 22.9 = 0$$

In other words, if the sample mean is subtracted from each measurement in the sample, some of the resulting differences will be positive and some will be negative and their algebraic sum will always be equal to zero. Since the sum of these deviations is zero, it is also true that their average is equal to zero.

Actually, the foregoing sample result is only a special case of formula (4.7), when $c = -\mu_x$. We therefore have the general result: *The average of algebraic deviations about the mean of a distribution is always equal to zero.* This result holds for samples, finite populations, or infinite populations and for both discrete and continuous variables. For example, consider the distribution shown in Table 4.4 which refers to the distribution of a discrete variable in an infinite population. The mean μ of this distribution is 1.59, and the algebraic deviations from the mean are $(0 - 1.59)$, $(1 - 1.59)$, $(2 - 1.59)$, and $(3 - 1.59)$. A fraction .04 of the population has a deviation of $(0 - 1.59)$, a fraction .41 has the deviation $(1 - 1.59)$, . . . , and a fraction .08 has a deviation of $(3 - 1.59)$. Therefore the average algebraic deviation about the arithmetic mean of the distribution is

$$[(-1.59)(.04) + (-.59)(.41) + (+.41)(.47) + (+1.41)(.08)] = 0$$

The geometrical or physical significance of the fact that the average of algebraic deviations about the mean of a distribution is always equal to zero is that the mean is the center of gravity or point of balance of the distribution. Thus if we could imagine the cells of the histograms of Fig. 4.5 as being made of blocks of wood or metal and placed on a thin board (the x axis), then this arrangement would be just balanced by holding a knife-edge under the board at the mean. It is for this reason that the mean is referred to as the *first moment* of the distribution.

One further interesting observation concerning the mean may be made as a direct result of this "point-of-balance" property. If we have a perfectly symmetric distribution (see Sec. 3.4), then the mean must coincide with the point of symmetry since the right half of the distribution will then just balance the left half. A special case of this has already been examined in the discussion of Fig. 4.3 where we found that the mean of a rectangular distribution on the interval 0–60 minutes was exactly equal to 30 minutes.

4.3. Locating a Distribution by Its Percentiles

4.3.1. *Infinite Populations and Continuous Variables.* The arithmetic mean is not the only measure of location which has the desirable property that shifting a distribution to the right or left by a specified amount also shifts the measure of location by exactly the same amount. In this section we shall consider several closely related measures which also possess this property. These measures are easy to define and comprehend for the distribution of a continuous variable in an infinite population but are frequently a bit troublesome to define for sample or finite-population distributions whether the variable be discrete or continuous. Consequently, we shall reverse our usual procedure and treat infinite populations and continuous variables first.

The simplest way to introduce this group of measures is by means of the cumulative percentage distribution. In Figs. 3.10, 3.11, and 3.12 are given three hypothetical cumulative percentage distributions for occurrence times of births (in minutes after the hour), duration of marriages terminated through divorce or annulment (in years), and time of arrival at a Sunday-morning church service (in minutes before or after the start of the service). We are assuming that these distributions are for continuous variables and infinite populations. The three cumulative percentage distributions are reproduced in Fig. 4.6.

From these cumulative distributions we can readily obtain many related quantities which will serve as measures of location for the original distributions. For example, we can ask for that duration of a terminated marriage such that 90 per cent of the population of terminated marriages lasted for less than this value and 10 per cent lasted for more than this value. This value can be obtained from the second distribution in Fig. 4.6 by drawing (as indicated) a horizontal line from 90 per cent on the cumulative percentage scale to the curve and then dropping a vertical line to the scale of the variable. The resulting value of the variable is equal to approximately 19.5 years. That is, if we have a hypothetical distribution of terminated marriages as pictured in Fig. 3.11, then 90 per cent of the terminated marriages in the population lasted for less than 19.5 years and 10 per cent lasted for more than 19.5 years. This value of the variable is called the 90th percentile of the distribution.

The 90 per cent used in the preceding paragraph is clearly an arbitrary number, and we might have used any other value between 0 and 100 per cent. In general, therefore, we define the *pth percentile* of a distribution as that *value of the variable* such that p per cent of the elements in the population have values of the variable less than or equal to the value of the pth percentile. A number of different percentiles are indicated on the cumulative distributions of Fig. 4.6, and their approximate (i.e., as read from the graphs) values are shown in the table on page 93.

It is easy to see from an examination of the cumulative distribution of Fig. 4.6 that the effect of adding a constant amount to every observation in one of these populations will be to add or subtract the same amount from any percentile. The distributions are simply shifted bodily to the right or left by this amount, and the percentiles maintain their same relative positions in the distribution. Thus any of the percentiles may be

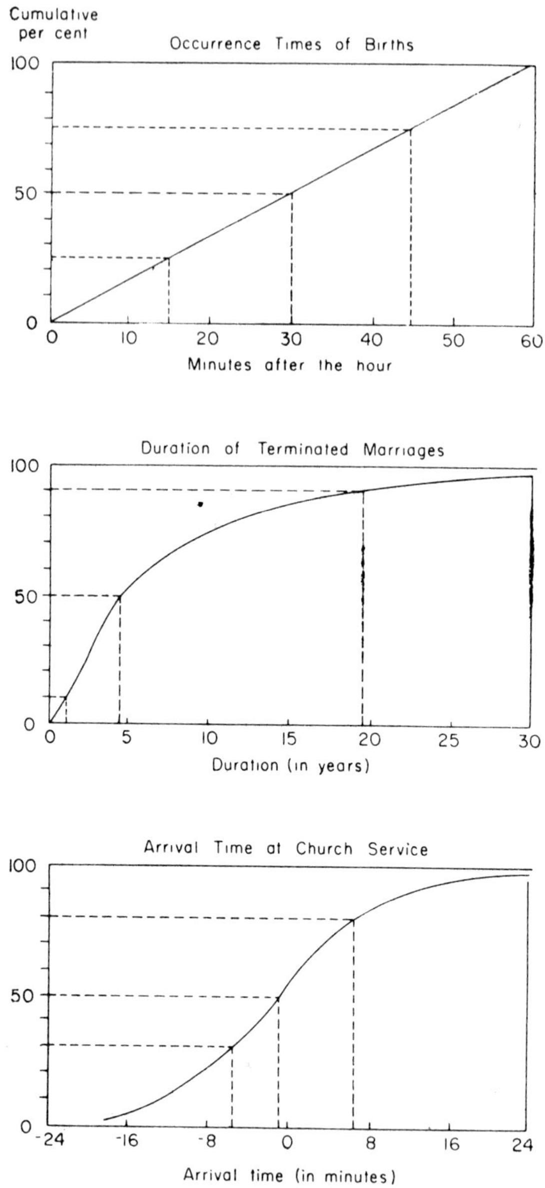

FIG. 4.6. Hypothetical cumulative percentage distributions for infinite populations of continuous variables.

used to compare the locations of two distributions having roughly similar shapes, just as was done with the arithmetic mean.

Some of the percentiles are customarily assigned special names. Thus the 50th percentile is called the *median* of a distribution, the 25th percentile is called the *lower quartile*, and the 75th percentile the *upper quartile* of a distribution; and the 10th, 20th, . . . , 90th percentiles are referred to as the *deciles* of a distribution.

Distribution	Percentile	Value of percentile
Occurrence times of births........................	25th	15 min
	50th	30 min
	75th	45 min
Duration of terminated marriages................	10th	1.8 years
	50th	4.5 years
	90th	19.5 years
Arrival time at church service....................	30th	−5.3 min
	50th	−1.0 min
	80th	+6.2 min

Although it is possible that one may be interested in almost any set of percentiles in connection with the study of some specific distribution, the median is usually preferred over the others as a general measure of location. This is primarily because it will always fall near the "center" of a distribution instead of in the tail of a distribution, as would be true, for example, of the 10th or 90th percentiles. Since half of the elements in a population have values of the variable smaller than the median and half have values larger than the median, we can make the following two observations concerning the median's geometrical interpretation:

1. If a vertical line is drawn at the median of a distribution curve (see Figs. 3.9, Form C, 3.11, and 3.12), the area under the curve will be divided into two equal parts since areas under such curves represent fractions of the population. This is in contrast to the arithmetic mean of the distribution which locates the center of gravity or point of balance of the distribution.

2. For a perfectly symmetric distribution the arithmetic mean and the median must coincide. Half the area must lie on each side of the point of symmetry, and this must also be the "balance" point of the distribution. Thus it was shown in Sec. 4.2.3 that the mean occurrence time for the rectangular distribution of births was 30 minutes, while Fig. 4.6 shows that the median occurrence time is also equal to 30 minutes.

4.3.2. *The Median for Distributions of Discrete Variables.* The difficulties involved in defining a median in certain situations are well illus-

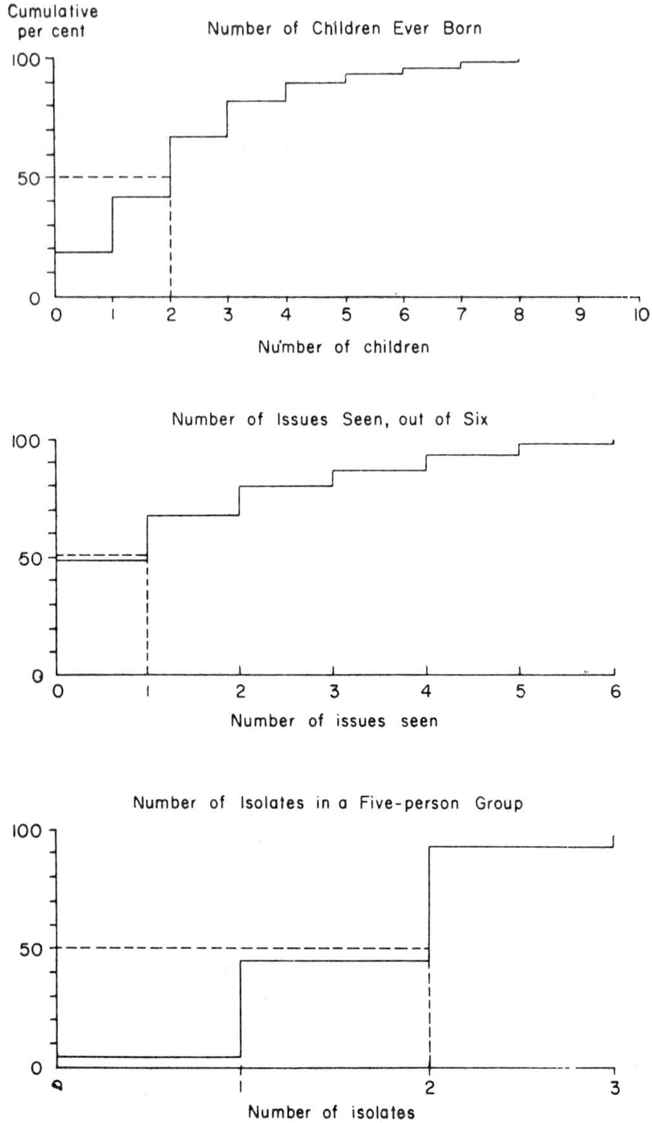

Fig. 4.7. Cumulative percentage distributions for discrete variables. Source: See data of Tables 3.4, 3.5, and 3.6.

trated if we attempt to apply the graphical procedure of the preceding section to distributions of discrete variables. Figure 4.7 presents the cumulative percentage distributions for three situations that we have already examined in some detail: (1) the distribution of native-white wives in unbroken first marriages by number of children ever born (see

Table 3.4), (2) the distribution of a sample of individuals in the United States by number of issues seen, out of six, of *Life* magazine (see Table 3.5), and (3) the distribution of five-member groups according to number of isolates on the basis of random choice (see Table 3.6). The first of these refers to a finite population, the second to a sample from a finite population, and the third to an infinite population.

We can use these cumulative percentage distributions to define a median or any other percentile in a manner similar to that demonstrated with the distributions of Fig. 4.6, provided that we regard the vertical portions, or "jumps," in the distributions as part of the curve. Thus in the first distribution we draw a horizontal line to the curve at 50 per cent, drop a vertical line to the scale of the variable, and find that the median value for this distribution is 2 children; for the second distribution this procedure leads to a median of 1 issue; and for the third it provides a median value of 2 isolates. We have therefore defined a median for these distributions and for distributions of discrete variables in general, although these medians do not have the clear-cut interpretation of the medians obtained in the preceding section. For example, the median number of children ever born is 2. We can observe from the distribution of Fig. 4.7 (or from Table 3.4) that 42.0 per cent of the finite population under study had values of the variable *less than* 2, while 65.6 per cent had values of the variable *less than or equal to* 2. Similarly, for the second distribution, 49.4 per cent of the sample had seen no issue, while 67.0 per cent had seen either one issue or no issue. In other words, it will be frequently impossible to find a value of a discrete variable such that half the sample or population has values *less than* or *less than or equal to* this value of the variable. This does not mean that the median is never used in such situations. It does mean that we must recognize this limitation of the median.

If a discrete variable has many possible values instead of only a few, as for the distributions of Fig. 4.7, it is ordinarily possible to come much closer to the ideal situation that existed for the distributions of continuous variables in infinite populations. Thus Fig. 4.8 shows a portion of the cumulative percentage distributions for the data of Table 4.1 on the number of days of disability associated with cases of acute respiratory illnesses and nonoccupational accidents. It will be observed that the medians are 5 and 8 days, respectively, and that the jumps occurring at the median are much smaller than in Fig. 4.7.

It should be observed that it is not necessary to draw a cumulative percentage distribution in order to determine a median. One has only to write down the cumulative per cents for a discrete variable distribution and take as the median the *first value* of the variable which has a cumulative per cent of over 50. In the event that the cumulative per cent associated with one value of the variable is exactly 50 per cent, then the horizontal line drawn from 50 per cent coincides with one of the flat portions of the cumulative percentage distribution and it is not clear what value of the variable should be used as the median. Under these circumstances, it is customary to arbitrarily take the median as being halfway between the value of the variable having a cumulative per cent

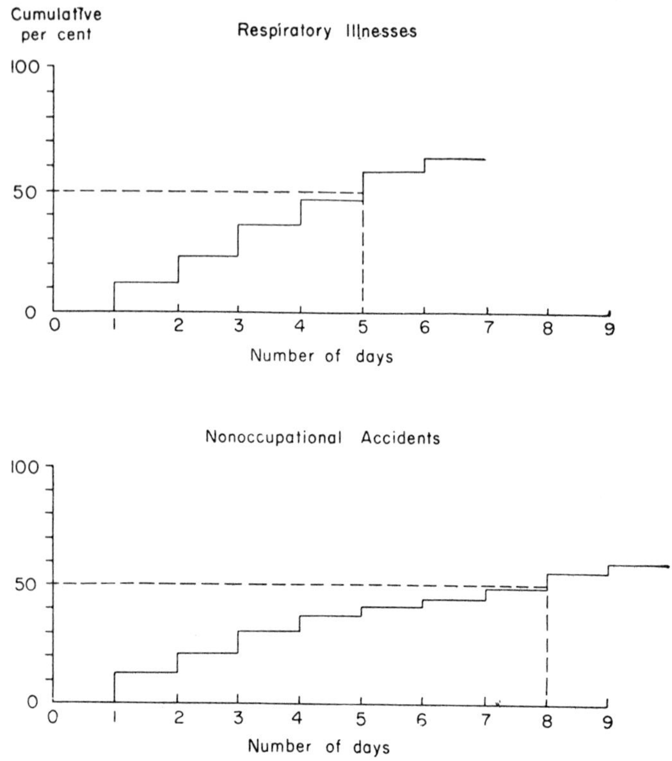

FIG. 4.8. Cumulative percentage distributions of number of days of disability for acute respiratory illnesses and nonoccupational accidents. SOURCE: See data of Table 4.1.

of 50 and the next greater value of the variable. Thus in Fig. 4.8, if the horizontal line from 50 per cent coincided with the flat portion of the diagram between 4 and 5 (for respiratory illnesses), the median would be taken as 4.5 days.

As a final point in this discussion, we note that with a small sample we may not wish to go to the trouble of determining cumulative percentages and may wish to obtain the median by direct observation. Under these circumstances the sample median may be defined as *the middle value of the variable when the observed values are arranged in order of magnitude.*

Thus our random sample of 10 nonoccupational accidents (see Sec. 4.2.1) showed the following numbers of days of disability:

8, 12, 7, 56, 1, 5, 119, 63, 3, 5

Arranged in order of magnitude, these become

1, 3, 5, 5, 7, 8, 12, 56, 63, 119

Since there is an even number of observations there are two middle values, corresponding to the fifth and sixth entries, and *we arbitrarily take the median as being halfway between these two middle values, namely,* 7.5 *days.* If there had been 11 observations, we would have taken the value of the variable corresponding to the sixth one since this is now exactly in the middle. The extension to other sample sizes is immediate. The observations are arranged in order of magnitude, and we count up to the $[(n + 1)/2]$th observation and use that value of the variable as the median. If n is odd, this gives us the middle value, and if n is even, it places us halfway between the two middle values.

4.3.3. *Samples and Finite Populations: Continuous Variables.* The median, or any other percentile, of a class-interval distribution may be obtained graphically from the cumulative-percentage polygon of the preceding chapter. Let us apply this procedure to the distribution pre-

TABLE 4.5. DURATION OF WORK STOPPAGES ENDING IN 1950

Duration of stoppages, days	Number of stoppages	Per cent of stoppages	Cumulative number of stoppages	Cumulative per cent of stoppages
0 up to 1	584	12.6	584	12.6
1 up to 4	838	18.0	1,422	30.6
4 up to 7	739	15.9	2,161	46.5
7 up to 15	1,045	22.5	3,206	69.0
15 up to 30	727	15.6	3,933	84.6
30 up to 60	545	11.7	4,478	96.3
60 up to 90	170	3.7	4,648	100.0
Total. . .	4,648*	100.0		

* There were 164 stoppages which lasted for 90 days and over. These have been omitted. It is therefore possible to compute the arithmetic mean of the cutoff, or "truncated," distribution.

SOURCE: *Analysis of Work Stoppages during 1950,* Bulletin 1035, table 13, U.S. Department of Labor.

sented in Table 4.5 which relates to the finite population of all work stoppages (strikes) ending in the United States in the year 1950 (excluding the 3.4 per cent of this population which had lasted for 90 days or more). We shall regard the duration of a work stoppage as a continuous variable, although recording procedures might be such that it is really discrete. Actually, if a discrete variable can and does assume many values, it introduces little inaccuracy to use the histogram and cumulative-percentage polygon for graphical presentation.

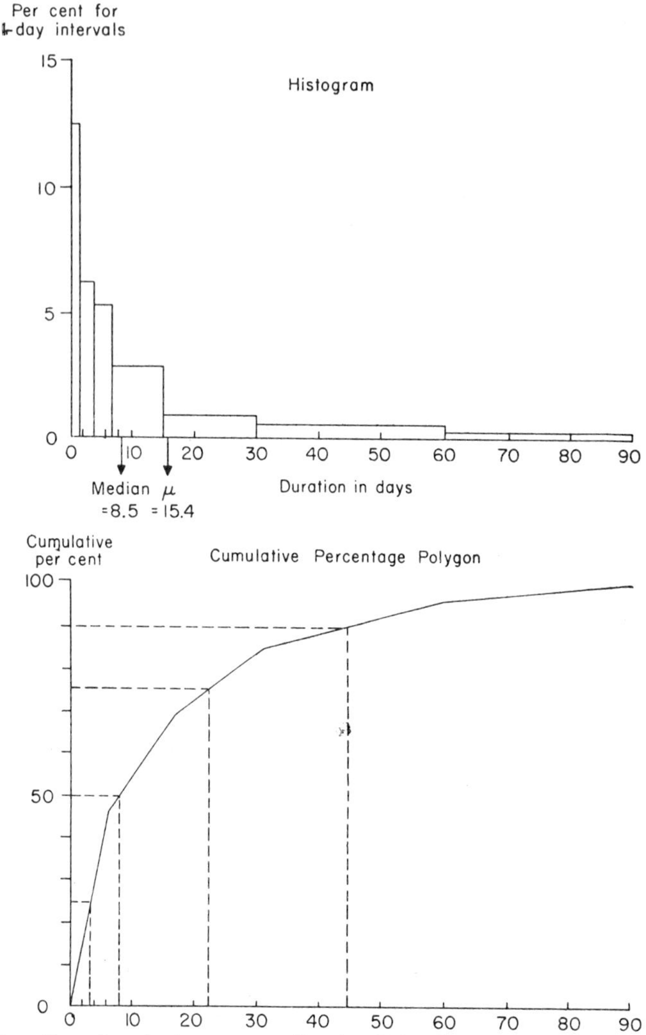

FIG. 4.9. Duration of work stoppages ending in 1950. SOURCE: See data of Table 4.5.

The histogram and cumulative-percentage polygon for this distribution are presented in Fig. 4.9. Notice that the class intervals are of unequal length and that this has been taken into account in drawing the histogram. The value of any percentile may be immediately obtained from the polygon of Fig. 4.9. Thus the lower quartile is 3.1 days, the median is 8.2 days, the upper quartile is 20.8 days, and the 90th percentile is 43.8

days. By way of comparison, we find the arithmetic mean of this distribution to be

$$(.5)(.126) + (2.5)(.180) + (5.5)(.159) + (11.0)(.225) + (22.5)(.156) + (45.0)(.117) + (75.0)(.037) = 15.4 \text{ days}$$

Thus the mean is almost twice as large as the median for this distribution.

It might be observed that the graphical procedure of Fig. 4.9 for determining the median is essentially an "interpolation" procedure that can also be carried out numerically. From the cumulative percentage distribution of Table 4.5 we see that the 50th percentile falls in the interval "7 up to 15." We must therefore add to 7 (the lower boundary of this "median" interval) some fraction of 8 (the length of the interval). This fraction may be determined in the following manner. A total of 46.5 per cent of the observations have values of the variable less than 7, and it is necessary to take an additional 3.5 per cent (50 − 46.5) of observations to attain the desired 50 per cent. These will be the lower 3.5 per cent out of the 22.5 per cent of all observations falling in the interval "7 up to 15." Thus *for this median interval*, we are taking a fraction 3.5/22.5 = .156 of the observations in the interval. Assuming that observations are evenly spread over the interval, we shall also add this fraction of the length of the interval to 7. Therefore the median is approximated as

$$7 + 8\left(\frac{50 - 46.5}{22.5}\right) = 7 + 8(.156) = 8.2 \text{ days}$$

Any other percentile of the distribution may be approximated in the same manner.

The foregoing numerical method for obtaining the median of a class-interval distribution is essentially another version of the ordering procedure set forth in the last section. If individual observations are available, then the argument given there is equally applicable to distributions of continuous variables in samples and finite populations. It will prove convenient to use the individual observations if the sample size is small or if the observations can be efficiently ordered and counted, as is the case when punched-card equipment is used for processing data.

4.4. A Comparison of the Mean and Median

There is no single measure of location that is "ideal" for all purposes. Even the very concept of location is somewhat ambiguous unless we speak only of shifting a given distribution from one spot to another and look for a parameter whose change will mirror this shift. The best that one can hope to do is to use a measure that is in some sense or other satisfactory for a given purpose. It will usually be sufficient to restrict one's choice to the mean and median although there are other measures of location which are useful in particular circumstances. A few comments on some of these will be found in the next section.

No attempt has been made to contrast the mean and median. However, we can abstract from our definitions, examples, and discussion to bring out some of the salient points. These are:

1. Although the notion of a median is quite precise, namely, that value of the variable such that half of a population or sample have smaller values and half have larger values, this ideal can seldom be realized unless we are dealing with the theoretical distribution of a continuous variable in an infinite population. Some arbitrariness is involved in defining a median for other situations, and the resulting value does not have the clear 50-50 property. The arithmetic mean, on the other hand, is clearly defined under all except very rare circumstances.

2. A bodily shift in the location of a distribution will effect a corresponding shift in the mean or median (and other percentiles as well) if we are consistent in handling the median when some arbitrariness is necessary.

3. The mean and median of any perfectly symmetric distribution are identical.

4. The mean and median will not coincide for a skewed distribution. Since the mean is the center of gravity of a distribution, it will be "pulled" away from the median in the direction of the long tail of the distribution. Thus for the right-skewed distribution of days of disability for nonoccupational accidents (see Table 4.1 and Fig. 4.8) the mean is 18.3 days and the median is 8 days. Other examples of this same type can be found among the illustrations of Secs. 4.2 and 4.3.

5. The arithmetic mean ordinarily involves more computation than is required for the median. Also, the median may be readily obtained by graphical methods or through the use of very simple punched-card equipment.

Point 4 is particularly important if one is presenting highly skewed distributions to statistically uninformed audiences where the measure of location may be interpreted as a "typical" value of the variable. It is primarily for this reason that one usually speaks of median incomes (see Fig. 4.1), median number of school years completed, and so on.

The final criterion which may be applied to the choice of a measure of location is one which we are not yet in a position to discuss critically. It concerns the degree to which we expect a sample value of the mean or median (a *statistic*) to approximate the population mean or median (a *parameter*). Thus for the distribution of days of disability for nonoccupational accidents, we have just observed that the mean is 18.3 days and the median 8 days. The random sample of 10 accidents discussed in Sec. 4.2.1 had a mean of 27.9 days and a median of 7.5 days. The sample median is therefore closer to the population median than the sample mean is to the population mean. However, if another sample were drawn, these positions might well be reversed. We need to develop procedures which will allow us to distinguish between statistics on the basis of their expected behavior under sampling operations.

4.5. The Mode

There are situations in which we wish to have simple terminology to distinguish between distributions having a single maximum and those

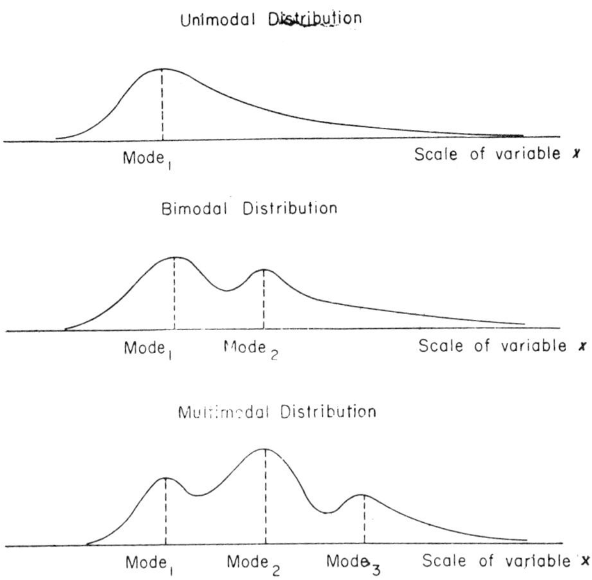

FIG. 4.10. Hypothetical distributions having different numbers of maxima. (Continuous variables and infinite populations.)

having two or more maxima. Figure 4.10 pictures three distributions of a continuous variable in an infinite population which will serve as examples of such distributions.

The first distribution in Fig. 4.10 has a single maximum. The value of the variable at which this maximum occurs is called the *mode* of the distribution, and this type of distribution is referred to as a *unimodal* distribution. The second distribution of Fig. 4.10 has two maxima and two modes and is called *bimodal.* Notice that the maxima in the second distribution are relative to the behavior of the distribution in a neighborhood about the value of the variable at which they occur. In other words, modes are defined at all *relative* maxima and not just at the *absolute* maximum of a distribution. The third distribution has three relative maxima and would therefore be called *multimodal.*

The most direct way of thinking about how a multimodal distribution might arise is in terms of the mixing together of two populations, each population having a unimodal distribution but with the modes occurring at different values of the variable. Thus if we mixed the Form A income distribution of Fig. 4.1 (farm laborers and foremen) with the Form C distribution (professional, technical, and kindred workers), we would obtain a bimodal distribution.

One can consider the possibility of using the mode of a distribution as a measure of the distribution's location. However, even the brief discussion which we have just given emphasizes two major shortcomings of the mode. These are (1) a distribution may have more than one mode and we would have the problem of choosing arbitrarily from among them; and (2) a distribution may not have a mode. The second point is illustrated by the rectangular distribution (see Fig. 4.3). In addition, many other difficulties are encountered in dealing with the properties of modes as determined from samples. Therefore, we shall, in dealing with distributions of quantitative variables, use the mode only in connection with describing the general shape of a distribution and not as a measure of location. It should be noted that the mean, median, and mode of a perfectly symmetric distribution with a central maximum do coincide (see Form A of Fig. 3.13).

EXERCISES

4.1. The arithmetic mean, median, and mode are sometimes called measures of "central tendency." What does the phrase "central tendency" mean to you? Cite distributions (from among those presented in Chaps. 3 and 4) for which you think this terminology is appropriate. Also cite distributions for which it would seem inappropriate.

4.2. The following table shows the distribution of the variable "miles to first major motor failure" for 189 buses operated by a large city bus company.

DATA FOR EXERCISE 4.2

Distance traveled to first major motor failure (thousands of miles)	*Number of failures*
0 up to 20	6
20 up to 40	11
40 up to 60	16
60 up to 80	25
80 up to 100	34
100 up to 120	46
120 up to 140	33
140 up to 160	16
160 up to 180	2
Total*	189

* We have here omitted two failures classified in the open-ended interval "180 and up."

SOURCE: D. J. Davis, "An Analysis of Some Failure Data," *Jour. Amer. Stat. Assoc.*, **47**:113–150 (1952).

a. Each failure falling in one of the above intervals was arbitrarily assigned the value of the mid-point of the interval, and a random sample of 15 failures was selected from the distribution (see Sec. 4.2.1). The following values were obtained: 90, 50, 150, 110, 90, 90, 110, 90, 50, 110, 90, 70, 50, 70, 150. Determine the arithmetic mean and median of this sample.

b. Compute the arithmetic mean and median for the entire distribution and compare your results with those obtained in *a.*

c. Plot on the same diagram (1) the cumulative-percentage polygon for the distribution as a whole, (2) the cumulative percentage distribution for the sample (use step-function form of Fig. 3.5), and (3) the smooth curve that you think might result for an indefinitely large number of the same type of buses operating under similar conditions.

4.3. Compute the arithmetic means of the three distributions given in Fig. 4.4. Note that the values of these means are indicated on the figure.

4.4. Determine the median of the distribution given in Exercise 3.3. Would the arithmetic mean of this distribution be larger or smaller than the median? Why?

4.5. One of the factors which is often studied in assessing the morale of a group of workers or in determining where to locate a plant is the length of time which workers must spend in traveling to and from work. One such study, made in a New England shipyard during World War II, reported the following distribution:

Traveling time to and from work, hr	*Per cent of workers*
Under 1	54.4
1 up to 2	30.9
2 up to 3	5.9
3 up to 4	2.9
4 up to 6	5.9
Total	100.0

Compute the arithmetic mean of this distribution.

4.6. Suppose that a book has 10 chapters and that the average number of pages per chapter is 40.

a. What is the total number of pages in this book?

b. Can you think of practical situations where one would wish to estimate the sum of values of a variable for all elements in a population rather than the mean per element?

4.7. Five individuals make the following errors in reporting their hourly earnings:

Reported earnings	True earnings	Error in reported earnings
\$2.06	\$2.11	−\$.05
2.40	2.33	+ .07
2.20	2.09	+ .11
1.92	2.17	− .25
2.12	2.35	− .23

Check numerically that the average reported earnings is equal to the average true earnings plus the average error (see Sec. 4.2.4).

4.8. Suppose one is trying to estimate the average age in a specified population of individuals. Each individual is asked to give his age, and the computation proceeds as though each gave his correct age. However, completely unknown to the investigator, one-quarter of the individuals underestimate their ages by 1 year, one-half give their correct ages, and one-quarter overestimate their ages by 2 years. Determine the numerical relationship between the true average age and the average age which the investigator obtains.

OTHER READING

Yule, G. Udny, and M. G. Kendall: *An Introduction to the Theory of Statistics*, 14th ed., Hafner Publishing Company, New York, 1950. Chapters 4 and 5 provide a thorough and reliable discussion of distributions and measures of location.

CHAPTER 5

THE SPREAD OF A DISTRIBUTION

5.1. Introduction

Distributions may differ not only in location but also in the extent to which they *spread* out along the scale of the variable. Thus we see from Fig. 4.1 that the income distribution for "professional, technical, and kindred workers" spreads out more on the income scale than does the distribution for "farm laborers and foremen," or conversely, the distribution for "farm laborers . . . " is more concentrated than is the distribution for "professional . . . workers." Other synonyms for spread are obtained if we think about the individual observations making up a distribution. A distribution is spread out if there is great variability or heterogeneity among the elements with respect to their associated values of the variable. Similarly, a concentrated distribution implies homogeneity among the elements of the sample or population.

This chapter will be devoted to a discussion of several measures that are commonly used to quantify the characteristic spread. No one of these measures is "ideal" for all purposes, and the presentation will therefore be arbitrary at times. This is not unlike the situation that existed for measures of location.

Some of the arbitrariness required in an introductory treatment of spread arises because the defined measures are used for many different purposes. Although we cannot consider these purposes in detail in this chapter, some illustrations are:

1. The spread of a distribution may be of direct interest. For example, consider the distribution of weights for a population of normal, healthy males of stated age and height. In specifying normal weight, it is as important to know the spread of this distribution around a measure of location as it is to know the location. Otherwise it is impossible to assign even crude meanings to the terms "overweight" and "underweight."

2. The spread of a distribution may be of interest only in so far as it aids in gauging the precision with which a sample mean estimates the corresponding population mean. For example, suppose we wish to estimate the average annual income of all families living in a community. If every family has the same income, the distribution of incomes will have

a spread of zero for any reasonable measure and a sample of one family will provide a perfect estimate of the average income. On the other hand, if incomes are spread between $0 and $25,000, a sample of considerable size may be needed to provide an "adequate" estimate of the average family income for this community. The spread of the population distribution is one of the major determinants of the precision of the sample mean.

5.2. The Variance and Standard Deviation

5.2.1. *Samples and Finite Populations.* Table 5.1 presents two comparable frequency distributions which may be used to introduce the variance and standard deviation. These are distributions of straight-time hourly earnings for machine-tool operators and tool-and-die makers

TABLE 5.1. DISTRIBUTIONS OF STRAIGHT-TIME HOURLY EARNINGS FOR MACHINE-TOOL OPERATORS AND TOOL-AND-DIE MAKERS
(St. Louis, Mo., January, 1954)

Straight-time hourly earnings, dollars	Machine-tool operators		Tool-and-die makers	
	Number*	Per cent	Number*	Per cent
1.50 up to 1.60	2	.2		
1.60 up to 1.70	12	1.2		
1.70 up to 1.80	40	4.0		
1.80 up to 1.90	119	11.7		
1.90 up to 2.00	120	11.9		
2.00 up to 2.10	129	12.7		
2.10 up to 2.20	163	16.1	52	5.0
2.20 up to 2.30	99	9.8	34	3.3
2.30 up to 2.40	318	31.4	71	6.9
2.40 up to 2.50	8	.8	206	19.9
2.50 up to 2.60	2	.2	629	60.9
2.60 up to 2.70			41	4.0
Total..............	1,012	100.0	1,033	100.0

* These are *estimated* population numbers. See text.

SOURCE: U.S. Department of Labor, Bureau of Labor Statistics, *Occupational Wage Survey of St. Louis, Mo.*, table A-3, January, 1954.

in the St. Louis, Mo., metropolitan area in January, 1954. The numbers of workers falling in the class intervals are estimates of the corresponding population numbers, but we shall treat them as though they were obtained from complete enumerations of these populations.

Data similar to these are collected for many different areas, and from time to time for the same area, by the Bureau of Labor Statistics. Although not important for present purposes, it should be recognized that such undertakings involve many problems of definition and sample selection. Great care must be used in defining the occupations of machine-tool operator and tool-and-die maker, and any other occupations for which data are to be obtained, in order to have comparable data for different establishments and areas. Also, it is necessary to obtain a sample of individual workers by first choosing a sample of establishments which hire workers in a given occupation and then collecting data on individual workers from these establishments. The data of Table 5.1 refer to individuals working in manufacturing establishments employing more than 100 workers at the time of the survey.

A discussion of all of the foregoing points will be found in the quoted source for Table 5.1 and in a paper by Cohen [1]. The general importance of these studies is emphasized in the following quotation from Cohen:

> The findings of all of these studies are used in wage determination through collective bargaining or employer personnel action. They are also used for comparison of wage levels in various parts of the country, and by public agencies in making wage determinations for employees who are paid on the basis of prevailing rates. In addition, they provide necessary information for formulation of public policy on wages, as in minimum wage legislation, and for the analysis of trends in economic developments.

One final observation may be made concerning these distributions before discussing their spreads. They are not quite as "regular" as were the distributions previously treated. Thus the distribution for machine-tool operators has two modal class intervals, "2.10 up to 2.20" and "2.30 up to 2.40." The original report used mainly 5-cent class intervals, and the resulting distributions showed even more irregularities. These irregularities occur because of a group effect in wage determination. All workers in a given establishment or union may be subject to the same uniform wage scale, although there may of course still be variations among individual workers within a group. As a result of this, we shall have a "mixing" effect such as was referred to in Sec. 4.5. This in no way means that it is inappropriate to examine the distribution of wages for all workers, although for some purposes it would be important to study and compare the distributions for individual unions and establishments.

In order to simplify the opening treatment of spread, a random sample of 10 hourly earnings was drawn from each of these distributions (see Sec. 4.2.1). The following results were obtained:

Machine-tool operators:	2.07, 1.72, 2.13, 2.36, 2.35,
	2.39, 2.14, 2.28, 1.71, 1.75
Tool-and-die makers:	2.51, 2.58, 2.57, 2.57, 2.12,
	2.53, 2.57, 2.57, 2.31, 2.57

The arithmetic mean $\bar{x}_M$ of the 10 hourly earnings for machine-tool operators is \$2.09, and $\bar{x}_T$ of the 10 hourly earnings for tool-and-die makers is \$2.49. Using the entire distributions of Table 5.1 we obtain $\mu_M = \$2.13$, and $\mu_T = \$2.49$. (The procedure used in calculating the mean of a class-interval distribution has been discussed in conjunction with Table 4.3.)

Let us now look at the two random samples of hourly earnings and attempt to define a quantity that will measure the extent to which these earnings are spread out (or differ among themselves). First of all, we must recognize that the actual location of a distribution should not be allowed to influence the measure of spread obtained. That is, if we increase the hourly earnings of each machine-tool operator by 10 cents, the sample distribution is shifted to the right by 10 cents, the mean changes from \$2.09 to \$2.19, and yet the spread of the distribution is unchanged. It is therefore appropriate that we measure spread about a measure of location, preferably the mean of the distribution.

The spreads of the two sample distributions about their respective means can be readily grasped from Fig. 5.1. Values of the variable are shown on the two vertical scales, and the points represent the separate observations, each set of 10 observations being numbered from smallest to largest. The horizontal lines are drawn at the arithmetic means of the two distributions. The vertical lines represent the deviations of the individual observations in each distribution from the mean. Those lines extending below the mean represent negative deviations, and those extending above the mean represent positive deviations. It is clear that the deviations are generally larger for the first distribution than for the second and that an appropriate summary of each set of deviations into a single quantity will provide a measure of spread.

The 10 deviations for each of the two samples are presented in Table 5.2. As in the opening discussion of the arithmetic mean (Sec. 4.2.1), x_i is the value of the variable for the ith sample element and $(x_i - \bar{x})$ is its deviation from the appropriate mean. Thus for the machine-tool operator sample, if we take $i = 2$, then $x_2 = 1.72$ and

$$x_2 - \bar{x}_M = 1.72 - 2.09 = -.37$$

The first obvious approach in summarizing the deviations of Table 5.2 is to obtain their average. This, however, is fruitless since it has already been demonstrated that the sum of algebraic deviations about the sample mean is always equal to zero. Nevertheless, we wish the magnitudes of these deviations to determine our measure of spread. There are two ways out of this dilemma. The first of these is simply to ignore all minus signs and treat every deviation as though it were positive, and the second is to square each deviation and thereby eliminate the minus signs, a negative number multiplied by itself giving a positive product. At this point we shall have to be quite arbitrary in this matter and simply state that the use of squared deviations leads to a more valuable measure of spread. We shall now define and compute this measure.

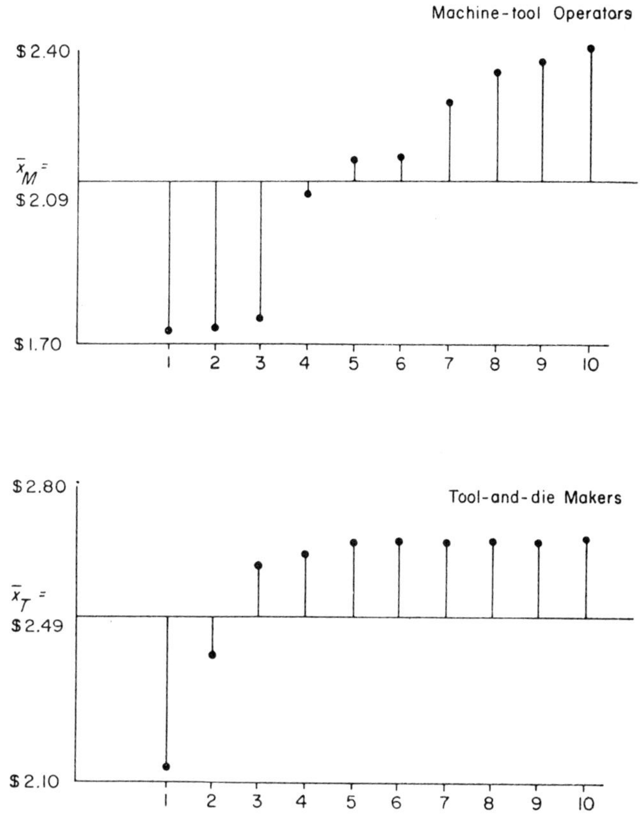

FIG. 5.1. The spread of hourly earnings around the sample mean.

Table 5.2 gives the value of each squared deviation in the two sample distributions. We take as a measure of spread the average of these squared deviations about the mean, using a divisor* of one less than the number of elements in the sample. For the machine-tool operator sample this measure of spread is .6680/9 = .0742 and for the tool-and-die maker sample it is .2114/9 = .0235. These quantities are called

* Obtaining the average in terms of $(n - 1)$ rather than in terms of n simplifies some of the expressions that are used in more advanced statistical work. Reference will be made to this fact in Chaps. 9 and 12.

sample variances and will henceforth be designated by the symbol s^2. Thus $s_M^2 = .0742$ and $s_T^2 = .0235$. The value of s_M^2 is larger than that of s_T^2, and we therefore have a quantitative verification of the impression given in Fig. 5.1 that the hourly earnings of machine-tool operators are more variable from operator to operator than are the hourly earnings of tool-and-die makers. For the moment this is all we shall ask of a measure

TABLE 5.2. COMPUTATION OF THE SAMPLE VARIANCE FOR THE TWO RANDOM SAMPLES OF HOURLY EARNINGS

Machine-tool operators			Tool-and-die makers		
Hourly earnings, dollars x_i	Deviations from the mean $(x_i - \bar{x}_M)$	Squared deviations from the mean $(x_i - \bar{x}_M)^2$	Hourly earnings in dollars x_i	Deviations from the mean $(x_i - \bar{x}_T)$	Squared deviations from the mean $(x_i - \bar{x}_T)^2$
2.07	− .02	.0004	2.51	+ .02	.0004
1.72	− .37	.1369	2.58	+ .09	.0081
2.13	+ .04	.0016	2.57	+ .08	.0064
2.36	+ .27	.0729	2.57	+ .08	.0064
2.35	+ .26	.0676	2.12	− .37	.1369
2.39	+ .30	.0900	2.53	+ .04	.0016
2.14	+ .05	.0025	2.57	+ .08	.0064
2.28	+ .19	.0361	2.57	+ .08	.0064
1.71	− .38	.1444	2.31	− .18	.0324
1.75	− .34	.1156	2.57	+ .08	.0064
20.90	0.00	.6680	24.90	0.00	.2114

$$\bar{x}_M = 2.09 \qquad \bar{x}_T = 2.49$$

$$s_M^2 = \frac{\sum_{i=1}^{10} (x_i - \bar{x}_M)^2}{9} = \frac{.6680}{9} = .0742 \qquad s_T^2 = \frac{\sum_{i=1}^{10} (x_i - \bar{x}_T)^2}{9} = \frac{.2114}{9} = .0235$$

of spread. It should provide a quantitative basis for saying that one distribution is more or less spread out than another, or alternatively, for establishing an order relationship among comparable distributions with respect to spread.

The considerations of the foregoing example can be formalized as follows. If we have a sample of n observations, where x_1 is the value of the variable for the first sample element, x_2 is the value for the second, . . . , and x_n is the value for the last, then the sample variance s^2 is defined as

$$s^2 = \frac{(x_1 - \bar{x})^2 + (x_2 - \bar{x})^2 + \cdots + (x_n - \bar{x})^2}{n - 1}$$

$$= \frac{\sum_{i=1}^{n} (x_i - \bar{x})^2}{n - 1} \tag{5.1}$$

Similarly, if we have a finite population containing N elements, we define the population variance σ^2 (read sigma squared) to be

$$\sigma^2 = \frac{(x_1 - \mu)^2 + (x_2 - \mu)^2 + \cdots + (x_N - \mu)^2}{N - 1}$$

$$= \frac{\sum_{i=1}^{N} (x_i - \mu)^2}{N - 1} \tag{5.2}$$

Notice that we are continuing to observe the convention introduced in the last chapter that the italic letter s^2 is used to denote a sample value or statistic while the corresponding Greek letter σ^2 is used for the population value or parameter. After a few preliminaries, we shall compute σ_M^2 and σ_T^2 for the two distributions of Table 5.1 and compare these with the values obtained for s_M^2 and s_T^2.

Although s^2 and σ^2 can be used to establish an order relationship among distributions with respect to spread, their units of measurement are not convenient for many purposes. The original observations of Tables 5.1 and 5.2 were expressed in dollars. The mean hourly earnings and deviations from the means are also in dollars. However, when we square deviations to obtain the variance, we also square the units of measurement. Thus $s_M^2 = .0742$ squared dollars. This measure can be converted back to dollars by taking its positive square root. We therefore obtain $s_M = \sqrt{s_M^2} = \sqrt{.0742 \text{ squared dollars}} = \sqrt{.0742}$ dollars $= .27$ dollars. Similarly $s_T = \sqrt{.0235 \text{ squared dollars}} = .15$ dollars. The positive square root s of the sample variance is called the *sample standard deviation*, and the square root σ of the population variance is the *population standard deviation*. These standard deviations have the same units as the original measurements. Distributions are given the same ordering whether variances or standard deviations are used. Formulas (5.1) and (5.2) serve as definitions of the sample and population standard deviations if the square root is taken of both sides of each expression.

Formulas (5.1) and (5.2) are not convenient for computation, nor are they suited for generalizing to an infinite population. However, there are many different ways in which they can be written so that they will serve these other purposes. For example, algebraic manipulation leads to the following form of (5.1):

$$s^2 = \frac{1}{n-1}\left(\sum_{i=1}^{n} x_i^2 - \frac{\left(\sum_{i=1}^{n} x_i\right)^2}{n}\right) \tag{5.3}$$

The application of this formula to finding s_M^2 (see Table 5.2 for data) will illustrate its use. Thus,

$$\begin{aligned} s_M^2 &= \frac{1}{10-1}\left[(2.07^2 + 1.72^2 + \cdots + 1.75^2) - \frac{(20.90)^2}{10}\right] \\ &= \frac{1}{9}\left[(4.2849 + 2.9584 + \cdots + 3.0625) - \frac{436.81}{10}\right] \\ &= \frac{1}{9}(44.3490 - 43.6810) = \frac{1}{9}(.668) = .0742 \end{aligned}$$

This may appear to involve more computational work than does the use of squared deviations from the mean, but the entire operation can be performed very efficiently with an automatic desk calculating machine. Moreover, this computational procedure offers fewer possibilities for rounding and computational errors. Formula (5.3) can also refer to σ^2 if n is replaced by N.

Still further variations of (5.1) and (5.2) are obtained if we have a distribution which specifies the number or fraction of elements having each possible value of the variable. For example, if N_1 elements in a finite population have the first value of the variable x_1, each of these elements will contribute an amount $(x_1 - \mu)^2$ to the numerator of (5.2). Their total contribution to the sum of squared deviations will be $N_1(x_1 - \mu)^2$. Similarly, those having the second value will contribute an amount $N_2(x_2 - \mu)^2$, and so on. Therefore we have the relation

$$\sum_{i=1}^{N} (x_i - \mu)^2 = \sum_{i=1}^{k} N_i(x_i - \mu)^2$$

where x_i in the left-hand summation represents the value of the variable for the *ith population element* ($i = 1, 2, \ldots, N$) and x_i in the right-hand summation represents the *ith value of the variable* ($i = 1, 2, \ldots, k$). When the symbol x_i is given this latter meaning, then any of the following forms of σ^2 are equivalent to (5.2):

$$\sigma^2 = \frac{\sum_{i=1}^{k} N_i(x_i - \mu)^2}{N - 1}$$

$$= \frac{1}{N-1}\left(\sum_{i=1}^{k} N_i x_i^2 - \frac{\left(\sum_{i=1}^{k} N_i x_i\right)^2}{N}\right)$$

$$= \frac{N}{N-1}\left(\sum_{i=1}^{k} p_i(x_i - \mu)^2\right)$$

$$= \frac{N}{N-1}\left(\sum_{i=1}^{k} p_i x_i^2 - \mu^2\right) \tag{5.4}$$

The second form of (5.4) is most convenient for computation, while the last two forms are most convenient for extending the concept of variance

TABLE 5.3. COMPUTATION OF VARIANCE FOR DISTRIBUTION OF STRAIGHT-TIME HOURLY EARNINGS OF TOOL-AND-DIE MAKERS
(See Table 5.1.)

Straight-time hourly earnings, dollars	Mid-points of intervals x_i	Number of workers N_i	Computations for variance		
			x_i^2	$N_i x_i$	$N_i x_i^2$
2.10–2.20	2.15	52	4.6225	111.80	240.3700
2.20–2.30	2.25	34	5.0625	76.50	172.1250
2.30–2.40	2.35	71	5.5225	166.85	392.0975
2.40–2.50	2.45	206	6.0025	504.70	1,236.5150
2.50–2.60	2.55	629	6.5025	1,603.95	4,090.0725
2.60–2.70	2.65	41	7.0225	108.65	287.9225
Total...........	....	1,033	......	2,572.45	6,419.1025

$$\mu_T = \frac{\sum_{i=1}^{6} N_i x_i}{N} = \frac{2{,}572.45}{1{,}033} = 2.49$$

$$\sigma_T^2 = \frac{1}{N-1}\left(\sum_{i=1}^{6} N_i x_i^2 - \frac{\left(\sum_{i=1}^{6} N_i x_i\right)^2}{N}\right) = \frac{1}{1{,}032}\left[6{,}419.1025 - \frac{(2{,}572.45)^2}{1{,}033}\right]$$

$$= \frac{1}{1{,}032}(6{,}419.1025 - 6{,}406.0978) = \frac{13.0047}{1{,}032} = .0126$$

to an infinite population. The transition from the first form to the third form is carried out in the manner used to obtain formula (4.4).

We shall illustrate the use of the second form of (5.4) by computing the variance σ_T^2 of the distribution of straight-time hourly earnings of tool-and-die makers (Table 5.1). These computations are summarized in Table 5.3. Since we are dealing with a class-interval distribution, each hourly earning in an interval is assigned a value corresponding to the mid-point of that interval. That is, $x_1 = 2.15$, $x_2 = 2.25$, . . . , and $x_6 = 2.65$. The k of formula (5.4) is here equal to 6, since there are six class intervals. Every individual quantity used during the course of the computation is entered in Table 5.3. However, actual computation with an automatic calculating machine does not require that each of these be written down. One must write down the x_i, the N_i, and the x_i^2, but such a machine will perform the multiplication of N_i by x_i and N_i by x_i^2 and give the required sums of these quantities, $\sum_{i=1}^{k} N_i x_i$ and $\sum_{i=1}^{k} N_i x_i^2$, without requiring that each individual product be written down. The final result of these computations is that $\sigma_T^2 = .0126$ squared dollars. In a similar fashion we find that $\sigma_M^2 = .0409$ squared dollars.

We can now summarize the results of our computations for the distributions of Table 5.1 and for the random samples that were drawn from these distributions. These results are:

	Machine-tool operators	Tool-and-die makers
Sample:		
$\bar{x}$	2.09 dollars	2.49 dollars
s^2	.0742 squared dollars	.0235 squared dollars
s	.27 dollars	.15 dollars
Population:		
μ	2.13 dollars	2.49 dollars
σ^2	.0409 squared dollars	.0126 squared dollars
σ	.20 dollars	.11 dollars

The quantities entered under the heading "sample" are statistics, and these are estimates of the corresponding population quantities, parameters. It will be observed that both sets of quantities give the same relative picture of the two distributions, namely, that the tool-and-die makers have a higher wage level than do the machine-tool operators and that there is less variability among the earnings of tool-and-die makers than there is among the earnings of machine-tool operators. The sample mean for tool-and-die makers happens to be equal to the population mean, but if another random sample were drawn this would not necessarily be the case.

Formula (5.4) may also be used for the computation of sample variances if we make the following change of notation: N is replaced by n, N_i is replaced by n_i, μ is replaced by $\bar{x}$, and p_i is replaced by n_i/n.

5.2.2. *Infinite Populations and Discrete Variables.* The variance of the distribution of a discrete variable in an infinite population may be defined in the following manner. Consider the third form of (5.4) for the variance of a finite population. This could be carried over immediately to an infinite population, as was done for (4.4) in the case of the arithmetic mean, if it were not for the multiplying factor $N/(N - 1)$. However, this factor behaves in a very "regular" manner as the size of a finite population is increased. Thus $10/9 = 1.1111$, $100/99 = 1.0101$, $1{,}000/999 = 1.0010$, and $10{,}000/9{,}999 = 1.0001$. That is, the factor $N/(N - 1)$ approaches closer and closer to 1 as N increases. Since we can think of obtaining an infinite population from a finite population by letting N become indefinitely large, we take the factor $N/(N - 1)$ to be 1 and define the variance of a discrete variable in an infinite population to be

$$\begin{aligned} \sigma^2 &= \sum_{i=1}^{k} p_i(x_i - \mu)^2 \\ &= \sum_{i=1}^{k} p_i x_i^2 - \mu^2 \end{aligned} \tag{5.5}$$

Thus the first form of (5.5) says that σ^2 is equal to the average of squared deviations about the mean, while the second form says that it is equal to the average of the squared values of the variable less the square of the mean.

The Form B distribution of Fig. 3.13 will serve to illustrate the use of (5.5). This is the distribution of the variable "number of correct 'guesses' out of five" in an indefinitely large population of individuals, each of whom is given five "true"-"false" statements which must be checked in the absence of any knowledge about their content. The computations are given in Table 5.4 where σ^2 is found to be 1.25. The standard deviation σ is equal to $\sqrt{1.25} = 1.12$ correct guesses. Notice that this is a perfectly symmetric distribution and that the mean is therefore at the exact center of the distribution. Here again we interpret the variance and standard deviation simply as measures of spread with which to compare distributions.

The variance of an infinite population, being the average of squared deviations about the mean of the population, is frequently referred to as the expected value of the squared deviations and is written as $E[(x - \mu)^2]$. Another notation which we shall sometimes use for the variance is $V(x)$. Thus we have the notational equivalence:

$$\sigma^2 = E[(x - \mu)^2] = V(x)$$

5.2.3. *Infinite Populations and Continuous Variables.* The variance of a distribution of a continuous variable in an infinite population may be defined by using the limiting process that was applied in the case of the mean (see Sec. 4.2.3). Let us suppose we wish to define the variance of the Form A distribution of Fig. 4.2. Using the four-equal-interval division of Form B of this figure and applying formula (5.5) we may take

TABLE 5.4. COMPUTATION OF VARIANCE FOR HYPOTHETICAL DISTRIBUTION OF NUMBER OF CORRECT GUESSES OUT OF FIVE FOR A TWO-CHOICE GUESSING EXPERIMENT

Number of correct guesses x_i	Fraction of infinite population expected to obtain specified number of correct guesses p_i	Computations for variance		
		x_i^2	$p_i x_i$	$p_i x_i^2$
0	$1/32$	0	0	0
1	$5/32$	1	$5/32$	$5/32$
2	$10/32$	4	$20/32$	$40/32$
3	$10/32$	9	$30/32$	$90/32$
4	$5/32$	16	$20/32$	$80/32$
5	$1/32$	25	$5/32$	$25/32$
Total.....	1	...	$80/32$	$240/32$

$$\mu = \sum_{i=1}^{6} p_i x_i = 80/32 = 2.5$$

$$\sigma^2 = \sum_{i=1}^{6} p_i x_i^2 - \mu^2 = 240/32 - (2.5)^2 = 7.5 - 6.25 = 1.25$$

as an approximation to σ^2

$$\sigma^2_{\text{approx}} = (x_1 - \mu)^2 \frac{h(x_1)(b-a)}{4} + (x_2 - \mu)^2 \frac{h(x_2)(b-a)}{4} + (x_3 - \mu)^2 \frac{h(x_3)(b-a)}{4} + (x_4 - \mu)^2 \frac{h(x_4)(b-a)}{4}$$

$$= \sum_{i=1}^{4} (x_i - \mu)^2 \frac{h(x_i)(b-a)}{4}$$

It will be recalled that $h(x_i)$ represents the height of the curve at the point x_i and that $h(x_i)(b - a)/4$ is therefore the area of a rectangular cell centered around x_i.

If we use 16 intervals (as in Form D of Fig. 4.2), then we obtain a better approximation to σ^2, namely,

$$\sigma^2_{\text{approx}} = \sum_{i=1}^{16} (x_i - \mu)^2 \frac{h(x_i)(b - a)}{16}$$

We can think of continuing this subdivision process without end, and the variance σ^2 of the distribution is defined to be the result finally obtained.

This process may be applied to the rectangular distribution of birth times pictured in Fig. 4.3, although with considerably more difficulty than was encountered in obtaining the mean of this distribution. Using the four-interval subdivision of Fig. 4.3, where

$$\frac{h(x_i)(b - a)}{4} = \frac{(1/60)(60 - 0)}{4} = 1/4$$

for each x_i, we have

$$\sigma^2_{\text{approx}} = (7.5 - 30)^2(1/4) + (22.5 - 30)^2(1/4) + (37.5 - 30)^2(1/4) + (52.5 - 30)^2(1/4)$$
$$= \frac{60^2}{12}\left(1 - \frac{1}{4^2}\right)$$

The final form of this expression can be obtained by arithmetical manipulation which is unimportant for present purposes. It can be shown that if we divide the interval 0 to 60 into k equal divisions instead of the four used above, then

$$\sigma^2_{\text{approx}} = \frac{60^2}{12}\left(1 - \frac{1}{k^2}\right)$$

Now σ^2 was defined to be the result obtained if the subdivision process were continued without end. That is, we are supposed to let the number k of equal intervals increase indefinitely. The only part of σ^2_{approx} which depends on k is the factor $1/k^2$. As k assumes successively larger values, for example, 10, 100, 1,000, and so on, this quantity becomes successively smaller—.01, .0001, .000001, . . . —and approaches closer and closer to zero. Therefore $[1 - (1/k^2)]$ approaches closer and closer to 1, and we have for indefinitely large k

$$\sigma^2 = \frac{60^2}{12} = 300 \text{ min}^2$$
$$\sigma = 17.3 \text{ min}$$

The foregoing numerical argument for the rectangular distribution has been presented merely to illustrate that one can obtain the variance of a distribution of a continuous variable in an infinite population. It is not amenable to general application. The determination of the variances of most such distributions requires the use of more advanced mathematical methods than is expected of the readers of this book.

5.2.4. *The Effects of Simple Changes in a Distribution on the Variance and Standard Deviation.* In this chapter we are using the variance only as a quantity with which to order distributions with respect to spread. (Other uses of the variance will be given in Chap. 9.) Even with this restricted interpretation, there are a number of useful observations which can be made concerning the effect which certain simple changes in a distribution will have on the variance of the distribution.

In the preceding chapter we noted that adding a constant amount c to each observation in a distribution merely shifts the entire distribution by this same amount and that $\mu_{x+c} = \mu_x + c$, where μ_x is the mean of the original distribution and μ_{x+c} is the mean of the distribution after c has been added to each x. Since the bodily shift of a distribution does not change its shape and since the variance measures spread about the arithmetic mean, such a shift does not affect the variance of the distribution. That is, $\sigma^2_{x+c} = \sigma^2_x$. This result follows formally by noting that $(x_i + c) - (\mu + c)$, a deviation from the mean after the shift has been made, is equal to $(x_i - \mu)$, the corresponding deviation before the shift is made. If we put this result in terms of the hourly-wage distributions of Table 5.1, it says that giving each worker an increase or decrease of 10 cents an hour (or any other fixed amount) does not change the spread of the distribution, although it does of course change the location of the distribution.

Another type of change in a distribution is obtained if we consider the effect of multiplying each value of the variable by the same amount, say b. In the case of a wage distribution this is equivalent to giving each worker the same percentage increase in wages. For example, suppose all workers are given a 10 per cent increase in hourly earnings. Then a machine-tool operator earning \$2.07 per hour will receive \$2.07 + (.10)(\$2.07), or (1.10)(\$2.07) = \$2.277. Thus a new rate is obtained from an old rate by multiplying by $b = 1.10$. It is clear that this change will affect not only the location of a wage distribution but also its spread. This latter follows from the fact that a low rate will not be increased by as many dollars as will a high rate. A worker originally receiving \$1.71 will be given an increase of (.10)(\$1.71) = \$.171, while one originally receiving \$2.39 will be given an increase of (.10)(\$2.39) = \$.239.

The effects on the mean and variance of a finite population of multiplying each observation by a constant b may be evaluated from formulas (4.3) and (5.2). Thus

$$\begin{aligned}\mu_{bx} &= \frac{\sum_{i=1}^{N} bx_i}{N} = b\left(\frac{\sum_{i=1}^{N} x_i}{N}\right)\\ &= b\mu_x \qquad (5.6)\end{aligned}$$

That is, multiplying each observation in a distribution by a constant also multiplies the mean by this constant. Similarly,

$$\sigma_{bx}^2 = \frac{\sum_{i=1}^{N} (bx_i - b\mu_x)^2}{N-1} = \frac{\sum_{i=1}^{N} b^2(x_i - \mu_x)^2}{N-1}$$

$$= b^2 \left(\frac{\sum_{i=1}^{N} (x_i - \mu_x)^2}{N-1} \right) = b^2\sigma_x^2 \tag{5.7}$$

From (5.7) we immediately have $\sigma_{bx} = b\sigma_x$. That is, multiplying each observation in a distribution by a constant multiplies the variance by the square of the constant and the standard deviation by the constant itself.

Although (5.6) and (5.7) were obtained for a finite population, they also hold for samples and infinite populations and may be illustrated numerically with the data of Table 5.2. Thus if each hourly wage for machine-tool operators is multiplied by 1.10, the new mean wage is (1.10)(2.09) = 2.30 dollars, the new sample variance is

$$(1.10)^2(.0742) = .0898 \text{ squared dollars}$$

and the new sample standard deviation is (1.10)(.27) = .30 dollars.

Formulas (5.6) and (5.7) are very helpful when we wish to change the units of a variable. Many changes in units are made by multiplying by an appropriate conversion factor, and a few familiar illustrations are:

Original units of variable x_i	Multiplying factor b	Units of bx_i
Dollars.	100	cents
Cents.	1/100	dollars
Minutes.	1/60	hours
Hours.	60	minutes

The application of these considerations to the distribution of hourly earnings of machine-tool operators (Table 5.1) is:

Distribution parameter	Original computations		Conversion factor	New parameter
μ	2.13 dollars	×	100	= 213 cents
σ^2	.0409 (dollars)^2	×	$(100)^2$	= 409 (cents)^2
σ	.20 dollars	×	100	= 20 cents

Their application to the distribution of occurrence times of births (Sec. 5.2.3) is:

μ	30 min	$\times$	$(1/60)$	$= .5$ hr
σ^2	300 (min)^2	$\times$	$(1/60)^2$	$= .0833$ (hr)^2
σ	17.3 min	$\times$	$(1/60)$	$= .29$ hr

This ability to change easily the units of the mean, variance, and standard deviation can also be helpful computationally. Thus computations for an income distribution may be made in terms of "thousands of dollars," thereby doing away with many zeros. The factor 1,000 may then be multiplied back in at the end of the computation—$(1{,}000)^2$ in the case of the variance—if the results are to be expressed in dollars.

5.3. Measures of Spread Related to the Percentiles of a Distribution

5.3.1. *The Range.* In Sec. 4.3 we discussed the percentiles of a distribution as measures of location. It is also possible to use these percentiles in defining a number of measures of spread which are easily computed and readily understood. These measures are quite useful for describing in simple terms the spread of a distribution but, with some few exceptions, are not as valuable for most work in statistics as are the variance and standard deviation.

The *range* is one measure of spread belonging to this group. It is defined to be the largest value of the variable minus the smallest value. For the random sample of the hourly earnings of 10 machine-tool operators from the distribution of Table 5.1 (2.07, 1.72, 2.13, 2.36, 2.35, 2.39, 2.14, 2.28, 1.71, 1.75) the sample range is equal to $2.39 - 1.71 = \$.68$. The corresponding population range is $2.59 - 1.50 = \$1.09$. The sample and population ranges of the distribution of earnings of tool-and-die makers (see Tables 5.1 and 5.2) are $2.58 - 2.12 = \$.46$ and $2.69 - 2.10 = \$.59$, respectively.

Although the range orders these wage distributions with respect to spread in the same manner as did the variance and standard deviation, it has a number of shortcomings for general use. Two of these are reasonably apparent from general considerations. They are:

1. The value of the sample range will depend on the size of the sample. It will tend to be "small" for "small" samples and to increase with increasing sample size. This is because there are apt to be few observations (or a small fraction of the population) in the tails of a distribution and these "extreme" observations are not likely to be drawn into "small" samples. The range is a valuable measure of spread in industrial-quality-control applications of statistics where samples of approximately four or five observations can be effectively and consistently used. Under these circumstances—"small" samples of constant size—the foregoing shortcoming is not important.

2. The range depends only upon the "extremes" in a distribution and does not distinguish between distributions with respect to the manner in which observations are spread out within this range. As an illustration, consider distributions D and E in Fig. 4.5. These distributions have identical ranges, $6 - (-10) = 16$ for D and $8 - (-8) = 16$ for E, and yet intuitively we feel that D is more spread than E. The sample variance does show this difference since $s_D^2 = 32.5$ and $s_E^2 = 9.9$.

The other shortcomings of the range are of a more theoretical nature than the foregoing. As one example we may cite the fact that there are some very useful theoretical distributions for infinite populations for which the range cannot be defined. Thus one might consider the following experiment. A trial of the experiment consists of repeatedly tossing a coin until the *first* head appears, and the "number of tosses required" to obtain this first head is a variable associated with this trial. The possible values of the variable are 1, 2, 3, . . . , and it is impossible to state the largest value of the variable since a trial might conceivably require an indefinitely large number of tosses. Since the largest value of the variable is indefinitely large and therefore unknown, it is impossible to define the population range for the distribution of the variable.

5.3.2. *The Interquartile Range.* One of the liabilities of the range, namely, its failure to take cognizance of what goes on within a distribution, may be overcome to some extent by considering the distance between designated percentiles of a distribution. The most frequently used measure of this type is the *interquartile range*, defined to be the upper quartile (75th percentile) of the distribution minus the lower quartile (25th percentile) of the distribution. The interquartile range is therefore the range within which the *middle* 50 per cent of observations are concentrated. For the distribution of "duration of work stoppages ending in 1950"—see Table 4.5 and Fig. 4.9—we obtained an upper quartile of 21.5 days and a lower quartile of 3.1 days. Hence the interquartile range of this distribution is $21.5 - 3.1 = 18.4$ days.

The interquartile range is an easily understood measure of spread. Consequently it is of some value for comparing the spread of distributions for statistically uninformed audiences. It is, however, subject to the same features of arbitrariness that characterize the median and the other percentiles of a distribution when one is dealing with small samples or discrete variables (see Secs. 4.3 and 4.4). Moreover, the interquartile range cannot be easily adapted to many of the problems where a measure of spread is required.

Other measures of spread can also be defined from the percentiles. For example, one might employ the 90th percentile minus the 10th percentile or the 80th percentile minus the 20th percentile. There are situations—large samples from a particular form of distribution for a continuous variable—where the use of such measures is advantageous for computational reasons, but we shall not consider such problems in this book.

EXERCISES

5.1. There are many approaches to statistical and mathematical analyses of literary style. Some of the differences in approach arise from differences in the fundamental element of analysis (e.g., the word, the sentence, etc.). The following two distributions refer to the population of *words* in each of two novels, and the variable is the number of syllables in a word. The first book is in English, and the second in German.

Number of syllables in a word	Huxley, *Brave New World*, per cent of words	Mann, *Zauberberg* (*Magic Mountain*), per cent of words
1	69.82	52.91
2	19.64	29.00
3	6.96	11.62
4	2.43	4.38
5	.....	1.37
6	.....	0.37
Total.......	98.85	99.65

SOURCE: Wilhelm Fucks, "On Mathematical Analysis of Style," *Biometrika*, **39** (pts. 1, 2):122–129 (1952).

It will be observed that the per cents do not add to exactly 100. The source presumably omitted a few extreme observations, and this fact should be ignored in answering the following questions.

a. A random sample of 10 words was drawn from each of these two distributions, and the number of syllables in these words were:

Brave New World: 1, 1, 2, 1, 1, 2, 2, 4, 1, 1
Zauberberg: 2, 2, 1, 3, 2, 1, 1, 1, 2, 3

For each of these two sample distributions compute (1) the mean $\bar{x}$, (2) the variance and standard deviation, s^2 and s, and (3) the range.

b. For each of the two population distributions compute (1) the mean μ, (2) the variance and standard deviation, σ^2 and σ, and (3) the range. Note that in computing σ^2 you will be using the fourth form of (5.4). Assume that the factor $N/(N - 1)$ is close to 1 and can therefore be ignored.

c. What per cent of the words in *Brave New World* have numbers of syllables which are more than one standard deviation larger than the mean (i.e., larger than $\mu + 1\sigma$)? What per cent have numbers of syllables more than two standard deviations larger than the mean (i.e., larger than $\mu + 2\sigma$)? What are the corresponding per cents for *Zauberberg*? Draw a sketch of the two distributions and indicate these fractions on the sketch.

5.2. The following table shows the distribution of ages at which each of 241 famous authors published their "one best book."

Age at which "one best book" was published	*Number of authors*
20 up to 30	20
30 up to 40	73
40 up to 50	80
50 up to 60	44
60 up to 70	22
70 up to 80	2
Total	241

SOURCE: Harvey C. Lehman, *Age and Achievement*, chap. 15, Princeton University Press, Princeton, N.J., 1953.

Determine the median, mean, variance, and standard deviation of this distribution.

5.3. The following table gives the distribution of the variable "distance moved" for 12,292 changes in place of residence made within Cleveland, Ohio, during the period 1933 to 1935.

Distance moved (thousands of feet)	Number of changes of residence	Per cent of changes
0 up to 3	5,585	45.4
3 up to 6	2,471	20.1
6 up to 9	1,313	10.7
9 up to 15	1,168	9.5
15 up to 21	537	4.4
21 up to 33	612	5.0
33 up to 45	395	3.2
45 up to 63	186	1.5
63 up to 93	25	.2
Total..............	12,292	100.0

SOURCE: Samuel A. Stouffer, "Intervening Opportunities: A Theory Relating Mobility and Distance," *Amer. Sociol. Rev.*, **5**:845–867 (1940).

This distribution was used in testing the hypothesis that "the number of persons going a given distance is directly proportional to the number of opportunities at that distance and inversely proportional to the number of intervening opportunities."

a. Compute the mean, variance, and standard deviation of this distribution. Note that the class intervals are not of equal length.

b. Determine the interquartile range of the distribution.

c. Suppose that all "movements" entering into this distribution were expressed in miles rather than in thousands of feet. What would be the mean, variance, and standard deviation of this distribution?

5.4. Suppose you are purchasing an automatic packaging machine for a company and have to choose between machines *A* and *B*. The machine purchased is to be used in putting up 5-pound 1-ounce packages of flour (which will be labeled as containing 5 pounds). However, you do not expect any such machine to put exactly 5 pounds and 1 ounce in each package. Sometimes more will be put in, and sometimes less. You observe these machines in operation and find that, on the average, each puts 5 pounds and 1 ounce in a package. The distributions of package weights are symmetric around this mean for both machines. However, the distribution of package weights for machine *A* has an interquartile range of 1 ounce, while the distribution for machine *B* has an interquartile range of 2 ounces. What effect would this difference in spread have on your choice of a machine—ignoring differences in cost and in maintenance costs?

REFERENCE CITED

1. Cohen, Samuel E.: "Studies of Occupational Wages and Supplementary Benefits," *Monthly Labor Rev.*, **77**:292–297 (1954).

OTHER READING

Wilks, S. S.: *Elementary Statistical Analysis*, Princeton University Press, Princeton, N.J., 1948. Chapter 3 provides a precise discussion of the mean and variance.

Yule, G. Udny, and M. G. Kendall: *An Introduction to the Theory of Statistics*, 14th ed., Hafner Publishing Company, New York, 1950. Chapter 6 gives an extensive treatment of measures of spread (referred to as dispersion).

CHAPTER 6

SIMPLE RANDOM SAMPLING

6.1. The Problem of Sample Selection

We have repeatedly stressed that it is both desirable and necessary to use samples in statistical investigations. Furthermore, we have provided several examples in which it was possible to compare sample results with results for the entire population from which the samples were drawn. Thus we have used sample distributions to approximate population distributions, sample means and medians to approximate population means and medians, and sample variances and ranges to approximate population variances and ranges. These comparisons were artificial in that the extent to which the sample values deviated from the corresponding population values was known exactly. This will not be the situation in practice. Ordinarily, only the sample characteristics will be available. From these, and from our knowledge of the method by which the sample was selected, we must be able to make a quantitative evaluation of the sample "representativeness." This chapter will be devoted primarily to a basic selection procedure known as simple random sampling which enables us to make this quantitative evaluation. Before doing this, however, we shall examine briefly a few appealing procedures which lead to samples whose representativeness cannot be assessed quantitatively.

Any individual or organization will always be tempted to select the most readily accessible portion of a population for study. Thus an inspector of coal shipments would find it most convenient to sample a carload of coal by taking a sample from the top of the car; a university researcher would find it most convenient to conduct his experiments on members of his classes; a market research organization would find it most convenient to interview individuals living in the community where the home offices of the organization are located; or a doctor would find it most convenient to study the effects of a new drug by limiting his observations to individuals undergoing treatment in the hospital where he works.

Each of the foregoing may be regarded as an example of convenience selection. The difficulty of evaluating the representativeness of a sample selected by such a procedure may be illustrated by examining more closely the case of the doctor who tries his new treatment or drug only on the patients in one hospital. For the sake of simplicity, let us assume that the treatment either brings about a complete cure or has no noticeable effect. Consequently the doctor records the percentage of individuals who are cured. He is interested in saying how close this percentage is to that percentage which would be obtained by treating every individual having the disease at that particular time.

The difficulty associated with the use of such a sampling procedure arises from the fact that patients having a particular disease in a specific hospital usually cannot be regarded as representative of the entire population of individuals having the disease. The mere fact that they are in a specified hospital may mean that they have had prior treatment for the disease, that they are more seriously ill than those who are not in this hospital, or that they come from a higher social class than those not in this hospital. If any one of these factors, or of the many others that have not been mentioned, are in some way related to the effect of the treatment (e.g., the treatment works well on "mild" cases but is ineffective on "serious" cases), the results of the experiment may be suspect when they are extended to the entire population of individuals having the disease. And yet nothing in the selection procedure or in the sample results allows one to evaluate the effects of these possibilities in any precise manner. The sole responsibility for generalizing to the entire population rests on the intuition and judgment of the experimenter. If it were extremely important that one know the "cure rate" for the entire population, then the single-hospital study would have to be regarded as a pretest and the treatment would next be tried on a sample from the entire population. This is similar to the developments leading to the tests of the Salk vaccine for poliomyelitis during the spring and summer of 1954.

Another simple selection procedure, somewhat related to convenience selection, is that of judgment selection. It is natural for an "expert" in a particular field to reason as follows: "I have intimate knowledge concerning the characteristics of the population from which I desire a sample. I can also maintain objectivity in selecting a sample. Combining this objectivity and knowledge, I can, on the basis of judgment alone, choose a sample to represent the population." Putting this in its simplest terms, such a person would array all the population elements before him and would indicate by personal judgment whether or not any given population element was to be included in the sample.

The characteristics of a sample chosen in such a manner will clearly depend upon the following three factors: (1) the actual knowledge which the expert possesses of the over-all characteristics of the population, (2) the actual knowledge which he has of the characteristics of each element or group of elements, and (3) the stated or unstated criteria with which the expert combines 1 and 2 to decide whether or not any element (or group of elements) is to be included in the sample. Unfortunately, there is usually no way of assessing these points quantitatively. In other words, we may not quarrel with the fact that the expert knows his population, but there is no way of measuring how he uses this information in selecting the sample. Consequently we cannot specify the degree of representativeness of samples obtained solely by judgment.

These comments should not be interpreted as meaning that a judgment sample can never be used. In some instances, especially when the sample is an extremely small portion of the population, judgment selection is the only possibility. As a simple illustration, suppose that one wishes to make a community study in a United States city belonging to a specified size class. It will obviously be impossible to make precise statements about the entire population of such cities from the study of a single city. Consequently the city must be selected on a judgment basis, taking into account as many as possible of those considerations which will have an important influence on the study. The discussion accompanying Table 3.4 treats an actual example of this type of selection.

Convenience and judgment selection suffer from the fault that the decision whether or not a population element is to be included in the sample depends upon the element's characteristics and upon the subjective judgment of an individual. Neither of these effects is amenable to measurement, and there is therefore no way of assessing quantitatively the representativeness of the resulting samples. Furthermore, one can cite situations in which samples chosen according to these and related procedures have given definitely misleading results. Possibly the most notorious of these is the *Literary Digest* poll. The *Digest* conducted its poll by mailing out to some 10 million persons a ballot accompanied by a subscription form. More than 2 million ballots were returned in each poll. In spite of the great size of this sample and its rather close prediction of the Presidential election of 1932, it produced a tremendous error in the 1936 election. Apart from the minor party candidates, the ballots showed approximately 40 per cent of the sample favoring Roosevelt and 60 per cent favoring Landon. The results of the actual election were almost precisely the reverse, 60 per cent for Roosevelt and 40 per cent for Landon. Although there were factors other than sampling which contributed to this error, e.g., last-minute shifts of opinion, the major portion of it can undoubtedly be traced to sampling. The original preparation of the mailing lists (subscription lists, telephone directories, car registrations, and the like) and the differential response rate (possibly the "better-educated" people were more apt to return a ballot than were the "less well-educated" people) among the various groups in the sample could account for a large portion of the error.

The foregoing considerations immediately lead one to ask whether there are selection procedures which do not depend upon nonmeasurable factors like subjective judgment. Such selection procedures do exist, and in this chapter we shall consider the most basic of these, namely, simple random sampling.

6.2. Definition of Simple Random Sampling: The Model

Let us suppose that we wish to draw a sample of size 3 from a finite population containing the six elements A, B, C, D, E, and F. There are 20 possible samples, any one of which might be chosen. These 20 samples are:

$$ABC, ABD, ABE, ABF, ACD, ACE, ACF, ADE, ADF, AEF,$$
$$BCD, BCE, BCF, BDE, BDF, BEF, CDE, CDF, CEF, DEF$$

A sampling procedure must specify how to select one of these samples for study.

Simple random sampling is by definition that procedure which gives each of these 20 samples an equal chance of being chosen. The phrase "equal chance" will for present purposes be interpreted as follows. Before a sample is drawn, we have no way of predicting exactly which one of the 20 it will be. However, if we think of drawing an indefinitely large number of samples—where any one of the 20 may be chosen on any of the drawings—then our definition of simple random sampling requires that sample ABC occur on $1/20$ of all drawings, ABD on $1/20$ of all drawings, . . . , and DEF on $1/20$ of all drawings.

Notice particularly that the foregoing definition of simple random sampling does not tell how actually to select a sample. Rather, it specifies the characteristics of a *model* to which the mechanical selection process is supposed to conform. (Mechanical procedures for the actual selection of simple random samples will be discussed in the following section.) Moreover, this model refers to what would happen if an infinite number of samples were chosen instead of the single sample to be used for a particular investigation. This may seem like an unnecessary complication, but we shall see later that a particular sample result must be evaluated with reference to this infinite set of drawings. As a final point, we see that no extraneous factors, such as were operative in convenience and judgment selection, are allowed to influence the choice of a sample. Every population element receives the same treatment, irrespective of its own characteristics or of the judgment of an expert.

One can think of simple random sampling as an experiment. A single trial of this experiment consists of drawing a sample, and the trial is categorized according to the actual sample (one of the 20) drawn. Then we can think of performing an infinite number of trials of the experiment.

The simple-random-sampling model says that the result (sample drawn) on any one trial should have no influence on the result of any other trial and that each possible sample should occur an equal fraction of the time in the infinite set of trials. Thus we have, by definition, the distribution given by the first and third columns of Table 6.1 for the variable "sample drawn" in an infinite set of drawings. This distribution is similar to those already discussed (for infinite populations) in the preceding chapters.

TABLE 6.1. DISTRIBUTION OF THE SAMPLE MEAN IN AN INFINITE SET OF SIMPLE RANDOM SAMPLES
(Simple random samples of size 3 drawn from a finite population of six elements)

Sample drawn	$\bar{x}$	Fraction of samples	Sampling distribution of $\bar{x}$: $\bar{x}$	Sampling distribution of $\bar{x}$: Fraction of samples
ABC	12/3	1/20		
ABD	13/3	1/20		
ABE	14/3	1/20		
ABF	15/3	1/20		
ACD	14/3	1/20	12/3	1/20
ACE	15/3	1/20	13/3	1/20
ACF	16/3	1/20	14/3	2/20
ADE	16/3	1/20	15/3	3/20
ADF	17/3	1/20	16/3	3/20
AEF	18/3	1/20	17/3	3/20
BCD	15/3	1/20	18/3	3/20
BCE	16/3	1/20	19/3	2/20
BCF	17/3	1/20	20/3	1/20
BDE	17/3	1/20	21/3	1/20
BDF	18/3	1/20		
BEF	19/3	1/20		
CDE	18/3	1/20		
CDF	19/3	1/20		
CEF	20/3	1/20		
DEF	21/3	1/20		
Total.		1		1

The distribution given by the first and third columns of Table 6.1 may be immediately used to obtain distributions (in an infinite set of drawings) for other variables that can be associated with any given random sample. For example, let us suppose that the elements *A*, *B*, *C*, *D*, *E*, and *F* are actually six families and that their respective incomes for a specified year are (in thousands of dollars) 3, 4, 5, 6, 7, and 8. The mean income μ for this finite population of six families is 5.5 (thousands of dollars). There is also a sample $\bar{x}$ for each of the 20 possible samples. Thus if one should happen to draw sample ABC, $\bar{x}$ is equal to

$$\frac{(3 + 4 + 5)}{3} = \frac{12}{3}$$

or if one draws CDE, $\bar{x}$ is equal to $(5 + 6 + 7)/3 = 18/3$. The correspondence between the sample drawn and the sample mean income is given in the first two columns of Table 6.1. Thus we see that sample ABC occurs on $1/20$ of all drawings. Therefore its mean, $\bar{x} = 12/3$, also occurs on $1/20$ of all drawings.

If we are interested in the sample mean as an approximation to the population mean, the individual elements that make up a sample are of no concern. The mean of the sample is the only quantity of interest. Therefore the left side of Table 6.1 has been "collapsed" into the right side. For example, sample ABE has a mean of 14/3 and occurs on $1/20$ of all drawings; sample ACD also has a mean of 14/3 and occurs on *another* $1/20$ of drawings. No other possible samples have means of 14/3. As a consequence, a sample mean of 14/3 will occur on $2/20$ of all drawings, those on which either ABE or ACD occurs. Thus the right-hand side represents the distribution of the sample mean $\bar{x}$ in an infinite set of drawings of simple random samples. This distribution is called the *sampling distribution* of $\bar{x}$.

It should be observed that we can compute the mean and variance of this sampling distribution just as we have done in the preceding chapters where we were dealing with infinite populations. Designating the mean of this distribution as $\mu_{\bar{x}}$, or $E(\bar{x})$, we have through the use of formula (4.6)

$$\mu_{\bar{x}} = E(\bar{x}) = (1/20)(12/3) + (1/20)(13/3) + (2/20)(14/3) + \cdots + (1/20)(21/3) = 330/60 = 5.5$$

A minor notational point deserves mention in connection with this computation. The variable is now the sample mean which can assume 10 possible values: $\bar{x}_1 = 12/3$, $\bar{x}_2 = 13/3$, . . . , $\bar{x}_{10} = 21/3$. Therefore in applying (4.6), we have the notational correspondence $x_1 = \bar{x}_1$, $x_2 = \bar{x}_2$, . . . , $x_{10} = \bar{x}_{10}$.

In a manner similar to that used for the mean, the variance of the sampling distribution of $\bar{x}$, designated as $\sigma_{\bar{x}}^2$ or $V(\bar{x})$, may be obtained from an application of formula (5.5):

$$\begin{aligned}\sigma_{\bar{x}}^2 = V(\bar{x}) &= [(1/20)(12/3)^2 + (1/20)(13/3)^2 + (2/20)(14/3)^2 + \cdots + (1/20)(21/3)^2] - \mu_{\bar{x}}^2 \\ &= (5{,}550/180) - (5.5)^2 = 7/12 = .583\end{aligned}$$

The distinguishing characteristic of *simple* random sampling is that *each possible sample* of size n from a population of N elements has an equal chance of being selected. There are circumstances under which it is desirable to maintain the "equal-chance" feature of simple random sampling but to exclude some of the possible samples from consideration. Some of the ways in which this can be done will be discussed in detail in Chap. 10. The following example is presented to illustrate this point and thereby to emphasize the need for the adjective *simple.*

Let us suppose that the homes of six families (A, B, C, D, E, and F) are spread out on a very long city street and that an interviewer must walk to the homes of the three families drawn into the sample. It will be assumed that family A lives at one end of the street, B next, C next, and so on. Under these circumstances, it would require less walking to interview three "adjacent" families than it would to interview a family at one end of the street, a second family in the middle, and a third family at the other end of the street. Therefore we might decide to choose one of the two samples ABC and DEF (i.e., samples which contain "adjacent" families) and to exclude from consideration the other 18 possible samples. Furthermore, we might decide to choose one of these two samples in such a manner that each has an equal chance of being selected. This selection procedure has the "equal chance" characteristic of simple random sampling, but it excludes certain possible samples from consideration and therefore is not the same as simple random sampling.

The model for the foregoing selection procedure states that if an indefinitely large number of samples is drawn, then sample ABC will occur on ½ of the drawings and sample DEF will occur on ½ of the drawings. Therefore the means of samples ABC and DEF will occur on corresponding fractions of the drawings. The sampling distribution of $\bar{x}$ for this selection procedure is given in Table 6.2. It will be observed that the mean

TABLE 6.2. DISTRIBUTION OF THE SAMPLE MEAN IN AN INFINITE SET OF DRAWINGS WHERE SAMPLES ABC AND DEF OCCUR WITH EQUAL FRACTIONS

Sample drawn	$\bar{x}$	Fraction of samples
ABC	12/3	½
DEF	21/3	½

$$\mu_{\bar{x}} = E(\bar{x}) = (\tfrac{1}{2})(12/3) + (\tfrac{1}{2})(21/3) = 5.5$$
$$\sigma_{\bar{x}}^2 = V(\bar{x}) = (\tfrac{1}{2})(12/3)^2 + (\tfrac{1}{2})(21/3)^2 - (5.5)^2$$
$$= \tfrac{9}{4} = 2.25$$

of this sampling distribution is the same as the mean for the simple-random-sampling distribution of Table 6.1. However, the variance of this distribution is almost four times larger than the variance of the distribution of Table 6.1. If our only concern is the precision with which a sample mean estimates a population mean, then this difference in spread leads us to prefer simple random sampling over the modified procedure. On the other hand, the modified procedure has been introduced because it involves less travel cost. This will be the ordinary situation when samples are to be drawn from finite populations. Some procedures will give more precise estimates than others and at the same time may "cost" more in application. The practical problem therefore is to balance precision against cost and to decide what is best for a particular set of circumstances.

The foregoing discussion may seem artificial in two respects: (1) we have talked about an infinite set of samples when in practice we shall have only a single sample; and (2) we have learned about the sampling distribution of $\bar{x}$ by assuming a complete knowledge of the finite population from which the samples were drawn. Actually these are not serious limitations, and much of the remaining portion of this book will be devoted to the chain of reasoning and the resulting techniques which will allow us to deal with such problems.

6.3. Selection of Simple Random Samples

6.3.1. *Translation of the Model into a Mechanical Procedure.* The discussion of the preceding section did not tell how to select simple random samples. Rather, the model specified what is expected if a "simple-random-sampling procedure" is used to choose an infinite number of samples. We must now devise mechanical selection procedures which will, we think, give results agreeing with those required by the model. The phrase "we think" has been introduced to emphasize that it is impossible to select an infinite number of samples and thereby obtain an exact check on a procedure.

The most direct way of selecting one or more simple random samples has already been discussed and applied in the preceding chapters and may be easily illustrated in terms of the 20 samples of Table 6.1. Suppose we prepare 20 identical cards, labeling each card with the designation for a different one of the 20 samples. These cards are shuffled thoroughly, and one card, or one sample, is drawn "blindly." This mechanical procedure can clearly be applied repetitively. That is, we draw a single card and mark down the sample selected; this card is then returned to the deck, the deck of 20 cards is again shuffled thoroughly, a card is drawn blindly, and the sample selected is marked down; and so on. There is no reason for thinking that any one card—and therefore any one sample—will be preferred over any other in the application of this procedure, and hence we expect each possible sample to occur the same fraction of the time in an infinite set of such drawings. Thus we have a mechanical selection procedure which is expected to produce results in accordance with the model for simple random sampling.

Although the foregoing procedure provides a way of drawing simple random samples, it is not convenient for most situations that one must face in practice. First, there will ordinarily be so many samples, even in small finite populations, that it would be impossible to enumerate them all; and second, the preparation and shuffling of cards would be extremely time-consuming. To illustrate the first difficulty, we shall now obtain an expression for the total number of different samples of size n that may be drawn from a finite population of N elements. Then we shall consider a mechanical selection procedure that requires less work than the preparation and shuffling of cards.

6.3.2. *The Number of Possible Samples of Size n from a Finite Population of N Elements.* A general expression for the total number of samples of size n that may be drawn from a finite population of N elements can be obtained by a systematic enumeration argument. This argument can be illustrated in terms of drawing samples of size 3 from a population containing the six elements A, B, C, D, E, and F. We know from the results of Sec. 6.2 that there are 20 such samples.

A systematic enumeration of the possible samples can proceed as in the following diagram, where it is assumed that the first sample element is A. Similar diagrams can be started with B as the first element, or with C, . . . , or with F, but it is not necessary to put these other five down in order to see the argument. The diagram is:

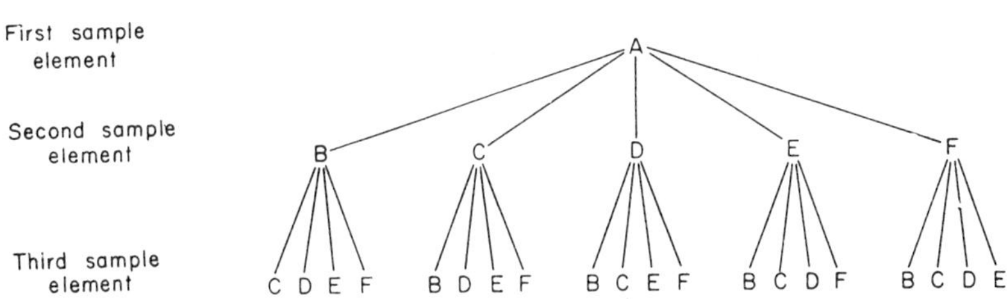

When A is the first sample element, we see that the second sample element can be any one of the remaining five elements and that the third sample element can then be any one of the four elements left. Thus this diagram represents $5 \cdot 4$ arrangements of the six elements taken three at a time. The diagram with B as the first element would also contain $5 \cdot 4$ arrangements, as would each of the diagrams for C, . . . , and F. Therefore the total number of such arrangements is $6 \cdot (5 \cdot 4) = 120$.

Actually, we are not interested in all 120 arrangements. These arrangements distinguish between samples not only with respect to the elements contained in the samples but also with respect to the *order* in which the elements occur. For our purposes it makes no difference whether we have sample ABC or BCA. Each contains the same elements and will have the same mean and the same variance. Therefore we must eliminate this duplication.

The argument for counting duplicates is the same no matter which sample of size 3 we examine, so let us focus attention on sample ABC. The following diagram picks the duplicates out of the six larger diagrams:

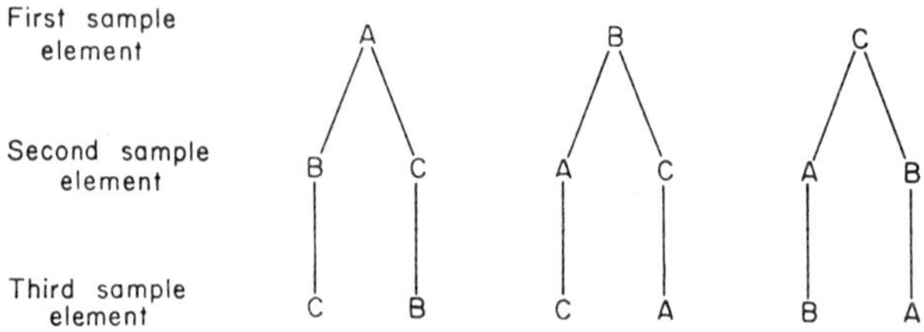

Thus we see that sample ABC is counted $3 \cdot 2 \cdot 1 = 6$ times in the 120 arrangements, and we really wish to count it only once. Similarly, any other sample we may name is counted six times instead of once. From this it is apparent that 120 is six times as large as the desired number. Thus the number of samples of size 3 that can be drawn from a finite population of six elements is $(6 \cdot 5 \cdot 4)/(3 \cdot 2 \cdot 1) = 20$.

This result can be written in a more conventional form by multiplying both numerator and denominator by the factor $3 \cdot 2 \cdot 1$, leading to

$$\frac{(6 \cdot 5 \cdot 4) \cdot (3 \cdot 2 \cdot 1)}{(3 \cdot 2 \cdot 1) \cdot (3 \cdot 2 \cdot 1)} = \frac{6!}{3!3!}$$

The quantity $6!$ is read as "six factorial," and the factorial symbol ! directs us to take the integer preceding the symbol and to form the product of it with each smaller integer, all the way down to 1. Thus

$$5! = 5 \cdot 4 \cdot 3 \cdot 2 \cdot 1 = 120$$

and, in general,

$$N! = N \cdot (N - 1) \cdot (N - 2) \cdot \cdot \cdot \cdot \cdot 3 \cdot 2 \cdot 1$$

The quantity $0!$, if it occurs, is defined to be 1.

The foregoing argument could have been carried out in general terms, and we would have obtained the result that the total number of samples of size n that can be drawn from a population containing N elements is equal to

$$\frac{N!}{n!(N - n)!} \tag{6.1}$$

This expression is frequently referred to as the "number of combinations of N things taken n at a time" and is abbreviated ${}_NC_n$, or ${}^NC_n$, or $\binom{N}{n}$.

With expression (6.1) at our disposal, it is now easy to demonstrate that the enumeration, card-shuffling procedure for selecting random samples is of little practical value. For example, suppose we have a "small" finite population of 100 elements and wish to choose a sample of size 5. The number of samples to be considered is

$$\frac{100!}{5!95!} = \frac{\overset{5}{\cancel{100}} \cdot 99 \cdot 98 \cdot 97 \cdot \overset{16}{\cancel{96}} \cdot \cancel{(95!)}}{\cancel{5} \cdot \cancel{4} \cdot \cancel{3} \cdot \cancel{2} \cdot 1 \cdot \cancel{(95!)}} = 75{,}287{,}520$$

It is clearly ridiculous to think of enumerating such a large number of samples, to say nothing of preparing and shuffling cards.

6.3.3. *Random Digit Tables.* We may do two things to make it easier to select random samples from a finite population. The first of these is to change the manner of attack. That is, instead of using the entire sample as a selection unit, elements are drawn *one by one* from the population until the desired sample size is attained. The only requirement on this procedure is that at each stage in the drawing all elements that have not yet been placed in the sample have an "equal chance" of being selected. It can be shown that this requirement is equivalent to the one contained in the original definition of simple random sampling.

Again it is reasonable to expect that this form of the simple-random-sampling model can be approximated mechanically by a card-shuffling technique. Thus a random sample of size 5 from a population of 100 elements would be chosen in the following manner. One hundred identical cards would be prepared, and on a single card would be placed the designation for a single population element. These cards would be thoroughly shuffled, and five cards drawn out "blindly." The five elements thus drawn would constitute a simple random sample. One has only to compare the labor involved in preparing and shuffling 100 cards with that involved in preparing and shuffling cards for the 75,287,520 possible samples noted at the end of the preceding section to appreciate the gains which this change in technique has effected.

Even the one by one technique becomes quite overwhelming if the population is of appreciable size. The work required to prepare and shuffle a large number of cards would frequently be prohibitive. However, even this difficulty can be overcome by a very simple expedient. Returning to the population of six families (A, B, C, D, E, and F), let us distinguish among these families by assigning to each a *number* rather than a name or letter. Thus we have family 1, family 2, . . . , and family 6. Then the six cards we prepare have numbers recorded on them, and a sample of size 3 would be drawn by selecting three of these cards, i.e., three of the numbers between 1 and 6. The operation of drawing a random sample of three numbers between 1 and 6 is the same whether we are talking about six families, six electric-light bulbs, six students, six time intervals, six pages of a book, and so on. Therefore it is reasonable to perform an extremely large number of drawings with numbers alone, record the results of these drawings, and use these drawings whenever it is necessary to select random samples. In other words, some central agency does the "dirty work" of preparing, shuffling, and drawing cards, and the sampler must only assign numbers to his population elements and make use of this catalogue of drawings. This procedure has been carried out on an extensive basis, and the results are known as *random-number*, or *random-digit*, tables.

We can think of obtaining random digits in the following manner. Ten identical cards are prepared, on each of which is noted one of the 10 digits 1, 2, . . . , 9, 0. These cards are thoroughly shuffled, a single card is drawn out blindly, and the digit appearing on the card is written down; the card drawn is *replaced* in the deck, the cards are reshuffled, and a second drawing is made; and so on. Actually there are more efficient ways of carrying out this drawing operation (e.g., through the use of high-speed electronic calculating equipment), but it is not necessary for us to consider these procedures. The end result is a table of random digits similar to that exhibited in Table 6.3. The random digits in this table are grouped in sets of 5 and 25, but this has been done solely to facilitate reading the table.

We shall first illustrate the use of such a table in some simple situations and then apply it to a few more complex problems. Suppose we have a group of eight individuals from which we wish to select a random sample of size 4. A digit is assigned to each individual, thereby using up the digits 1, 2, . . . , 8. A starting point is arbitrarily chosen in a random-digit table. For ease of reference, let us start at the first digit on line 507 of Table 6.3. We shall then use each successive digit as we go across this line. The first digit is 0 and is ignored since it has not been assigned to any of the population elements; the second digit is 6, and we therefore have drawn individual 6 into our sample; the third digit is also 6, and this is ignored since individual 6 has already been chosen; the fourth digit "chooses" individual 7; the fifth digit "chooses" individual 1; and the sixth digit "chooses" individual 4. Therefore our sample of size 4 consists of individuals 6, 7, 1, and 4. With the starting point designated above, we could also have gone down the first column of the table until four individuals were chosen. The reader can check that this would give the sample consisting of individuals 2, 3, 1, and 4. The only requirements in using this table are that an arbitrary starting point be taken and that consecutive digits be read in a *consistent* (down or across the table) manner. Those digits that do not apply, either because one or

TABLE 6.3. 4,000 RANDOM DIGITS

Line No.	Random digits									
500	67885	99295	47271	38655	59513	96960	31718	08974	16122	20535
501	52380	29769	70660	57425	50891	75044	84257	73315	38181	28673
502	93140	26307	82265	78382	19681	56585	08975	76764	39956	83450
503	84663	89963	71584	57696	30829	60527	64947	34899	28805	28397
504	91830	51842	99838	39839	66971	67177	74219	35637	35634	93581
505	81746	29991	81096	94279	02968	62561	02479	82126	25702	67953
506	88088	50293	83423	86206	39935	23253	43041	48941	85787	08388
507	06671	43574	84908	67295	33623	55060	28174	48415	02529	22009
508	24524	05283	30460	32399	80423	56929	40852	69969	88541	05979
509	91496	64730	57198	83145	39750	03568	54669	98679	04297	51047
510	31492	47734	31343	31180	00232	19707	24823	75079	73943	17997
511	08446	91252	39879	58682	82972	18417	39203	36681	42895	08459
512	15618	17941	52594	43277	16530	40052	91100	87422	47230	95699
513	49794	50492	87439	86354	04546	65333	11057	77727	19748	38722
514	91821	18107	42125	89239	28847	54623	38783	47803	31414	38450
515	03697	89186	30579	44188	26532	08420	80723	48100	60748	76330
516	45832	08311	16051	04475	13400	48527	46073	17439	56498	94632
517	09021	16871	83366	14896	04219	38375	87890	90217	42370	61028
518	85101	76771	83715	94737	69973	74187	01958	59691	86712	86570
519	60984	76342	13648	85250	28323	48379	45141	36277	51845	29039
520	03553	05128	59866	51281	68124	17007	24729	29710	41439	40574
521	11774	86746	89698	56020	37810	88972	11361	95583	70786	00589
522	74473	87513	17690	61427	72914	32517	01804	97910	06327	30246
523	33049	02622	41026	80875	41293	16752	84225	84414	37137	68956
524	08095	64981	28180	38629	76962	23840	17477	75268	48297	70340
525	57888	13938	38554	86836	02195	30270	55484	53364	54705	41380
526	56316	37723	00234	21424	26664	63804	75139	36534	18579	09833
527	98849	72762	59767	52497	24227	83152	71794	21398	99456	89215
528	51632	54799	27973	68568	68465	98500	28681	18369	24279	96335
529	12874	82160	67202	85199	27908	67022	49810	77929	96212	81153
530	77884	07032	01671	53362	28119	56786	30883	28540	76029	03774
531	64611	19736	25589	46569	45206	48215	69523	17423	91807	90039
532	30393	58319	85098	66519	57571	24541	03562	14400	62731	82534
533	61477	89731	18421	29861	52829	00838	78040	43350	74323	82892
534	84746	28302	13264	07595	00134	12933	46831	24864	47275	20527
535	09110	28485	30326	99826	64005	99308	65779	42760	90066	03974
536	38688	39968	32604	11694	46262	73262	45405	43923	67397	88228
537	56405	17839	92073	57622	93328	15442	50186	07570	58001	31000
538	08915	11467	14793	82691	51238	12485	51745	18192	05985	36826
539	89434	38669	91592	88799	65621	67237	59541	19657	93402	58705

TABLE 6.3. 4,000 RANDOM DIGITS (*Continued*)

Line No.	Random digits									
540	73553	78280	69125	95591	81168	91927	25976	89077	71690	19404
541	64603	59752	74698	44233	67602	38615	31303	28650	53700	89819
542	07783	04351	77451	47350	21234	16016	41532	76508	23063	44993
543	43983	33356	61715	96485	22121	78004	06316	87896	99289	93981
544	37850	66128	92735	45064	50924	24204	58816	65290	34392	55567
545	66416	72353	45775	68590	85685	72683	60090	37149	85347	57414
546	72336	12979	05720	92754	76911	96883	74420	05220	85815	23557
547	80567	44365	70254	50864	36619	51479	23281	76428	18580	34240
548	59289	49076	18439	29522	42541	04024	84446	92434	90407	77241
549	19690	78143	65919	13699	91844	91241	38361	67171	90551	05709
550	03474	76025	97043	33834	44638	54040	82797	00545	38159	16089
551	35870	89158	55864	98078	50563	36492	10994	85909	09018	19252
552	73887	67928	60045	70782	11937	04074	53814	46621	52577	94853
553	45968	73667	65062	73306	76045	78649	91654	53958	96537	95542
554	67622	54579	17279	67440	56441	20681	64011	52226	96618	32831
555	60664	67547	39523	02043	59748	01887	69229	94653	99271	98164
556	62155	09234	47367	13047	06364	35064	10073	06793	80248	29009
557	44969	11129	17139	79630	89772	26921	56949	23465	30036	17173
558	82459	96218	60768	76417	24405	18710	68887	82394	69729	82503
559	40873	41590	67255	30757	09657	91881	34578	09511	05417	58953
560	18532	10721	22029	48524	47778	00881	83489	03464	57462	97459
561	86689	39755	39547	00740	36666	07993	31671	86304	12970	73402
562	52849	31652	79655	11250	18463	57518	20306	25301	01374	51208
563	33298	87662	61849	60923	68685	69411	39266	80320	34844	89416
564	81569	83651	35795	40168	33501	01042	58931	03892	85188	74740
565	85476	23790	33842	89565	53359	25579	59049	62394	72435	12457
566	21904	18370	97035	57905	09581	91227	92754	37760	01411	07440
567	87175	88318	63242	85960	56690	12618	30493	11569	73723	07448
568	58830	00157	65814	21118	22140	73793	57855	81830	06795	13183
569	12625	30635	56429	73216	12342	36722	83886	96828	82870	90954
570	97614	02370	42160	73370	11944	49067	59452	80495	43911	46712
571	17033	68037	41963	03874	44856	82985	57453	84358	16120	04454
572	76624	00405	62369	55080	61880	51270	87807	10653	36894	70850
573	35660	00234	14705	93418	94084	82856	25384	71555	56754	78315
574	18291	91656	98079	52384	43306	65205	75903	58701	99496	50048
575	33557	87793	90857	10143	46726	84284	43635	41213	83845	70986
576	91408	80220	05728	68890	46577	21152	43759	43301	93661	97252
577	50106	10099	13722	18572	44024	00351	18173	23717	85114	85998
578	57782	63951	53723	86853	63851	79430	49181	46386	69666	55743
579	76162	71724	40028	94786	34457	16906	90040	30789	40281	94697

SOURCE: The Rand Corporation, *A Million Random Digits*, The Free Press, Publishers, Glencoe, Ill., 1955.

more digits were not assigned to population elements or because a digit has already been drawn into the sample, are simply ignored.

If a population consists of 100 or fewer elements—but more than 10—then it is necessary to distinguish between elements by using a two-digit number. The numbering scheme would then go 01, 02, 03, . . . , 10, . . . , 98, 99, and 00 (= 100). A sample of the required size would then be chosen by taking *consecutive pairs* of digits from the table, ignoring those pairs already used in the same sample and those pairs which were not assigned to elements. If a population contains 1,000 or fewer elements—but more than 100—then one uses the triplets 001, 002, . . . , 010, . . . , 100, . . . , 998, 999, and 000 and considers consecutive triplets of digits in the random-digit table. This process can be extended to finite populations of any size.

Random-digit tables can easily be used to select samples from any frequency distribution, and the results of such selections have been used in the preceding chapters to illustrate the difference between samples and populations. The only trick in this type of application is to make a judicious assignment of numbers to the population elements. One convenient way of accomplishing this is illustrated in Table 6.4 with the data of Exercise 4.2. There are 189 motor failures represented in this table, and a triplet of digits is therefore needed for identification. The first interval "0 up to 20" contains 6 failures, and these are assigned triplets 001, 002, . . . , 006. The second interval contains 11 failures, and these are assigned triplets 007, 008, . . . , 017. This process is easily carried out in terms of the cumulative number of failures.

Now suppose one wishes to draw a simple random sample of five failures from the distribution of Table 6.4. Triplets of random digits will be taken from a random-digit table. Thus starting at the left of line 520 of Table 6.3, we find that the first random triplet is 035. From Table 6.4 we see that 035 is the order number of a motor failure in the interval "60 up to 80." All elements in an interval are assumed to have a value of the variable corresponding to the mid-point of the interval, and this is 70 for the random triplet 035. The second random triplet on line 520 is 530, and this does not apply in the present example. The process is continued until a sample of size 5 is obtained, the complete details being given at the foot of Table 6.4.

It might be observed that this process is quite "wasteful" of triplets, all those over 189 having to be discarded. An equivalent mechanical procedure is obtained if we assign more than one triplet to an element, each element receiving the *same number of triplets* in order to preserve the "equal chance" feature. The term *equivalent* is used in this instance to mean that an infinite number of applications of this mechanical-selection procedure is expected to give results in conformance with the simple-

TABLE 6.4. SELECTION OF A RANDOM SAMPLE OF SIZE 5 FROM DISTRIBUTION OF MILES TO FIRST MAJOR FAILURE OF 189 CITY BUSES
(See Exercise 4.2.)

Distance traveled to first major motor failure (thousands of miles)	Number of failures	Cumulative number of failures	Numbers assigned to individual failures
0 up to 20	6	6	001, . . . , 006
20 up to 40	11	17	007, . . . , 017
40 up to 60	16	33	018, . . . , 033
60 up to 80	25	58	034, . . . , 058
80 up to 100	34	92	059, . . . , 092
100 up to 120	46	138	093, . . . , 138
120 up to 140	33	171	139, . . . , 171
140 up to 160	16	187	172, . . . , 187
160 up to 180	2	189	188, . . . , 189
Total.	189		

Random numbers drawn (starting at the first digit on line number 520 of Table 6.3):

035, ~~530~~, ~~512~~, ~~859~~, ~~866~~, ~~512~~, ~~816~~, ~~812~~, ~~417~~, *007*, ~~247~~, ~~292~~, ~~971~~, *041*,
~~439~~, ~~405~~, ~~741~~, *177*, ~~486~~, ~~746~~, ~~896~~, ~~985~~, ~~602~~, *037*

Values of the variable corresponding to the chosen random numbers (using midpoints of intervals):

70(*035*), 30(*007*), 70(*041*), 150(*177*), 70(*037*)

random-sampling model. An easy way of accomplishing this in the present situation is to select element 001 if any of the triplets 001, 201, 401, 601, or 801 are drawn; element 010 if any of 010, 210, 410, 610, or 810 are drawn; and element 189 if any of 189, 389, 589, 789, or 989 are drawn.

6.3.4. *Some Timesaving Ways of Assigning Serial Numbers to Population Elements.* There are many situations in which it would be an almost impossible task to list each population element for the purpose of assigning serial numbers. And yet often one can, by an ingenious analysis of the problem, avoid this listing operation. We shall discuss one example to illustrate a useful approach.

Suppose that an individual is studying the relative emphasis which one or more daily newspapers place on various types of news stories, for example, stories on national politics, state politics, foreign relations, trade unions, and so on. Attention might be focused on the front-page headlines, and each studied headline would be placed in one of the defined categories. Such studies may also be concerned with the content of radio programs, the content of books, and the like, and may, for example, provide clues to changes in a government's policy. Descriptions of a number of specific studies carried out along these lines will be found in a book on content analysis by Lasswell, Leites, and associates [1].

In order to be more definite concerning the newspaper-headline example, suppose that the study is limited to front-page headlines appearing in either *The New York Times* or *The New York Herald Tribune* during the course of a year (365 days). We therefore have a finite population consisting of all front-page headlines appearing in either of these two newspapers on any one of the 365 days. A single element of this population is a single headline, and a simple random sample of headlines is desired.

It would clearly be a very time-consuming task to go through these 700 issues and number each headline. As a matter of fact this would almost constitute a complete enumeration of the population since it would require little additional work to classify a headline once it is made available for numbering. This difficulty can be avoided in the following manner. Number the days as 001, 002, . . . , 365 and assign a digit 1 to the *Times* and a digit 2 to the *Herald Tribune*. Then by examining a few issues of each newspaper we can determine the *maximum* number of front-page headlines that appear in any single issue. As an illustration, let us take this maximum number as nine. A simple random sample of headlines is then selected by taking sets of five consecutive digits from a random-digit table, 000-0-0, where the first three digits determine the day, the fourth digit determines the newspaper, and the fifth digit determines the headline. The headlines need be numbered only for those day-newspaper combinations that are drawn into the sample. As in previous examples, five-digit sets which do not apply are ignored and a given headline is allowed to appear in the sample only once.

There are two features of the foregoing procedure that require brief mention. The selection of a digit for picking a headline from within a day-newspaper combination is made before the exact number of headlines is known for that combination and is based on the maximum number of headlines that could occur. Thus if digit 8 is drawn and there are only four headlines for that combination, the entire five-digit set is ignored and the next five-digit set is taken. This procedure is necessary to ensure the "equal chance" feature of simple random sampling. If one headline were always picked from each day-newspaper combination drawn, then headlines appearing on day-newspaper combinations with a "small" number of headlines would have larger chances of being drawn into the sample than those appearing on combinations with "large" numbers of headlines. The second feature is that the serial numbering of headlines within a day-newspaper combination must be done in a consistent manner without looking at the digit already drawn. Otherwise, personal biases in favor of certain types of content might destroy the basic requirements of simple random sampling. That is, an individual might, perhaps even unconsciously, note the random digit drawn and then number the headlines so as to make this digit correspond to certain preferred types.

The foregoing procedure may be viewed in terms of selecting a sample in three *stages*. At the first stage, a day is chosen; at the second stage, a newspaper is selected; and at the third and final stage, a headline (the ultimate population element) is drawn.

This device of assigning serial numbers in successive stages is applicable to a wide variety of problems. As illustrations we may cite the following:

1. Serial numbers could be assigned to *listings* in a telephone directory by using consecutive sets of random digits to refer to page-column-line combinations. Line numbers would have to be given only for page-column combinations drawn into the sample. Again one would use the *maximum* number of lines in selecting a random line number in order to avoid excessive emphasis on listings contained in short columns.

2. Serial numbers could be assigned to *workers* in a plant by using consecutive sets of random digits to refer to department-worker combinations.

3. Serial numbers could be assigned to *words* in a book by using consecutive sets of random digits to refer to page-line-word combinations.

4. Serial numbers could be assigned to *homes* in a city by using consecutive sets of random digits to refer to block-home combinations.

Efficiency is achieved in all these applications because serial numbers have to be assigned only to those population elements which are included in the prior combinations drawn into the sample. The only feature about which one must be careful, in addition to the "maximum number" problem mentioned earlier, is that a completely new "combination set" of random digits must be used to select *each* population element that is to be included in the sample. The sampling procedure leading to the distribution presented in Table 6.2 provides an illustration of this type of deviation from simple random sampling. In that situation we had six families living along a street, families A, B, and C being "close" to one another and families D, E, and F being "close" to one another. The first stage of this procedure consisted of choosing one of the two sets ABC or DEF, this being done by drawing one of the digits 1 (for ABC, say) or 2 (DEF). However, once a set was chosen, every family in the set was automatically included in the sample. Thus the first-stage digit is used to choose three families rather than a single family, and this violates the requirement for obtaining a simple random sample of families.

6.4. Obtaining Information from the Elements of a Simple Random Sample

A simple random sample, or a sample selected by any other procedure for that matter, is never the end product of a statistical study. Rather, it is the beginning of the study. Once the sample has been chosen, it is next necessary to obtain information from, or concerning, the elements in the sample and to use this information in describing the characteristics of the population. Illustrations of this have been given in the preceding chapters. We shall say that the "effective" sample consists of those sample elements from which usable information is obtained. Any deductions drawn from the model for simple random sampling require that the sample drawn and the "effective" sample be identical. Therefore, differences between the two may invalidate to some unknown extent statements made on the basis of the model.

If one draws a sample from a finite population of inanimate objects, then the actual sample and the effective sample will ordinarily be the same. A newspaper headline cannot refuse to answer questions, and an electric-light bulb drawn from a production line will not be at the movies when the inspector calls. However, when one is dealing with human populations, as is so often the case in social-science research, this agreement between actual and effective samples cannot be expected as a matter of course. In survey work, for example, interviewers may go to wrong addresses; some individuals may not be at home when the interviewer calls; other individuals may refuse to talk with an interviewer; or a substantial fraction of a sample may not return a mail questionnaire even after follow-up appeals have been used. Deductions made from the model must be qualified in terms of these sample losses.

In conclusion, it may be observed that such losses are not peculiar to samples. They also occur with complete enumerations. This is one of the reasons why a sample survey is frequently preferred over a census. It may be possible to reduce these losses in a "small-scale" sampling operation to such an extent that the total error of the sampling results—sampling error combined with "loss" error—is less than the "loss" error of the census. Decisions on a matter like this depend upon a large number of factors such as available resources (time, money, personnel, etc.), the accuracy of measurement, and the uses to which the obtained data are to be put.

6.5. The Limitations of Simple Random Sampling

If it were possible to ignore questions of feasibility and costs, simple random sampling would supply a solution to most finite-population sampling problems. However, every statistical study has practical and economic aspects that must be incorporated into the study design. This means that simple random sampling must frequently be modified in the interests of efficiency. For example, if an experimenter has a given amount of money to spend on the collection of data, he will attempt to devise a best, or "optimum," sampling procedure. That is, all sampling procedures costing this fixed amount of money will be examined; a comparison will be made among them with respect to the expected sampling error of estimates made from samples selected by each procedure; and that procedure whose sample estimates have the smallest sampling error will be chosen.

Two ways in which simple random sampling may be modified in the interest of feasibility and costs are:

1. The sample elements may be "clustered" into compact groups. Thus, if a population is spread over a wide geographic area, the elements in a simple random sample will ordinarily be widely scattered and interviewing costs will be much greater than if the sample elements are clustered into compact geographic areas. A very simple form of this clustering was used in the discussion accompanying Table 6.2.

2. Known characteristics of a population may be used in sample design so as to produce more precise estimates than would be given by a simple random sample costing the same amount of money.

Both of the foregoing points will be discussed and applied in Chap. 10, but we shall first have to develop the theory for evaluating sample results. This development will be carried out in terms of simple random sampling because simple random sampling provides the basic theory for dealing with all other types of sampling procedures.

6.6. Sampling from Infinite Populations

Simple random sampling from a finite population has been defined in terms of all possible samples. An individual element may appear not more than once in any particular sample. If one thinks of drawing sample elements one by one, then elements are being drawn *without replacement.* That is, once an element has been chosen for a given sample, it is not replaced in the population before the succeeding elements of this sample are drawn.

Elements may also be drawn one by one *with replacement.* That is, after an element has been selected for a given sample, it is replaced and has a chance of being selected on the next draw, and also on each succeeding draw. If this operation were carried out with the six-family illustration of Sec. 6.2, we would have to consider such samples (in addition to those set forth in that section) as AAA, AAB, and so on. One would rarely expect to use sampling with replacement for a finite population because it seems to be "wasteful" of information to let an element appear in a sample more than once. Later on we shall obtain expressions which will allow us to evaluate this "waste." However, it is sometimes convenient to use sampling with replacement in developing theory. This theory may be used to approximate what would happen if the sampling were done without replacement.

There is another way of looking at sampling with replacement from a finite population. Returning to the illustration, we can think of having an infinite number, or inexhaustible supply, of cards on one-sixth of which are recorded A's, on one-sixth of which are B's, . . . , and on one-sixth of which are F's. These cards are thoroughly shuffled or mixed, and three cards are drawn out. Since the supply is infinite, the drawing of any number of cards does not change the composition of the supply. Hence on any drawing, each member of the population has an equal chance of being selected.

If we have a truly infinite population of elements, the definition which we have given for simple random sampling has no meaning. Since there are an infinite number of elements, there are also an infinite number of samples and the phrase "equal chance" in our definition of simple random sampling is not applicable. Nevertheless, we frequently have to conduct repeated trials of an experiment which could conceivably be repeated an infinite number of times, and it is convenient to regard a finite set of such trials as a sample from the infinite population of all trials. The basis for dealing with problems of this type will be set forth in the following chapter.

EXERCISES

6.1. The following procedure is used to draw a sample of size 500 from the population of persons 21 years of age and over who reside in a specified city at a particular time. A listing is prepared of all households in the city, and a serial number is assigned to each household. Five hundred different sets of consecutive random digits are drawn from a random-digit table, thus obtaining a simple random sample of households. For each of the 500 selected households, a listing is prepared of the adult members of the household, and exactly *one* adult is drawn into the sample from each household by using a random digit.

a. Is the resulting sample a simple random sample of adults? Why or why not?

b. If you do not think that it is a simple random sample of adults, how would you suggest modifying the selection procedure so that a simple random sample of adults would result?

c. Suppose that *every* adult living in the 500 selected households were drawn into the sample. Would this give a simple random sample of adults? Why or why not?

6.2. A sample is desired from the population of households (in a city) having one or more children in high school. A simple random sample of all households in the city is obtained as in Exercise 6.1. These households are visited. If a household has children in high school, it is kept in the sample; if it does not have children in high school, it is discarded and the "closest" household (geographically) which has children in high school is "substituted" in the sample. Is the resulting sample a simple random sample from the population of households having children in high school? Why or why not?

6.3. In response to a newspaper notice of a single job vacancy, five men appear simultaneously at the employment office of a large company. It is likely that one of the first two or three individuals interviewed will be hired, and the interviewer therefore wishes to show no preference in the order in which they are taken. How would you use a random-digit table to assign an impartial ordering?

6.4. Suppose that an individual wishes to study the quality of grain in a field of standing wheat. The field is square, and a side is 1,000 feet in length. It is decided that five points will be chosen at random in the field, that a circle of radius 3 feet will be marked off around each point, and that the grain growing in these five circles will be harvested and analyzed. Assuming that distances are measured in feet, describe how one might use a random-digit table to locate five points in the field.

6.5. One of the techniques used in the study of social behavior is *time sampling.* Arrington [2] defines time sampling as "a method of observing the behavior of individuals or groups under the ordinary conditions of everyday life in which observations are made in a series of short time periods so distributed as to afford a representative sampling of the behavior under observation." Suppose this method is to be applied to the study of the behavior of children in a nursery school. This behavior will be observed during 25 5-minute intervals spread out over the course of a specified week. If the school is in session for 5 days—from 9 to 11 A.M. and from 1 to 3 P.M. on each day—how would you use a random-digit table to choose a random sample of 25 5-minute intervals?

6.6. Table 4.1 gives the distribution of days of disability for acute respiratory illnesses among the employed males in the Eastern Health District of Baltimore, 1938 to 1943.

a. Following the technique discussed in connection with Table 6.4, use the random-digit table to draw 10 simple random samples, each of *size* 5 from the distribution. Notice that a given three-digit number may not appear more than once in any one sample, although it may appear in two or more different samples.

b. Compute the mean $\bar{x}$ of each sample and observe how these compare with the population mean. Compute the *mean of your* 10 *sample means* and compare this with the population mean.

c. Compute the variance among *your* 10 *sample means.*

REFERENCES CITED

1. Lasswell, Harold D., Nathan Leites, and associates: *Language of Politics*, George W. Stewart, Publisher, Inc., New York, 1949.
2. Arrington, Ruth E.: "Time Sampling in Studies of Social Behavior," *Psych. Bull.*, **40**:81–124 (1943).

OTHER READING

Hansen, Morris H., William N. Hurwitz, and William G. Madow: *Sample Survey Methods and Theory*, vol. I, John Wiley & Sons, Inc., New York, 1953. Chapter 1 of this book provides an over-all view of the modern approach to sample-survey design.

CHAPTER 7

EXPERIMENTS WITH UNCERTAIN OUTCOMES AND PROBABILITY

7.1. Experiments with Uncertain Outcomes

In the preceding chapter the drawing of a simple random sample from a finite population of elements has been regarded as a trial of an experiment where the outcome of the trial is uncertain. That is, before a sample is selected by using a random-digit table there is no way of predicting with certainty which of the $N!/[n!(N - n)!]$ possible samples, or possible outcomes of the experiment, will occur. However, the mechanical procedure for drawing a simple random sample has been set up so that it could conceivably be *repeated* an infinite number of times and so that each possible sample (or outcome) is expected to occur an equal fraction of the time in this infinite set of trials.

Other examples of "conceptually repeatable experiments where the outcome of a single trial is uncertain" were also discussed in preceding chapters. The phrase in quotes has been stated in so general a manner that it is pertinent to many of the illustrative examples already presented and to a host of applications of statistics that have not been mentioned. For example:

1. A single trial of an experiment consists of tossing a coin having two faces, head and tail. The outcome of the trial is either a "head" or a "tail," the possibility of the coin standing on edge being excluded. Before the coin is tossed it is impossible to predict with certainty which face will lie uppermost. Also, we can think of conducting an infinite number of trials of this experiment "under identical conditions."

2. A single trial of an experiment is the birth of a child; the outcome is either a "male" or a "female." Excluding the possibility of prebirth tests for sex, it is impossible to predict with certainty the outcome of a trial. By observing an indefinitely large number of births, one is conducting an infinite number of trials of the experiment. It will be recalled that the discussion accompanying Fig. 3.9 also concerned this experiment, but there the possible outcomes were related to "number of minutes after the hour at which the birth occurred." Thus we see that the result of a

trial of an experiment may be categorized with respect to different sets of outcomes or variables, depending upon the purposes of the experiment.

3. The personnel office of a large company selects workers to fill a particular position. A single trial of an experiment consists of hiring an individual for the position. A possible set of outcomes for this experiment contains the two members "success" and "failure." Even though interviews, aptitude tests, references, and the like, are used in making a selection (and these form part of the conditions of the experiment), the outcome of a single trial cannot be predicted with certainty. One can think of conducting an infinite number of trials of this experiment, although in practice only a very small number of trials is possible.

4. Exercise 5.4 was concerned with the behavior of an automatic packaging machine. A single package is filled by the machine, and this may be regarded as a single trial of an experiment. An outcome of the trial is the exact weight of the package. Again it is possible to conceive of an infinite number of trials, each trial being classified according to the uncertain outcome "weight of package."

5. The methods of time sampling as applied to the study of children's behavior in nursery schools was briefly described in Exercise 6.5. A single trial of an experiment might consist of observing such a group of children for a 5-minute interval, possibly classifying a trial according to the "number of times that any child strikes another child." The outcome of such a trial is uncertain, and one can think conceptually of conducting an infinite set of trials.

The foregoing examples indicate the general applicability of the phrase "experiments with uncertain outcomes" and should emphasize the necessity for having a formal framework of concepts and techniques within which to treat the observed results of such experiments. The *mathematical theory of probability* provides this framework. In the present chapter some aspects of the intuitive background of this theory will be presented, and the essentials required in the application of statistical method will be developed.

By restricting attention to conceptually repeatable experiments, we shall be concerned with what may be termed *statistical* probability. It is also possible to consider "judgmental," "personal," "subjective," "psychological," or "degree of conviction" approaches to probability. These approaches attempt to attach quantitative measures of belief to such statements as "This strike will be settled without wage increases being granted to the workers," "My best friend married a girl much older than he, and his marriage will terminate in divorce," or "The next car I buy will be a 'lemon'." "Personal" approaches to probability have always had strong intuitive appeal for many people and have been considered formally by both philosophers and mathematicians, but their value in practical situations is still open to question. The interested reader will find a reference to such an approach at the end of this chapter.

7.2. The Probability of an Elementary Event

The major prerequisite for dealing with experiments having uncertain outcomes is that the conditions under which repeated trials of the experiment are to be made must be stated exactly. This statement of conditions defines the experiment, or alternatively, in the terminology of Chaps. 1 to 5, it describes an infinite population of trials of the experiment. Care in setting forth these conditions is essential. If ambiguity exists in the definition of the experiment, then ambiguity will also exist in the results obtained on the basis of the definition. Problems encountered in this respect are very similar to those that arise in defining a finite population, although the fact that we are concerned with conceptual experiments may make these problems less tangible. As an illustration of these difficulties, consider the example of hiring workers to fill a particular position, a single trial consisting of the hiring of one individual and the outcome of the trial being either a "success" or "failure." If a new personnel director does the interviewing, or if a new set of aptitude tests is used in the selection process, or if economic conditions change so that a new type of individual is applying for the position, the conditions of the experiment are changed and this change has to be taken into account.

Once the conditions are defined under which we can at least think of carrying out repetitions of an experiment, it is next necessary to agree upon a set of *elementary*, or basic, outcomes for a single trial of the experiment. We shall here refer to an elementary outcome of an uncertain experiment as an elementary *event*, abbreviated as E, and shall distinguish between different elementary events by the use of subscripts E_1, E_2, ...

We shall not attempt to provide any formal definition of what constitutes an elementary event, since this depends upon the experiment and the purposes for which it is conducted, but will present a number of examples which will illustrate the use of the concept:

1. A single toss of a coin results in either a "head" or a "tail." Elementary event E_1 would be a "head" and E_2 a "tail," or vice versa.

2. Exercise 2.1 referred to the "checkout waiting experience" of a housewife who always shops at a supermarket between 4 and 5 P.M. on Wednesday afternoons. One way of describing this experience is to specify the number of persons ahead of her when she arrives at the checkout station with the shortest line. The number of persons can take on the values 0, 1, 2, . . . , and each of these is an elementary event, designated as $E_1(=0)$, $E_2(=1)$, $E_3(=2)$, ... Suppose, however, that the outcome classifications "no person in line" and "one or more persons in

line" were used. The first of these is an elementary event since it corresponds to E_1. The second is a *compound* event in that it occurs whenever any one of the elementary events E_2, E_3, . . . , happens.

3. An employee in a plant is asked the question "How closely does your foreman check up on his men?" The answer categories "much too closely," "somewhat too closely," "just about right," "not quite closely enough," "not nearly closely enough," and "no answer" provide a set of elementary events for a single trial of this experiment. Thus E_1 might denote "much too closely," and so on.

It should be observed that the term elementary event is synonymous with the phrase "separate and distinct category or value of a variable" used previously in talking about the infinite population of trials associated with a conceptual experiment. As in the opening discussion of variables, we require that the elementary events associated with a given experiment be mutually exclusive (i.e., only one of the set can occur on a single trial of the experiment) and that the set of elementary events be exhaustive (i.e., there is no possible outcome for a single trial which is not included in the set). The elementary events associated with an experiment will form the basis for the succeeding discussion, and it is with these events that we shall build the compound events that ordinarily arise in the application of statistical method.

The definition of the model for simple random sampling given in the preceding chapter specified not only an experiment and its associated set of elementary events (a possible sample is an elementary event, and the set contains all possible samples) but also the fraction of times that each elementary event is expected to occur if an infinite number of repetitions of the experiment were carried out. That is, even though it is impossible to predict with certainty which elementary event will occur on a *single trial* of the experiment, we assume that we can talk in a quantitative and precise fashion about what will happen for an *infinite set of trials*. The fraction of times that an elementary event E is expected to occur for an infinite set of trials is called the *probability* of that event and will be symbolized by $\Pr(E)$. Thus if repeated simple random samples of size 2 are drawn from a finite population containing the five elements A, B, C, D, and E, then there are 10 elementary events: $E_1 = AB$, $E_2 = AC$, $E_3 = AD$, $E_4 = AE$, $E_5 = BC$, $E_6 = BD$, $E_7 = BE$, $E_8 = CD$, $E_9 = CE$, and $E_{10} = DE$. Furthermore, the simple-random-sampling model requires that $\Pr(E_1) = \Pr(E_2) = \cdots = \Pr(E_{10}) = 1/10$. In general, if we are drawing simple random samples of size n from a finite population containing N elements, then

$$\Pr(\text{any particular sample}) = \frac{1}{\{N!/[n!(N-n)!]\}}$$

that is, one divided by the total number of possible samples. Events having equal probabilities are often referred to as "equally likely" events.

From this treatment of simple random sampling, we see that the elementary events are those most basic outcomes for which we attempt to specify probabilities. They are not necessarily the most basic outcomes which can be identified for an experiment. For example, if we consider drawing the elements of a random sample one by one, we could look at the order in which the elements are drawn and distinguish between sample AB and sample BA. However, our model for simple random sampling specifies the probabilities in terms of the elements in a sample without regard to the order in which they were drawn.

The elementary events (possible samples) and their associated probabilities form the basic "building blocks" with which we shall attempt to develop a theory for handling the practical questions that arise in the use of simple random sampling (e.g., what can be said about the extent of agreement between the mean of a single random sample and the mean of the finite population from which the sample is drawn?). Some uses of these building blocks, and similar ones that arise in connection with other uncertain experiments, will be set forth in the succeeding portions of this book. Before doing this it is appropriate to examine the relationship between the probabilities specified by the model and the use of random-digit tables for the mechanical selection of samples.

Since we think of probabilities as referring to a sequence of repetitions of an experiment, let us use a random-digit table (see Table 6.3) to draw repeated samples from a finite population. For ease of reference, consider an N of 5 and an n of 2. We then have the 10 possible samples that were previously enumerated, where digit 1 is assigned to sample AB, 2 to AC, . . . , and 0 to DE. The results of drawing 1,000 samples are given in Table 7.1. The fraction of times that each sample was drawn is presented separately for each of 10 consecutive sets of 100 random digits, for each of two consecutive sets of 500 digits, and for the entire 1,000 digits.

We observe from Table 7.1 that there is considerable variability within and between sets of 100 digits with respect to the fraction of times that any given sample occurs, although these fractions do "cluster," as expected, around the value of .10. For example, sample AD occurs on 17 per cent of the trials in the second set of 100, while CE occurs on only 4 per cent of the trials in the eighth set of 100. On the other hand, there is much less variability within and between the two sets of 500 digits—CE occurring on 7.4 per cent of the trials in the first set, and BE occurring on 11.4 per cent of the trials in the second set of 500. Finally we note that there is still less variability between these per cents for the complete set of 1,000 trials.

TABLE 7.1. RESULTS OF DRAWING 1,000 SIMPLE RANDOM SAMPLES OF SIZE 2 FROM A FINITE POPULATION OF FIVE ELEMENTS

Sample drawn	Sets of 100 digits, fraction of times each sample drawn for:					Fraction of times in first 500 digits
	First	Second	Third	Fourth	Fifth	
AB	.09	.06	.07	.14	.11	.094
AC	.10	.14	.09	.05	.14	.104
AD	.11	.17	.12	.06	.11	.114
AE	.13	.09	.10	.09	.09	.100
BC	.11	.05	.15	.12	.11	.108
BD	.11	.09	.11	.13	.10	.108
BE	.12	.10	.06	.13	.07	.096
CD	.07	.09	.12	.14	.11	.106
CE	.05	.12	.05	.06	.09	.074
DE	.11	.09	.13	.08	.07	.096
Total....	1.00	1.00	1.00	1.00	1.00	1.000

Sample drawn	Sets of 100 digits, fraction of times each sample drawn for:					Fraction of times in second 500	Fraction of times in 1,000
	Sixth	Seventh	Eighth	Ninth	Tenth		
AB	.08	.06	.10	.14	.16	.108	.101
AC	.10	.13	.09	.08	.09	.098	.101
AD	.15	.09	.09	.09	.07	.098	.106
AE	.12	.11	.08	.06	.11	.096	.098
BC	.13	.10	.10	.12	.09	.108	.108
BD	.11	.15	.10	.10	.09	.110	.109
BE	.08	.11	.15	.11	.12	.114	.105
CD	.07	.06	.13	.09	.11	.092	.099
CE	.06	.12	.04	.13	.11	.092	.083
DE	.10	.07	.12	.08	.05	.084	.090
Total....	1.00	1.00	1.00	1.00	1.00	1.000	1.000

From the foregoing example, it appears intuitively reasonable to expect that the fraction of trials on which any given sample occurs will approach $1/10$ as the number of trials is increased indefinitely. Thus we see that one can think of the probabilities assumed by the simple-random-sampling model as being the ultimate (i.e., as the number of trials is increased indefinitely) values of these empirical fractions. Of course, it is impossible to conduct an infinite number of trials of the experiment and thus obtain a check on the foregoing conjecture, but it can be shown, for example, that the results given in Table 7.1 are not in disagreement with what would be expected on the basis of the model. The procedures for making such judgments concerning agreement or disagreement will be developed in later chapters, especially Chap. 11.

Abstracting from this discussion of simple random sampling, we can describe as follows the framework within which one can treat experiments having uncertain outcomes:

1. A conceptually repeatable experiment is defined by specifying the conditions under which repeated trials are to be made. It must be possible to think of an infinite number of such repetitions.

2. The outcome of any single trial cannot be predicted with certainty. However, a set of possible elementary outcomes or events that may occur on a trial can be enumerated. One and only one of these elementary events, denoted by $E_1, E_2, \ldots$, can occur on any given trial.

3. Associated with each elementary event is a number which is referred to as the probability of the event, designated by $\Pr(E)$.

4. The probability of an event E may be interpreted intuitively as the fraction of times that E would occur in an infinite number of trials of the experiment. Since it is impossible to conduct an infinite number of trials of any experiment, exact values for probabilities cannot be determined by experiment.

5. Since a probability is interpreted as the fraction of times that an event occurs, any probability must lie between 0 and 1 inclusive. This requirement is written formally as

$$0 \leq \Pr(E) \leq 1 \tag{7.1}$$

where $0 \leq \Pr(E)$ is read "0 is less than or equal to $\Pr(E)$," or alternatively, "$\Pr(E)$ is greater than or equal to 0." Similarly, $\Pr(E) \leq 1$ is read "$\Pr(E)$ is less than or equal to 1." If we wish to write "$\Pr(E)$ is greater than 0," we would use $\Pr(E) > 0$ or $0 < \Pr(E)$. $\Pr(E) < 1$ would be read as "$\Pr(E)$ is less than 1."

Now it is quite clear that simple random sampling is a particularly "satisfying" type of conceptually repeatable experiment with which to deal because (1) the numerical values of the probabilities of the elementary events (i.e., the fraction of times that each event is expected to occur in an infinite set of trials) were established by definition; (2) a mechanical procedure which was expected to conform to the model was devised for conducting trials of the experiment; and (3) we were actually able to conduct a "large" number of trials of the experiment for a small finite population. Very frequently we shall not be in this fortunate position. True enough, we may have a conceptual experiment and its associated set of elementary events. However, it may not be possible to specify the fractions of times we expect each event to occur in an infinite set of trials; it may even be difficult to define the conditions under which repetitions of the experiment are to be performed; and it may be impossible to conduct more than a "small" number of trials.

Actually, it is these very difficulties which make it necessary to work within the abstract framework of mathematical probability and to see what kinds of results can be obtained within this framework. In other words, we shall try to form a model for any conceptual experiment in which we are interested, just as we have done for simple random sampling; we shall attempt to deduce results on the basis of this model; and we shall then try to use these results in treating practical problems. The remainder of this chapter will be devoted to obtaining some very simple results on the basis of this framework, and these results will then be used to solve some of the problems that arise continually in the applications of statistical method.

TABLE 7.2. DISTRIBUTION OF LENGTHS OF TELEGRAMS DISPATCHED
(The Netherlands; 10,984 telegrams dispatched on Feb. 8, 1949)

Number of words	*Fraction of telegrams*
3–10	.284
11–15	.424
16–20	.180
21–25	.057
26–30	.023
31–35	.015
36–40	.005
41–45	.005
46–50	.003
Over 50	.004
Total	1.000

SOURCE: J. H. C. Lisman, *Econometrics, Statistics and Thermodynamics*, The Netherlands Postal and Telecommunications Services, The Hague, 1949.

As a conclusion to these remarks concerning the probability framework, it may be well to remind the reader that we have not required that the probabilities associated with the elementary events of an experiment be equal. It so happens that they are equal for the simple-random-sampling model (where the elementary event is the sample drawn) and will also be equal for the events of many other experiments that will be considered, but this is not a necessary requirement. For example, suppose an experiment consists of observing an individual send a telegram and that the elementary events are defined by "the number of words in the telegram." We would not expect, as an illustration, that Pr(number of words = 10) would be equal to Pr(number of words = 50). Some numbers that might be taken as approximations to such probabilities are given in Table 7.2, and we see that they are far from equal. Notice that these empirical fractions are given, not for the elementary events, but for combinations of elementary events.

7.3. The Possible Outcomes of an Experiment: Compound Events

The elementary events of an experiment, and their associated probabilities, form the building blocks for developing the theory of probability. The need for further development is quite clear if it is recognized that many of the previously cited applications of statistics deal not with elementary events but with *compound* events, that is, events obtained by various types of combinations of elementary events. Some of the useful ways of obtaining compound events will be explored in this section. The procedures needed for treating such events on a probability basis will be developed in the following two sections.

One way of combining elementary events was illustrated in Table 6.1. There we were concerned with the behavior of the sample mean $\bar{x}$ when repeated simple random samples of size 3 were selected from a finite population of six elements (A, B, C, D, E, and F). The elementary events in the simple-random-sampling model are the separate and distinct samples that may be chosen. However, we observed in this connection that the individual population elements drawn into a sample are of no concern if attention is focused on the sample mean. Consequently, we wish to combine those elementary events that give rise to the same sample mean. A particular value of the sample mean is a compound event if two or more of the possible samples have this value for their means. The correspondence set up in Table 6.1 was:

Compound event	*Statement in terms of elementary events*
$\bar{x} = 12/3$	Sample ABC occurs
$\bar{x} = 13/3$	Sample ABD occurs
$\bar{x} = 14/3$	Either sample ABE or ACD occurs
$\bar{x} = 15/3$	Either ABF, or ACE, or BCD occurs
$\bar{x} = 16/3$	Either ACF, or ADE, or BCE occurs
$\bar{x} = 17/3$	Either ADF, or BCF, or BDE occurs
$\bar{x} = 18/3$	Either AEF, or BDF, or CDE occurs
$\bar{x} = 19/3$	Either BEF or CDF occurs
$\bar{x} = 20/3$	CEF occurs
$\bar{x} = 21/3$	DEF occurs

It will be observed that each of these compound events is determined by an "either . . . or . . . or . . . " combination of elementary events and that this combination may even be continued in terms of the sample means themselves. For example, we might feel "unhappy" for certain reasons if our sample mean were going to be less than 15/3 or greater than 18/3 and attempt to find the probability of this event. An analysis into elementary events would then be:

Event.	The sample mean $\bar{x}$ is either less than 15/3 or greater than 18/3.
Event in terms of different $\bar{x}$'s. . . .	The sample mean $\bar{x}$ is either 12/3, or 13/3, or 14/3, or 19/3, or 20/3, or 21/3.
Event in terms of possible samples.	Either *ABC*, or *ABD*, or *ABE*, or *ACD*, or *BEF*, or *CDF*, or *CEF*, or *DEF* occurs.

Events obtained by "either . . . or . . . or . . . " combinations of other events may be defined for almost every illustrative example that has been discussed thus far in this book, and such compound events play an important role in any statistical investigation. A few additional examples are:

1. A trial of an experiment consists of asking a single adult resident of the United States the question "How do you feel about the general idea of having an organization like the United Nations?" The elementary events, or separate response categories in this instance, are given in Table 3.3. The compound event "approve of the idea" is, in terms of the listed elementary events, "*either* the person approves 'without qualification' *or* he approves 'with qualifications or uncertainty'."

2. A single trial of an experiment consists of observing the duration of a marriage that terminated through divorce or annulment in Oregon in 1950 (see Table 3.7). If we choose to think of the duration of the terminated marriage as falling in one and only one of the 11 class intervals of Table 3.7, these 11 class intervals determine a set of elementary events for the experiment. The event "lasted for less than 3 years" is then compound since it corresponds to "*either* the marriage lasted for less than 1 year, *or* it lasted from 1 up to 2 years, *or* it lasted from 2 up to 3 years."

3. A single trial of an experiment consists of observing the number of job changes that a tool-and-die maker has made in the 11-year period preceding the time of observation (see Table 3.10). The compound event "one or more job changes" can be stated as "*either* one change, *or* two changes, *or* three changes, ..."

Another type of combination of elementary events may be required whenever we have a conceptual experiment, a single trial of which can be categorized simultaneously with respect to two or more different *sets* of elementary events. This situation has not been previously discussed, except in Sec. 2.3 where it was observed that one might wish to describe a population with respect to two or more variables simultaneously. However, such situations will become increasingly important in succeeding considerations and will therefore be introduced at this point.

It was noted in Sec. 7.1 that the birth of a child may be classified both with respect to the qualitative variable "sex" and with respect to the quantitative variable "number of minutes after the hour at which the birth occurred." Of course many other variables or sets of outcomes could also be used, but attention will be restricted to these two only. In order to emphasize the essentials of the argument, let us agree to consider "time of birth" in terms of the two elementary events "first half hour" and "second half hour." Then we have the following set of possible compound events with respect to which a single birth may be categorized: (1) the birth occurs in the first half hour *and* the child is male; (2) the birth occurs in the first half hour *and* the child is female; (3) the birth occurs in the second half hour *and* the child is male; and (4) the birth occurs in the second half hour *and* the child is female. Thus we have compound events defined on the basis of the simultaneous occurrence of the agreed-upon elementary events.

Several illustrations of the combination of events by an " . . . and . . . and . . . " relationship are:

1. A single trial of an experiment is obtained by observing an acute respiratory illness for an employed male (see Table 4.1 and accompanying discussion). The trial could be classified according to "number of days of disability associated with the illness" and to "age of the individual having the illness." One might then be concerned with such compound events as "the illness results in more than 10 days of disability, and the individual is more than 40 years of age." Compound events defined in this manner would be of great interest if, for example, one were attempting to determine differential rates according to age for disability insurance.

2. A single trial of an experiment consists of observing an individual employed as a tool-and-die maker and classifying the trial in relation to "straight-time hourly earnings" and "industry of employment." Interest might then be centered in compound events like "a worker earns between $2.50 and $2.60 per hour *and* is employed in the aircraft industry," or "a worker earns less than $2.30 per hour *and* is employed in the motor vehicle industry."

A special and extremely important case of the " . . . and . . . and . . . " combination of events arises when two or more trials of a single experiment are considered in conjunction with one another. For example, we have previously referred to the "hiring" experiment, a single trial of which consists of hiring an individual to fill a particular type of position, where the set of elementary outcomes has the two members "success" and "failure." Suppose individuals are hired in pairs, rather than singly. We can, if we like, look upon this as a new conceptual experiment. A single trial consists of hiring two individuals. The possible outcomes of this compound experiment can be related back to the original experiment as follows:

Possible outcomes of compound experiment in terms of original experiment	*An alternative set of possible events for compound experiment: number of "successful" hirings*
First individual is a "success" *and* second is a "success"	2
First individual is a "success" *and* second is a "failure"	1
First individual is a "failure" *and* second is a "success"	1
First individual is a "failure" *and* second is a "failure"	0

The following two sections will be devoted to the development and application of several basic rules of procedure that permit one to proceed formally from the probabilities for elementary events to the probabilities for compound events.

7.4. Probabilities for "Either . . . or . . . " Combinations of Events

Let us first consider an experiment and its associated set of elementary events E_1, E_2, ... Each event E is assigned a number $\Pr(E)$ lying between 0 and 1, inclusive. Although the mathematical theory of probability assumes that these numbers are given and is not ordinarily concerned with their values, we shall illustrate numerically the results of this and the following section by drawing simple random samples of *size* 1 from various finite populations. A sequence of repetitions of this experiment can therefore be obtained by drawing a single population element at random, noting the outcome, replacing the element in the population, making another random draw of a single element, and so on.

Some of the natural properties of the $\Pr(E)$'s have already been used. For example, since $\Pr(E)$ is interpreted as the fraction of times that event E occurs in an infinite number of trials of an experiment and since one and only one of the elementary events must occur on each trial, it is clear that we should require

$$\Pr(E_1) + \Pr(E_2) + \cdots = 1 \tag{7.2}$$

That is, for any experiment the sum of the probabilities for all the elementary events will be equal to 1. This relationship between probabilities has been noted for simple random sampling (see Table 6.1) and for the other conceptual experiments discussed in previous chapters (see Table 3.6 and Fig. 3.13, Form B, together with the accompanying descriptive material).

Another simple but important property of a probability can be obtained by an argument similar to that used in obtaining expression (7.2). Suppose we consider a particular elementary event E and its associated probability $\Pr(E)$ and ask for the probability that E does *not* occur, $\Pr(\text{not } E)$. If E occurs on a fraction $\Pr(E)$ of an infinite sequence of repetitions of the experiment, then the event "not E" will occur on the remaining fraction of trials. That is,

$$\Pr(\text{not } E) = 1 - \Pr(E) \tag{7.3}$$

A useful way of combining the $\Pr(E)$'s for obtaining probabilities of certain types of compound events may be derived from the concept of an elementary event and from the empirical interpretation of a probability. This combining property is: *the probability for an arbitrary event A is equal to the sum of the probabilities of the elementary events that make up the event A*, where the term "arbitrary event" refers to any event that can be defined in terms of "either . . . or . . . " combinations of elementary events. For example, let A be the event "either E_1 occurs or E_2 occurs." Then E_1 occurs on a fraction $\Pr(E_1)$ of an infinite sequence of trials; E_2 occurs on a fraction $\Pr(E_2)$ of all trials, but the trials on which E_2 occurs are different from those on which E_1 occurs; and therefore A occurs on a fraction $\Pr(E_1) + \Pr(E_2)$ of trials. Thus $\Pr(A) = \Pr(\text{either } E_1 \text{ occurs or } E_2 \text{ occurs}) = \Pr(E_1) + \Pr(E_2)$.

The foregoing combining property of the $\Pr(E)$'s is easily illustrated by simple random sampling. Table 7.3 gives the distribution of time intervals (in days) between successive compensable accidents in one district of a coal mine.

TABLE 7.3. DISTRIBUTION OF TIME INTERVALS BETWEEN SUCCESSIVE COMPENSABLE ACCIDENTS IN ONE DISTRICT OF A COAL MINE

Time interval, days	Number of intervals	Fraction of intervals
0	37	.378
1	15	.153
2	18	.184
3	10	.102
4	6	.061
5	2	.021
8	6	.061
10	1	.010
12	1	.010
14	1	.010
23	1	.010
Total........	98	1.000

SOURCE: B. A. Maguire, E. S. Pearson, and A. H. A. Wynn, "The Time Intervals between Industrial Accidents," *Biometrika*, **39**:168–180 (1952).

These 98 time intervals would ordinarily be regarded as 98 observations from an infinite population of time intervals, but here we shall treat them as a finite population from which to draw random samples of size 1. There are 98 elementary events for this experiment, since any one of the 98 time intervals may be drawn on a single trial, and the simple-random-sampling model requires that each have equal probability. That is Pr(of any specified E) $= 1/98$. Suppose we now ask for the probability that "an interval is drawn whose length is zero," the phrase in quotations defining an arbitrary event A in the sense of the preceding discussion. The drawing of any one of the 37 intervals having length zero will give rise to A, and therefore

$$\Pr(A) = \Pr(x = 0) = 37(1/98) = .378$$

Similarly, $\Pr(x = 1) = .153$, $\Pr(x = 2) = .184$, and so on. Probabilities for compound events are given the same empirical interpretation as were the probabilities for elementary events. That is, if repeated simple random samples of size 1 are drawn from the distribution of Table 7.3, then $\Pr(x = 0) = .378$ is the fraction of times an interval of length zero is expected to be drawn in an infinite number of trials.

The result obtained in this particular example is completely general and can be extended to any problem where *simple random samples of size* 1 are being drawn from a finite population. In such situations, *the probability that the element drawn has a specified characteristic* (or a specified value of the variable) *is always equal to the fraction of the population* having this characteristic. As a matter of fact, this is the very reason why the symbol p (standing for probability) was used in the preceding chapters to denote the fraction of a population having a stated value of a variable.

It is quite clear that in drawing simple random samples of size 1 from the distribution of Table 7.3, one is not really interested in which of the 98 intervals occurs on a given trial, except in so far as information is obtained about the value of the variable "length of interval." Therefore, once the probabilities for the events $x = 0$, $x = 1$, etc., have been obtained by applying the definition of simple random sampling, these events can be taken as a starting point for further considerations. That is, the events E_1 $(x = 0)$, E_2 $(x = 1)$, . . . , E_{11} $(x = 23)$ may be regarded as a set of elementary events for the drawing of a random sample of size 1 since one and only one of them can occur on any single trial of the experiment and since we know their respective probabilities. The associated probabilities are given in the right column of Table 7.3. These elementary events can be combined to obtain compound events in many ways, and the probabilities of these compound events may again be obtained by using the fundamental combining property for the probabilities of elementary events. For example:

1. Pr(x is less than 2) = Pr(either $x = 0$ or $x = 1$) $= \Pr(x = 0) + \Pr(x = 1) = .378 + .153 = .531$

2. Pr(x is greater than 8) = Pr(either $x = 10$, or $x = 12$, or $x = 14$, or $x = 23$) $= \Pr(x = 10) + \Pr(x = 12) + \Pr(x = 14) + \Pr(x = 23) = .04$

The property that the probability of any arbitrary event (obtained by "either . . . or . . . " combinations of elementary events) is equal to the sum of the probabilities of the elementary events that make up the arbitrary event can now be used to obtain very general rules for "either . . . or . . . " combinations of all types of events. Suppose we ask, with reference to Table 7.3, for the probability that "either x is less than 3 *or* x is greater than 10." The probability of this event may be obtained by the following analysis:

Arbitrary event	Elementary events making up arbitrary events	Probabilities
A_1 (x is less than 3)	Either $x = 2$ or $x = 1$ or $x = 0$	$\Pr(A_1) = .184 + .153 + .378 = .715$
A_2 (x is greater than 10)	Either $x = 12$ or $x = 14$ or $x = 23$	$\Pr(A_2) = .010 + .010 + .010 = .03$
Either A_1 or A_2	Either $x = 2$ or 1 or 0 or 12 or 14 or 23	$\Pr(\text{either } A_1 \text{ or } A_2) = (.184 + .153 + .378) + (.010 + .010 + .010) = .745 = \Pr(A_1) + \Pr(A_2)$

Thus we have the result that $\Pr(\text{either } A_1 \text{ or } A_2) = \Pr(A_1) + \Pr(A_2)$.

The only special characteristic of the events A_1 and A_2 used in carrying through the preceding analysis was that they had *no elementary events in common.* Two arbitrary events that have no elementary events in common are called *mutually exclusive* events; that is, if either one of them occurs on a given trial of an experiment, then the other cannot occur on the same trial. The foregoing analysis can be extended easily to any number of arbitrary mutually exclusive events, and we have the very useful result: If the arbitrary events $A_1, A_2, \ldots$, are mutually exclusive (i.e., no two of them have an elementary event in common), then

$$\Pr(\text{either } A_1 \text{ or } A_2 \text{ or } \ldots) = \Pr(A_1) + \Pr(A_2) + \cdots \tag{7.4}$$

General formulations of (7.4) are possible for "either . . . or . . . " combinations of events that do have elementary events in common, but such formulations will not be required in this book. Discussions of this point will be found in references cited at the end of this chapter (e.g., Feller, pp. 19, 61).

There is one type of outcome for a conceptual experiment which was emphasized in the preceding chapters but which has not yet been treated explicitly in this chapter. This is the situation where the outcome of the experiment is characterized by a value of a continuous variable. As was observed in Secs. 2.2 and 3.3.2, we do not speak of separate and distinct values of the variable (i.e., outcomes of the experiment) if we think in terms of perfect measurement. Nevertheless, the "elementary event" approach of this section can be used if attention is restricted to *intervals within which the variable may fall,* rather than to individual values of the variable.

Let us return to the convenient illustration of the rectangular distribution of the occurrence times (in minutes after the hour) of births. For ease of reference, this distribution is given in Fig. 7.1. It will be recalled that the total area under this distribution is 1 [corresponding to the requirement of (7.2) that the sum of probabilities for the elementary events be 1] and that the fraction of the population having values of the variable in any interval is equal to the area over that interval. This situation can be interpreted according to the probability terminology of this

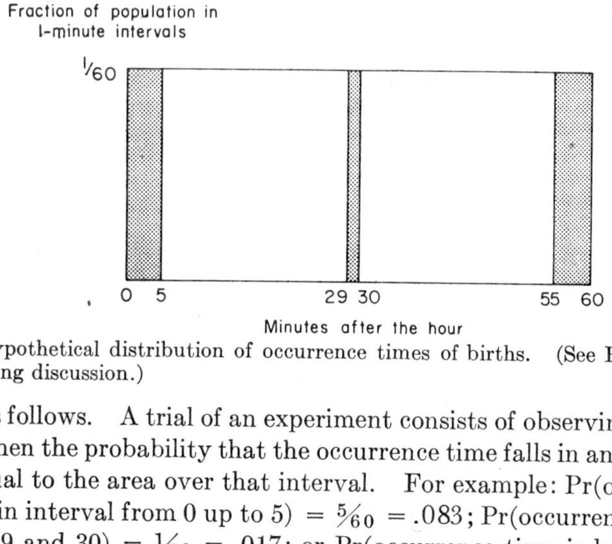

FIG. 7.1 Hypothetical distribution of occurrence times of births. (See Fig. 3.9 and accompanying discussion.)

chapter as follows. A trial of an experiment consists of observing a single birth. Then the probability that the occurrence time falls in any interval is also equal to the area over that interval. For example: Pr(occurrence time falls in interval from 0 up to 5) = $5/60$ = .083; Pr(occurrence time is between 29 and 30) = $1/60$ = .017; or Pr(occurrence time is between 55 and 60) = $5/60$ = .083.

Expression (7.4) may also be used to combine these probabilities if we recognize that two events A_1 and A_2, where each is defined in terms of an interval on the scale of the variable, are mutually exclusive if they contain no common values of the variable (i.e., they do not overlap). Thus we have such results as the following:

1. Pr(*either* x is in the interval 0 up to 5 *or* it is in the interval 5 up to 10) = Pr(x is in the interval 0 up to 5) + Pr(x is in the interval 5 up to 10) = .085 + .085 = .170.

2. Pr(*either* x is in the interval 0 up to 5 *or* it is in the interval 55 up to 60) = Pr(x is in the interval 0 up to 5) + Pr(x is in the interval 55 up to 60) = .085 + .085 = .170

7.5. Probabilities for "... and ... and ..." Combinations of Events

It was observed in Sec. 7.3 that the need frequently arises for considering experiments whose single trials may be classified simultaneously with respect to two or more variables or, alternatively, two or more sets of elementary outcomes. In this section the basic probability approach required for dealing with such experiments will be developed.

As an example of the foregoing type of situation, consider the data of Table 7.4. These data were obtained from a sample of 37,844 marriage licenses issued in Seattle during the period 1939 to 1946, and each license is classified according to the "previous marital status" of both the bride and the groom. The previous marital status of either the bride or the groom is one of three elementary events: E_1 (single), E_2 (divorced), and E_3 (widowed). However, when the previous marital status of the bride and the groom are considered jointly, there are nine possible categories into which a license may fall: bride single *and* groom single, bride single *and* groom divorced, . . . , bride widowed *and* groom widowed. These nine possibilities are represented by the nine cells in the interior of the table, and the number of licenses falling in each cell is indicated. The numbers in parentheses are the fractions of the 37,844 licenses that fall in each cell, and they add to 1.

TABLE 7.4. CLASSIFICATION OF MARRIAGE LICENSES ACCORDING TO THE PREVIOUS MARITAL STATUS OF THE BRIDE AND THE GROOM
(Data taken from a sample of licenses issued in Seattle, 1939–1946)

Previous marital status of the bride	Previous marital status of the groom			Total
	Single	Divorced	Widowed	
Single	24,756 (.654)	2,397 (.063)	459 (.012)	27,612 (.729)
Divorced	3,917 (.104)	3,057 (.081)	694 (.018)	7,668 (.203)
Widowed	1,046 (.028)	721 (.019)	797 (.021)	2,564 (.068)
Total	29,719 (.786)	6,175 (.163)	1,950 (.051)	37,844 (1.000)

SOURCE: Charles E. Bowerman, "Assortative Mating by Previous Marital Status: Seattle, 1939–1946," *Amer. Sociol. Rev.*, **18**:170–177 (1953).

For present purposes we shall regard these data as referring to a finite population containing $N = 37{,}844$ elements, and our considerations will again be in terms of drawing a simple random sample of size 1 from the population. One further preliminary point concerns notation. The number of elements falling in a cell will be denoted by an N with a double subscript. Thus $N_{SS} = 24{,}756$, the number of licenses for which both the bride and the groom were single. Similarly $N_{DW} = 694$, the number of licenses for which the bride was divorced and the groom widowed. The first subscript indicates the previous marital status of the bride (a row of the table), and the second subscript indicates the previous marital status of the groom (a column of the table). There will also be occasion to refer to the number of licenses for which the bride was single (without regard to the previous marital status of the groom), and these cases will be represented as $N_{S\cdot} = 27{,}612$. Similarly $N_{\cdot W} = 1{,}950$, the number of licenses for which the groom was widowed (without regard to previous marital status of the bride).

Now let us consider drawing repeated random samples of *size* 1 from this population of licenses. The results of the preceding section have shown us how to determine the probabilities of various outcomes. For example:

$$\Pr(SS) = \frac{N_{SS}}{N} = .654 \qquad \Pr(WD) = \frac{N_{WD}}{N} = .019$$

$$\Pr(DD) = \frac{N_{DD}}{N} = .081 \qquad \Pr(SW) = \frac{N_{SW}}{N} = .012$$

$$\Pr(\text{bride } S) = \frac{N_{S\cdot}}{N} = .729$$

$$\Pr(\text{groom } S) = \frac{N_{\cdot S}}{N} = .786$$

Notice that Pr(bride widowed *and* groom divorced) has been abbreviated as $\Pr(WD)$.

The foregoing probabilities describe the situation as presented in Table 7.4, but it is desirable to attempt to push the analysis back one step further. For example, one may ask, What is the relationship between a probability for a compound event, say $\Pr(SS)$, and the probabilities for the elementary events combined to form this compound event, Pr(bride single) and Pr(groom single)? This question can be answered by the following simple manipulation:

$$\Pr(SS) = \frac{N_{SS}}{N} = \frac{N_{S\cdot}}{N} \cdot \frac{N_{SS}}{N_{S\cdot}}$$

where the second form is obtained from the first by multiplying both numerator and denominator by $N_{S\cdot}$ (which does not change the value of the probability) and placing $N_{S\cdot}$ over N. Recognizing that $N_{S\cdot}/N$ is Pr(bride S), this becomes

$$\Pr(SS) = \Pr(\text{bride } S) \cdot \frac{N_{SS}}{N_{S\cdot}}$$

The numerator of the quantity $N_{SS}/N_{S\cdot}$ is the number of licenses for which the bride is single *and* the groom single, while the denominator is the number of licenses for which the bride is single (without regard to the previous marital status of the groom). $N_{SS}/N_{S\cdot}$ can be interpreted as follows. Suppose we take all licenses for which the bride is single $N_{S\cdot}$ and draw one of these at random. Then $N_{SS}/N_{S\cdot}$ is the probability of obtaining a license for which the groom is single, where we know already that the bride for this license is single. A probability such as this is usually called a *conditional probability*. Therefore we have

$$\begin{aligned}\Pr(SS) &= \Pr(\text{bride } S) \cdot \Pr(\text{groom single } \textit{given that} \text{ bride is single}) \\ &= \Pr(\text{bride } S) \cdot \Pr(\text{groom single} \mid \text{bride single})\end{aligned}$$

Notice that the phrase "given that" has been replaced by a vertical bar. This is reasonably standard notation and will be used throughout the remainder of this book.

This analysis can be applied to any of the cells of Table 7.4. Some representative applications are:

$$\begin{aligned}\Pr(DW) &= .018 = \Pr(\text{bride } D) \cdot \Pr(\text{groom } W \mid \text{bride } D) \\ &= (.203) \cdot (694/7{,}668) = (.203) \cdot (.091) \\ &= .018 \\ \Pr(DW) &= .018 = \Pr(\text{groom } W) \cdot \Pr(\text{bride } D \mid \text{groom } W) \\ &= (.051) \cdot (694/1{,}950) = (.051) \cdot (.356) \\ &= .018 \\ \Pr(SS) &= .654 = \Pr(\text{bride } S) \cdot \Pr(\text{groom } S \mid \text{bride } S) \\ &= (.729) \cdot (24{,}756/27{,}612) = (.729) \cdot (.897) \\ &= .654\end{aligned}$$

Notice that the first two of these use the fact that our analysis has symmetry, namely,

$$\begin{aligned}\Pr(DW) &= \Pr(\text{bride } D) \cdot \Pr(\text{groom } W \mid \text{bride } D) \\ &= \Pr(\text{groom } W) \cdot \Pr(\text{bride } D \mid \text{groom } W)\end{aligned}$$

The reader can check this symmetry in terms of the original data.

The drawing of simple random samples of size 1 from the population represented in Table 7.4 has provided direction for dealing with " . . . and . . . and . . . " combinations of events. Abstracting from this discussion, we define the following rule for such combinations: Given any two arbitrary events A and B, representing outcomes of conceptually repeatable experiments, the probability that both A and B occur is given as

$$\begin{aligned}\Pr(\text{both } A \text{ and } B \text{ occur}) &= \Pr(A) \cdot \Pr(B \mid A) \\ &= \Pr(B) \cdot \Pr(A \mid B) \qquad (7.5)\end{aligned}$$

It will be recalled that $\Pr(A)$ represents the probability that A occurs without regard to occurrence or nonoccurrence of B, $\Pr(B)$ represents the probability that B occurs without regard to the occurrence or nonoccurrence of A, $\Pr(B \mid A)$ is the probability that B occurs *given that* A has already occurred, and $\Pr(A \mid B)$ is the probability that A occurs *given that* B has already occurred. Expression (7.5)—together with a special case that will soon be discussed—and expression (7.4) will form the basis for our succeeding development of statistical method.

As an illustration of the joint application of expressions (7.4) and (7.5), let us return to the one-by-one procedure used in Chap. 6 for drawing a simple random sample from a finite population. Suppose we have a finite population containing the 10 elements $E_1, E_2, E_3, \ldots, E_{10}$ and wish to draw a simple random sample of size 2 from this population. Since the two elements are to be drawn one at a time, it is necessary to analyze the experiment in terms of the "element obtained on the first draw" *and* the "element obtained on the second draw." In order to be more definite, let us ask for the probability that the resulting sample contains the two elements E_1 and E_2. The event "sample contains elements E_1 and E_2" can be written as "either 'E_1 on first draw *and* E_2 on second' *or* 'E_2 on the first *and* E_1 on second'." The probability of this event can be obtained by a joint application of (7.4) and (7.5).

Let A be the event "E_1 is drawn on the first trial" and B the event "E_2 is drawn on the second trial," then (7.5) says

$$\Pr(E_1 \text{ on first } \textit{and } E_2 \text{ on second}) = \Pr(E_1 \text{ on first}) \cdot \Pr(E_2 \text{ on second} \mid E_1 \text{ on first})$$

On the first trial a single element is being drawn at random from the 10 elements. Therefore $\Pr(E_1 \text{ on first}) = 1/10$. Looking at the second trial, *assuming E_1 has been drawn on the first*, we are drawing a single element at random from the nine remaining elements $E_2, E_3, \ldots, E_{10}$. Therefore

$$\Pr(E_2 \text{ on second} \mid E_1 \text{ on first}) = 1/9, \text{ and}$$
$$\Pr(E_1 \text{ on first } \textit{and } E_2 \text{ on second}) = 1/10 \cdot 1/9 = 1/90$$

This analysis can be applied in a similar fashion to the event "E_2 on the first *and* E_1 on the second," and we obtain the same answer as before, namely, that $\Pr(E_2 \text{ on first } \textit{and } E_1 \text{ on second}) = 1/90$.

Finally, we observe that the events "E_1 on first and E_2 on second" and "E_2 on first and E_1 on second" are mutually exclusive since if E_1 is drawn on the first trial it cannot also be drawn on the second. Therefore, using expression (7.4),

$$\begin{aligned}&\Pr(\text{sample is composed of elements } E_1 \text{ and } E_2)\\&= \Pr(E_1 \text{ on first } \textit{and } E_2 \text{ on second}) + \Pr(E_2 \text{ on first } \textit{and } E_1 \text{ on second})\\&= 1/90 + 1/90 = 1/45\end{aligned}$$

This agrees with the analysis of Chap. 6 since there are $10!/(2!8!) = 45$ possible samples of size 2, and the "equal chance" feature of simple random sampling requires that each have the same probability, namely, $1/45$.

The quantity $\Pr(A \mid B)$ entering into (7.5) was defined to be the probability that event A occurs *given that* (or on the condition that) B has already occurred. It seems natural to compare this conditional probability with $\Pr(A)$, the probability that would be assigned to event A in the absence of any knowledge concerning the occurrence or nonoccurrence of event B. There are many combinations of events A and B for which these two probabilities cannot be expected to be equal. For example:

1. In drawing simple random samples of size 1 from the 37,844 marriage licenses of Table 7.4, we have Pr(bride single) = .729. However, we also have

$$\Pr(\text{bride single} \mid \text{groom single}) = \frac{24{,}756}{29{,}719} = .833$$

$$\Pr(\text{bride single} \mid \text{groom divorced}) = \frac{2{,}397}{6{,}175} = .388$$

and $$\Pr(\text{bride single} \mid \text{groom widowed}) = \frac{459}{1{,}950} = .235$$

Thus the probability of drawing a "single" bride *depends* to a considerable extent on the previous marital status of the groom.

2. Table 4.1 gives the distribution of days of disability for 1,090 acute disabling illnesses, separated according to the two causes "respiratory" and "nonoccupational accidents." If a simple random sample of one illness is drawn from these 1,090, then

$$\Pr(\text{illness lasts more than 21 days}) = \frac{84}{1{,}090} = .077$$

However,

$$\Pr(\text{illness lasts more than 21 days} \mid \text{disability due to respiratory illness}) = \frac{51}{968} = .053$$

while

$$\Pr(\text{illness lasts more than 21 days} \mid \text{disability due to nonoccupational accident}) = \frac{33}{122} = .270$$

Thus the probability of an event defined in terms of length of disability *depends* on the cause of disability.

If it happens that we are dealing with two events A and B for which $\Pr(A)$ does equal $\Pr(A \mid B)$, then we see that the occurrence or nonoccurrence of A does not depend on the occurrence or nonoccurrence of B. It is then customary to replace $\Pr(A \mid B)$ by $\Pr(A)$ in (7.5) and to state as a matter of definition that two events A and B are *independent* if

$$\Pr(\text{both } A \text{ and } B \text{ occur}) = \Pr(A) \cdot \Pr(B) \tag{7.6}$$

The illustration and application of (7.6) is most easily carried out in terms of repeated trials of a single conceptual experiment, a situation that was referred to at the end of Sec. 7.3. For example, suppose an experiment consists of drawing a single time interval at random from the distribution of Table 7.3, noting the length of the interval drawn, *replacing* the interval drawn into the population, and then drawing a second interval at random. Possible outcomes of this two-trial experiment are such events as "first trial 0 *and* second trial 0," "first trial 0 *and* second trial 2," "first trial 23 *and* second trial 3," and so on. It seems reasonable under these circumstances to assume that the outcome on the first trial has no influence on the outcome of the second trial, and vice versa. (Note that this would not be true if the interval obtained on the first draw were not replaced in the population before the second draw was made.) Therefore we say the first and second trial outcomes are independent. Some typical probabilities are:

$$\begin{aligned}
\Pr(\text{first trial 0 } \textit{and} \text{ second trial 0}) &= (.378) \cdot (.378) = .143\\
\Pr(\text{first trial 0 } \textit{and} \text{ second trial 3}) &= (.378) \cdot (.102) = .039\\
\Pr(\text{first trial 10 } \textit{and} \text{ second trial 2}) &= (.101) \cdot (.184) = .002\\
\Pr(\text{first trial 8 } \textit{and} \text{ second trial 23}) &= (.061) \cdot (.010) = .0006\\
\Pr(\text{first trial 23 } \textit{and} \text{ second trial 8}) &= (.010) \cdot (.061) = .0006
\end{aligned}$$

Observe that these probabilities are to be given the same empirical interpretation as before. That is, a single trial of an experiment now consists of making two independent random drawings of one element from the distribution of Table 7.3. Then .143, for example, is the fraction of all such pairs of drawings for which we expect the first element to be 0 and the second to be 0.

A second application of (7.6) can be phrased in terms of the distribution pictured in Fig. 7.1 of occurrence times of births. The occurrence time of a birth is observed and is classified into one of the three nonoverlapping intervals "first 5 minutes," "middle 50 minutes," and "last 5 minutes." Let us designate these three events by A_1, A_2, and A_3. The probabilities of these events are: $\Pr(A_1) = .085$, $\Pr(A_2) = .83$, and $\Pr(A_3) = .085$. Suppose we now make *independent* observations of the occurrence times of two births. Expression (7.6) can then be used to compute such probabilities as the following:

$$\Pr(\text{first birth } A_1 \textit{ and } \text{second birth } A_2) = (.085) \cdot (.830) = .071$$
$$\Pr(\text{first birth } A_1 \textit{ and } \text{second birth } A_3) = (.085) \cdot (.085) = .007$$
$$\Pr(\text{first birth } A_2 \textit{ and } \text{second birth } A_2) = (.830) \cdot (.830) = .689$$

Expression (7.6) can be extended to any number of independent events, and the general result is obtained that if events $A, B, C, D, \ldots,$ are independent, then

$$\Pr(A \textit{ and } B \textit{ and } C \ldots \text{ occur}) = \Pr(A) \cdot \Pr(B) \cdot \Pr(C) \cdot \cdots \quad (7.7)$$

Thus, in the example of the preceding paragraph, suppose we make independent observations on the occurrence times of three births. Then

$$\Pr(\text{first birth } A_2 \textit{ and } \text{second birth } A_3 \textit{ and } \text{third birth } A_1)$$
$$= \Pr(A_2) \cdot \Pr(A_3) \cdot \Pr(A_1) = (.83) \cdot (.085) \cdot (.085)$$
$$= .006$$

As a concluding point in this section, note that repeated and independent trials of an experiment provide an answer to a problem that was raised in Sec. 6.6. It was stated there that one cannot speak of drawing a simple random sample from an infinite population of elements. Nevertheless, one can speak of making n independent observations on elements from such a population. Therefore we see that "n independent trials" play a somewhat similar role in treating infinite populations as does "a simple random sample of size n" in treating a finite population.

7.6. Models Are Required for Treating Conceptually Repeatable Experiments

The preceding portions of this chapter have been concerned mainly with the statement and illustration of some basic definitions and manipulative rules that are fundamental to the mathematical theory of probability. These definitions and rules are necessary prerequisites for the use of a probability approach, but they tend to focus attention on one or more isolated outcomes of a conceptually repeatable experiment. Moreover, they are not sufficient, by themselves, to ensure the success of every attempt to apply probability theory. Therefore we wish to remind the reader that successful applications usually require a comprehensive or *model* view of a conceptually repeatable experiment. That is, the experiment must be viewed as an entity so that the relationships between the probabilities of the elementary events and the relationships between the outcomes of successive trials of the experiment are stated in unmistakable terms.

The model for simple random sampling, as described in Chap. 6, is one example of such a model. This particular model specified a complete set of elementary events (the possible samples of size n that may be drawn from a finite population containing N elements), the numerical values of the probabilities for the elementary events, and the requirement that repeated trials of the experiment be independent. From this model it is possible to deduce, by means of the manipulative rules set forth in this chapter, results that will hold in repeated and independent drawings of simple random samples. For example, Table 6.1 gives a specific description of the behavior to be expected of $\bar{x}$ in repeated trials. Generalizations and modifications of this description will permit the evaluation *from the sample itself* of the degree to which the mean of a simple random sample may be expected to approximate the mean of the population from which the sample was drawn.

It was pointed out in Sec. 6.4 that information is rarely obtained from all population elements drawn into the sample and that practice does not conform exactly to the model. As a consequence, the model may seem inappropriate or even worthless. This, however, is not true. Practice often does agree with the model closely enough for the purposes of certain studies. Often the model may be changed to conform more closely to practice. For example, the model may be changed to allow for the possible effects of individuals who refuse to be interviewed, or enlarged to make use of information available about the population. These are important considerations in the current theory of sample-survey design.

Every situation in which probability theory is to be applied meaningfully must be analyzed in a manner somewhat similar to that set forth for simple random sampling. That is:

1. It is first necessary to obtain a probability model. This model will specify the conceptual experiment involved, the elementary events, the probabilities of the elementary events (although not necessarily in numerical terms), and the relationships among the results of repeated trials of the experiment.

The simple-random-sampling model was presented as an accomplished fact. It can be derived from the following very basic requirement: "A sample of size n is to be drawn from a finite population containing N elements. No information is available concerning the population, and the cost to the sampler will be the same for all possible samples. What sampling procedure should be used?" Other sampling models can be obtained by assuming that certain information about the population is known or by imposing certain cost requirements on the undertaking. Some of these models will be discussed in Chap. 10.

We usually design sampling models to fulfill stated specifications, and then try to "fit" practice to the model. In many other situations, the procedure is reversed and we attempt to "fit" models to given observable phenomena. These situations require that one formulate the way in which one thinks the mechanism underlying a phenomenon operates, and proceed from this to the probability model. Such a development may require great care. For example, the study of industrial accidents was briefly treated in Sec. 2.3, and we noted that one could start from one of the three assumptions: (*a*) accidents occur "by chance"; (*b*) some individuals are more liable to accidents than others; or (*c*) all individuals are "equally" liable to accidents in the beginning, but once an individual has sustained an accident, he then becomes more or less liable to future accidents. The model obtained will of course depend on which of these three assumptions is used, and the probabilities of the elementary events will be symbolic and not numerical.

2. From the models and the rules for treating probabilities, certain consequences will be deduced. If the model is designed to fulfill stated requirements and practice is then "fitted" to the model, as in the case of simple random sampling, it is only necessary to make sure that practice and model agree and to use the deduced consequences. If, on the other hand, the model is being "fitted" to some phenomenon, as in the case of accidents, the derived results will be compared with actual data to see how well the model fits the phenomenon.

3. If the model fits "satisfactorily," it will be used in succeeding work; if it fits only "somewhat satisfactorily," then an attempt may be made to modify the model; and, finally, the fit may be "unsatisfactory" and a new attempt will have to be made.

The actual development of models, as described under step 1, is often a complex undertaking which involves deep understanding in a particular subject-matter field. Furthermore, the manipulations involved in 2 then require considerable ingenuity and technical skill. Such matters are beyond the scope of this book. In the following chapters we shall, however, describe and apply some of the simple probability models that have been found most useful in the field of statistics, and the emphasis will therefore be on step 2 above.

EXERCISES

7.1. The following table presents the distribution of length of life for a group of grocery wholesalers, where the length of life of a firm is the number of years it was in business. These data are adapted from a study of 169 grocery wholesalers who were in business at any time during the period 1893 to 1951 in the city of Seattle, Wash.

Length of life, years		*Fraction of wholesalers*
0 up to 5		.651
5 up to 10		.160
10 up to 15		.089
15 up to 25		.047
25 and over		.053
	Total	1.000

SOURCE: Richard B. Still, "Mortality of Seattle Grocery Wholesalers," *Jour. Marketing*, **18**:160–165 (1953).

a. A *single* wholesale grocery firm is to be drawn at random from this group. What is the probability: (1) that it was in business for less than 5 years? (2) That it was in business for 5 years or more? (3) That it was in business for less than 10 years? (4) That it was in business for 15 years or more? (5) That it was in business for at least 10 years? (6) That it was in business for either less than 5 years or for 25 years or more?

b. Suppose that a sample of size 2 is drawn *with replacement.* That is, a single firm is drawn at random and its length of life noted; the firm drawn is replaced in the population, and a second firm is drawn at random. What is the probability: (1) That both firms were in business for less than 10 years? (2) That both were in business for 25 or more years? (3) That the first firm drawn was in business for less than 5 years and the second was in business for 25 or more? (4) That one of the two firms was in business for less than 5 years and the other for 25 or more? (5) That at least one of the two was in business for 25 or more years?

7.2. Students of political behavior are often interested in the point in time at which a voter decides on the candidate for whom he is going to cast his ballot. Following the 1948 Presidential election, the Survey Research Center of the University of Michigan questioned a national sample of 390 individuals who had voted in the election. Each individual was asked, among other things, whether he had voted for Truman or Dewey and at what point he had made up his mind. The results were:

DATA FOR EXERCISE 7.2

When a decision was made on a candidate	Candidate voted for:		Total
	Truman	Dewey	
Knew all along	76	75	151
Decided at time of convention	47	60	107
Decided during campaign	30	23	53
Decided within 2 weeks of election	30	5	35
Decided on election day	6	4	10
Don't know when decision made	2	4	6
Time not ascertained	21	7	28
Total	212	178	390

SOURCE: Angus Campbell and Robert L. Kahn, *The People Elect a President*, University of Michigan, Survey Research Center, Ann Arbor, Mich., 1952.

For present purposes we shall regard these 390 individuals as constituting a finite population from which to draw simple random samples of size 1.

a. If a single individual is drawn at random from this population, what is the probability that the individual: (1) Knew all along for whom he was going to vote? (2) Decided on election day? (3) Decided before election day? (4) Voted for Truman? (5) Knew all along and voted for Truman? (6) Knew all along and voted for Dewey? (7) Decided before election day and voted for Truman? (8) Decided before election day and voted for Dewey?

b. If it is *known* that the individual drawn at random voted for Truman, what is the probability that he "knew all along" for whom he was going to vote? What is the corresponding probability if it is known that the individual voted for Dewey?

c. Compare Pr("Knew all along" and "voted for Truman") with Pr("Knew all along") · Pr("voted for Truman").

7.3. An adult who has had neither measles nor mumps suddenly discovers (too late for him to protect himself) that one of his children has mumps and that a second has measles. Suppose that the probability of his coming down with mumps after exposure is .15 and of coming down with measles after exposure is .20 and that it is possible to have both diseases at the same time. What is the probability that:

a. He has both diseases?
b. He has neither disease?
c. He has mumps and not measles?
d. He has measles and not mumps?

7.4. A survey is conducted to measure the readership of magazines A and B. It is found that 25 per cent of the families buy A and 40 per cent buy B. If there is no relationship between whether or not a family buys A and whether or not it buys B, what is the probability that a randomly selected family buys both A and B?

7.5. In a certain area 10 per cent of the homes have television and 25 per cent have automatic washers. A single home is drawn at random. Do you think that it would be appropriate to compute the probability of obtaining a home having *both* a television set *and* an automatic washer by multiplying .10 by .25? Why or why not?

OTHER READING

Feller, William: *An Introduction to Probability Theory and Its Applications*, vol. 1, John Wiley & Sons, Inc., New York, 1950. For those who want the best exposition of the subject available. The first chapters require no advanced mathematics but are quite condensed. Recommended only for those who intend to study statistics seriously.

Savage, Leonard J.: *The Foundations of Statistics*, John Wiley & Sons, Inc., New York, 1954. This book attempts to formalize a personal approach to probability. The casual reader can skip between the formulas and obtain the flavor of the argument.

Wilks, S. S.: *Elementary Statistical Analysis*, chap. 4, Princeton University Press, Princeton, N.J., 1948.

CHAPTER 8

THE BINOMIAL PROBABILITY MODEL AND STATISTICAL INFERENCE

Producers and consumers of empirical research must often make inferences and decisions in the absence of complete information. Some aspects of this problem have been emphasized in the preceding chapters by comparing numerical measures of certain characteristics of population distributions and the corresponding measures for samples drawn from these populations. This direct comparison will not be possible in practice. Only measures of the sample characteristics will be available, and inferences must be drawn from these to the entire population.

The purpose of this chapter is to show how the basic concepts of probability theory can be used to attack this problem. The discussion will center on one particularly simple but generally applicable type of situation where the need for inference arises. The simplicity of this situation will make it possible to develop the approach in complete detail, and the fundamental ideas of the approach can then be applied to other situations.

8.1. The Binomial Probability Model and Its Applicability

8.1.1. *Situations to Be Considered.* We are going to study problems that arise in describing populations, either finite or infinite, with respect to variables having only two possible values or categories. A variable having only two categories will be called a dichotomous variable or, more briefly, a dichotomy. There are many variables (and populations in which they can be studied) which are naturally dichotomous. In addition, almost any variable may be *reduced* to a dichotomy. A few examples will serve to illustrate these points.

Some qualitative variables which would ordinarily be regarded as having two natural categories are the following: (1) a coin is tossed, and the result is either a "head" or a "tail"; (2) a telephone number is dialed, and it is either "busy" or "not busy"; (3) a worker in a plant is either "absent" or "present" on a stated day; (4) an item produced by a machine is either "defective" or "nondefective"; (5) the answer to a true and false question is either "right" or "wrong"; (6) an individual is asked to designate the members of a group with whom he would be willing to work and each individual member is either "chosen" or "not chosen"; (7) a seed placed in the proper environment either "germinates" or "does not germinate"; and (8) a person is requested to give either a "yes" or "no" answer to a stated opinion question.

It is possible to study dichotomous variables in either finite or infinite populations. For example, one may wish to focus attention on the output of a machine during a stated period of time and thus obtain a finite population or "lot" of items, each item being "defective" or "nondefective"; or one may consider the machine as running continuously and thus conceive of an extremely large or even infinite population of items. The viewpoint adopted will depend upon the goals of a particular investigation. Thus the occasional buyer of items will be interested in the particular lot or lots which he receives, while the manufacturer will be more interested in the long-run quality of the product.

Even if a qualitative variable has more than two natural categories, it is always possible, although perhaps undesirable, to combine categories so that a dichotomous variable results. For example, Table 3.3 shows seven response categories for the question "How do you feel about the general idea of having an organization like the United Nations?" A number of ways in which these seven categories might be dichotomized are (1) "approve" and all other categories, (2) "approve or approve with qualifications" and all other categories, (3) "do not know what the United Nations is" and all other categories, and (4) "disapprove or disapprove with qualifications" and all other categories. These dichotomizations of the seven response categories illustrate the general line of approach that is always applied. One of the two reduced categories is a focal point for the analysis, while the other is a residual, or "all other," category. Each of the original categories can be made a focal point in its turn, but the approach of this chapter will not be directly applicable if one wishes to study all the original categories *simultaneously*.

Quantitative variables can also be dichotomized, the dichotomizations being based on the order relations among the different values of the variable. Any of the distributions of quantitative variables given in the earlier chapters suggest many ways of obtaining dichotomies. For example:

1. Exercise 3.3 gives the distribution of "number of persons" living in occupied dwelling units in the United States in 1950. Dichotomous reductions are (*a*) "one person" and "more than one person," and (*b*) "three persons or fewer" and "more than three persons."

2. Exercise 5.3 gives the distribution of the variable "distance moved," in thousands of feet, for changes in place of residence in Cleveland, Ohio, during the period 1933 to 1935. Dichotomous reductions are (*a*) "moved less than 3,000 feet" and "moved 3,000 feet or more," and (*b*) "moved

between 9,000 and 15,000 feet" and "moved less than 9,000 or more than 15,000 feet."

3. Table 6.1 shows the possible values of the sample mean $\bar{x}$ when simple random samples of size 3 are drawn from a small finite population. Dichotomous reductions are (*a*) the sample mean is "less than $^{14}\!/_3$" and the sample mean is "$^{14}\!/_3$ or greater," and (*b*) the sample mean is "less than $^{14}\!/_3$ or is greater than $^{19}\!/_3$" and the sample mean is "between $^{14}\!/_3$ and $^{19}\!/_3$, inclusive."

It should be clearly understood that the foregoing reductions of quantitative variables are given merely as illustrations and that there is no implication that one must use dichotomies in order to apply the ideas of inference to the study of distributions of quantitative variables. If the goals of a particular investigation make it appropriate to use a dichotomy, all well and good. If a dichotomy is not appropriate, there is no need to force the analysis into the framework of this chapter. Other methods of procedure are available, and some of these will be discussed in succeeding chapters.

The methods to be considered can be applied to any dichotomous variable. A framework of general notation and terminology is therefore needed. Attention will ordinarily be centered on one of the two categories, and this focal category will henceforth be referred to as "success," abbreviated S. The other category, frequently a residual category, will be called "failure," abbreviated F. Thus each element in the defined population is either an S or an F. Furthermore, this population is completely described with respect to the dichotomous variable if one specifies the fraction of the population falling in the S category. This population fraction will be denoted by the symbol p. The fraction of the population falling in the F category will always be equal to $(1 - p)$ since each population element must be in one and only one of the two categories S and F. It is customary to denote $(1 - p)$ by the symbol q, and it is clear that the sum of p and q is always 1. Under these circumstances, our problem is that of learning something about the value of p from observations made on a sample drawn from the population.

8.1.2. *The Binomial Model.* If a simple random sample of size 1 is drawn from a finite population having the properties just described, or if a single trial is conducted of an experiment which defines an infinite population having these properties, then the elements of mathematical probability can immediately be applied. That is, we have a conceptually repeatable experiment where each trial of the experiment must result in one of the two elementary events S or F. Furthermore, the probability of obtaining an S on a single trial is equal to p, that is, $\Pr(S) = p$.

Now it is quite clear that a single trial of this conceptually repeatable experiment is going to give very little information about the value of the unknown parameter p. However, the experiment is not only conceptually repeatable for an indefinitely large number of trials but is also practically repeatable for any selected finite number n of trials. It therefore seems intuitively reasonable that the number of S's occurring in a set of n trials will provide some information about the value of p. It will be assumed that the successive trials are independent of one another. That is, the occurrence of an S or F on any one or more of the set of n trials has no influence on whether S's or F's occur on any of the remaining trials.

Abstracting from this discussion, situations of the following type will be considered:

1. There exists a conceptually repeatable experiment, a single trial of which results in either an S or an F.
2. On any single trial of the experiment the probability of obtaining an S is equal to p. That is, $\Pr(S) = p$ and $\Pr(F) = 1 - p = q$.
3. n independent trials of the experiment will be performed, and the number of S's occurring will be recorded.

These three conditions define the *binomial model*. From this model it is possible to deduce, assuming a known value of p, how the number of S's in a sample of n will vary from sample to sample. This, however, is only a way station in attaining the real goal, namely, to infer from the number of S's in a sample something about an unknown value of p. Section 8.2 will treat the problem of deduction, while the remainder of the chapter will use these results to solve problems of statistical inference.

8.1.3. *The Approximate Nature of the Binomial Model.* The results that can be obtained from the binomial model provide exact descriptions of actual situations only if the requirements of the model hold. That is, we must be able to carry out n independent trials of the experiment, and we must know that the probability of obtaining an S on any trial is equal to p. It is rarely possible to fully satisfy these requirements. For example:

1. If the elements of a simple random sample are drawn one by one without replacement from a finite population, as described in Chap. 6, then the binomial model does not apply. The probability of obtaining an S on the tenth draw, for example, depends on the number of S's and F's that were obtained in the preceding nine draws. This point was illustrated in Sec. 7.5. On the other hand, if the elements are drawn one by one with replacement, then the constant probability and independence requirements of the model are satisfied and the theory can be used without reservation—assuming that no other difficulties are encountered in making measurements on the chosen sample elements, and so on.
2. Assume that the elements of a simple random sample are drawn one by one with replacement from a finite population but that a certain segment of the population refuses to give answers to the interviewers.

Under these circumstances, the binomial model describes the samples drawn from the *answer segment* of the population and ignores the existence of the *no-answer segment.*

3. An operator-machine combination is continually producing some small item. Each item is either "defective" or "nondefective," and interest is centered on the fraction of defectives. If the production process is examined throughout the course of a day, there may be a changing fraction of defectives because of increasing fatigue on the part of the operator; or if the process is studied over an extended period of time, such as a month, there may be a changing fraction of defectives because of a gradual shift in the adjustment of the machine. Either of these changes leads to a violation of the constant probability requirement of the binomial model.

4. A psychophysical experiment is performed by repeatedly presenting a sound of some critical intensity to an individual who either "hears" or "does not hear" each presentation. The individual remembers his previous responses, and these influence his future responses. Hence the results of successive trials may not be independent in the sense required by the binomial model.

Even though situations like the foregoing occur in practice, we may still be willing to use the binomial model, recognizing that it is only an approximation to the model we would like to use. It is not possible to attack the problem of inference without a model, and it may be impossible to devise an exact model for the actual experiment. Of course one would always like to know the degree of approximation involved in using an inexact model. This is possible under some circumstances. Thus we will be able to give some recommendations concerning sampling problems for which the binomial model may be used to describe results obtained by drawing simple random samples without replacement from a finite population.

8.2. Consequences Derived from the Binomial Model

8.2.1. *The Probability of Obtaining Any Stated Number of S's in a Set of n Binomial Trials.* A set (or sample) of n binomial trials can be characterized by the number of S's that occur in it, say u. The number of F's occurring in the same set of n trials will be equal to $(n - u)$. Now one can think of performing a set of n trials and observing u for this set; then performing a second set and again observing u; and so on indefinitely. The quantity u will vary from set to set, and its value for any given set cannot be predicted with certainty. Thus a set of n binomial trials can be viewed as a single trial of a compound experiment whose possible events are the possible values of u. We shall now obtain an expression for the probability that u assumes any stated value. This probability will depend upon n, the number of binomial trials, and upon p, the probability of obtaining an S on a single binomial trial.

The steps required in obtaining a general expression for the desired probabilities can be set forth with an example. Suppose that a simple random sample of size 3 is to be drawn from a population consisting of the professional personnel employed by a company. In order to ensure the applicability of the binomial model, the workers will be drawn one by one with replacement. Once a worker is drawn, he is asked, "If a professional friend of yours were offered a job with your company and you felt he had the proper qualifications for the job, would you advise him to take it or advise him to go somewhere else if he could?" A "take it" response will be denoted by S while any other response will be an F. We shall assume that a fraction p of the workers in the population would give a "take it" response and will determine the probability that u—the number of S's in the three drawings—is equal to 0, 1, 2, or 3.

The possible outcomes of this set of three drawings and their associated probabilities are presented in Table 8.1. The distinct outcomes are

TABLE 8.1. POSSIBLE OUTCOMES OF A SET OF THREE BINOMIAL TRIALS

Distinct outcomes	Identification of outcomes by:			Number of S's u	Probability of outcome	Probability of various values of u
	First drawing	Second drawing	Third drawing			
1	F	F	F	0	$qqq = q^3$	q^3
2	S	F	F	1	$pqq = pq^2$	
3	F	S	F	1	$qpq = pq^2$	$3pq^2$
4	F	F	S	1	$qqp = pq^2$	
5	S	S	F	2	$ppq = p^2q$	
6	S	F	S	2	$pqp = p^2q$	$3p^2q$
7	F	S	S	2	$qpp = p^2q$	
8	S	S	S	3	$ppp = p^3$	p^3

obtained by enumerating all sequences of S's and F's that might arise. These are given on the eight lines of the table. Each set of three drawings must result in one and only one of these sequences. The probability of obtaining a particular sequence, say FFS, follows from the application of (7.7). That is, the results of the separate drawings are independent in the probability sense, and therefore

$$\begin{aligned}&\Pr(\text{first drawing } F \textit{ and } \text{second drawing } F \textit{ and } \text{third drawing } S)\\&\quad = \Pr(\text{first drawing } F) \cdot \Pr(\text{second drawing } F) \cdot \Pr(\text{third drawing } S)\\&\qquad = qqp = pq^2\end{aligned}$$

A similar use of (7.7) gives the probabilities for the other seven sequences.

Actually, we are not interested in the order in which the S's and F's occur in a set of three trials. Rather, we are concerned only with the number u of S's which occur in the set. Table 8.1 shows that $u = 0$ only for the first sequence. Consequently $\Pr(u = 0) = q^3$. However, $u = 1$ for either the second, third, or fourth sequence. Since these sequences are mutually exclusive—only one can arise on any given set of three trials—$\Pr(u = 1)$ can be obtained through the use of (7.4). That is,

$$\begin{aligned}\Pr(u = 1) &= \Pr(\text{either } SFF \text{ or } FSF \text{ or } FFS) \\ &= \Pr(SFF) + \Pr(FSF) + \Pr(FFS) \\ &= pq^2 + pq^2 + pq^2 \\ &= 3pq^2\end{aligned}$$

The same argument applies to the case of $u = 2$ and $u = 3$, and the right-hand column of Table 8.1 gives the desired expressions, namely, $\Pr(u = 0) = q^3$, $\Pr(u = 1) = 3pq^2$, $\Pr(u = 2) = 3p^2q$, and $\Pr(u = 3) = p^3$.

As a numerical illustration of these results, suppose that 60 per cent of the professional workers in the plant would give a "take it" response. Then $p = .60$ and $q = (1 - p) = .40$. Substituting in the foregoing general expressions, we have

$$\begin{aligned}\Pr(u = 0) &= q^3 = (.40)^3 = .064 \\ \Pr(u = 1) &= 3pq^2 = 3(.60)(.40)^2 = .288 \\ \Pr(u = 2) &= 3p^2q = 3(.60)^2(.40) = .432 \\ \Pr(u = 3) &= p^3 = (.60)^3 = .216\end{aligned}$$

These values can be given the usual empirical interpretation of a probability. That is, if an indefinitely large number of samples of three workers were drawn in accordance with the described procedure from a population containing 60 per cent S's, then we would expect 6.4 per cent of these samples to have no S's, 28.8 per cent to have one S, 43.2 per cent to have two S's, and 21.6 per cent to have three S's. Observe that these probabilities add to 1, or the per cents to 100. This follows from the fact that one and only one of the values of u must occur on each set of three drawings.

The contents of Table 8.1 illustrate that there were two distinct steps involved in obtaining the probabilities of the various values of u for a set of three binomial trials: (1) the first step consisted of computing the probabilities associated with particular sequences of S's and F's, using the model requirement of trial-to-trial independence and the resulting multiplicative property of single-trial probabilities; and (2) noticing that all sequences with the same value of u have the same probability, we next took the second step of counting the number of sequences having a particular value of u and multiplying this number by the probability of any one of these sequences, thereby using the fact that the separate sequences are mutually exclusive. Thus Table 8.1 shows that the probability of any sequence having two S's ($u = 2$) is p^2q, that there are three such sequences, and that we must therefore have $\Pr(u = 2) = 3p^2q$.

These two steps can be applied to general situations with any number n of binomial trials. We first ask for the probability to be assigned to a particular sequence of n trials which has exactly u S's, where u can take on any of the values $0, 1, 2, \ldots, n$, and $(n - u)$ F's. Such a sequence will look like

$$SFFSS \cdots FSF$$

and its probability will be

$$pqqpp \cdots qpq$$

This probability is the product of u p's, since there are u S's and the probability of a single S is p, and $(n - u)$ q's, since there are $(n - u)$ F's and the probability of a single F is q. It can be written as p^uq^{n-u}, the order of multiplication of the p's and q's having no effect on the answer. Thus the result is obtained that any particular sequence of n binomial trials having u S's and $(n - u)$ F's has probability p^uq^{n-u} and this completes the first step in obtaining the general expression.

The second step requires counting the number of distinct sequences having exactly u S's and $(n - u)$ F's. An answer to this counting operation has already been obtained in another connection, and it is only necessary to state the problem so that the link is apparent. An ordered set of n trials has n positions

$$\underbrace{\_\ \_\ \_\ \cdots\ \_\ \_}_{n \text{ positions}}$$

in which S's and F's can be placed. We wish to know the number of ways in which u of these positions can be chosen to receive S's—the remaining $(n - u)$ automatically receiving F's. This question could just as well have been phrased, Given a finite population of n elements (positions), in how many ways can I choose a sample containing u elements (positions)? When the question is phrased in this manner, we see that the answer is given by expression (6.1), where N is replaced by n and n is replaced by u. That is, the number of sequences of length n that have exactly u S's and $(n - u)$ F's is

$$\frac{n!}{u!(n-u)!}$$

Putting these two expressions together, as was done in the special case of Table 8.1, the desired result is:

If n binomial trials are conducted, where p is the probability of an S on a single trial, then the probability of obtaining exactly u S's, where u can assume any of the values 0, 1, 2, . . . , n, is

$$\frac{n!}{u!(n-u)!} p^u q^{n-u} \tag{8.1}$$

The following example will illustrate the application of (8.1). Social psychologists have been increasingly concerned with studying the ability of individuals to predict the feelings, thoughts, and actions of other people and have therefore been attempting to identify and measure psychological processes variously referred to as "social perception," "social sensitivity," or "empathy." Thus Dymond [1], who carried out some of the early empirical research on this topic, defines empathy as "the imaginative transposing of oneself into the thinking, feeling, and acting of another and so structuring the work as he does."

One facet of Dymond's research was to bring together small groups of college students, to have the members of these groups work together on a class project, and then to request each member of a group to fill in a rating test. For example, each member of the group was asked to place *himself* in one of five ordered categories with respect to the trait "friendliness," where the two extreme categories were "very friendly" and "very unfriendly." Furthermore, each member of a group was asked to predict how he thought the other members of the group rated *themselves*.

Let us focus attention on a particular member of a group, say individual A, who is asked to predict how six other members of the group, say $B_1, B_2, \ldots, B_6$, rated themselves on friendliness. If A makes a correct prediction for B_1, then he scores a "success," and similarly for $B_2, \ldots, B_6$. As a starting point for an analysis of this type of experiment, one might question the very existence of any such process as empathy or social sensitivity and therefore assume that individual A merely guesses for each of $B_1, B_2, \ldots, B_6$. The question then is, What kind of score (number of S's) can we expect A to produce in this experiment if he is guessing? We shall here be concerned only with the answer to this specific question, recognizing however that this is actually only one step in the direction of deciding the more important question, Does empathy actually exist? This more general problem will be discussed in Sec. 8.4.

Since A must place each of $B_1, \ldots, B_6$ in one of the five categories on friendliness, his probability of a correct guess for a single individual is $\frac{1}{5}$. We therefore have a set of six binomial trials where the value of p is $\frac{1}{5}$. Expression (8.1) can be applied to give the probability of obtaining exactly u S's, where u can take on any of the values 0, 1, 2, . . . , 6. The computation of these probabilities is summarized in Table 8.2. The only special point to be noted in these computations is that $\Pr(u = 0)$

involves $(1/5)^0$ and $\Pr(u = 6)$ involves $(4/5)^0$. For present purposes, any number raised to the zeroth power is defined to be 1. That is, $(1/5)^0 = 1$, $(4/5)^0 = 1$, and similarly for all other such expressions.

TABLE 8.2. THE PROBABILITIES OF VARIOUS VALUES OF u IN A SET OF SIX BINOMIAL TRIALS
$(p = 1/5)$

Value of u	*Probability of* u [*from* (8.1)]
0	$\frac{6!}{0!6!}\left(\frac{1}{5}\right)^0\left(\frac{4}{5}\right)^6 = 1\,\frac{4{,}096}{15{,}625} = .2621$
1	$\frac{6!}{1!5!}\left(\frac{1}{5}\right)^1\left(\frac{4}{5}\right)^5 = 6\,\frac{1{,}024}{15{,}625} = .3932$
2	$\frac{6!}{2!4!}\left(\frac{1}{5}\right)^2\left(\frac{4}{5}\right)^4 = 15\,\frac{256}{15{,}625} = .2458$
3	$\frac{6!}{3!3!}\left(\frac{1}{5}\right)^3\left(\frac{4}{5}\right)^3 = 20\,\frac{64}{15{,}625} = .0819$
4	$\frac{6!}{4!2!}\left(\frac{1}{5}\right)^4\left(\frac{4}{5}\right)^2 = 15\,\frac{16}{15{,}625} = .0154$
5	$\frac{6!}{5!1!}\left(\frac{1}{5}\right)^5\left(\frac{4}{5}\right)^1 = 6\,\frac{4}{15{,}625} = .0015$
6	$\frac{6!}{6!0!}\left(\frac{1}{5}\right)^6\left(\frac{4}{5}\right)^0 = 1\,\frac{1}{15{,}625} = .0001$
	Total 1.0000

The probabilities in Table 8.2 immediately tell how well an individual can be expected to do on this experiment by guessing alone. Thus the probability of obtaining a score of 3 or more is equal to

$$.0819 + .0154 + .0015 + .0001 = .0989$$

while the probability of a score of 4 or more is

$$.0154 + .0015 + .0001 = .0170$$

Under these circumstances it appears reasonable to conclude that individuals are not "guessing" if they consistently obtain scores of 4 or more in this type of experiment. The rationale of such an argument will be treated in some detail in Sec. 8.4.

As a concluding point in this section, we note that extensive tabulations of the values of expression (8.1) have been prepared and are available in [2] and [3]. The bibliography at the end of this chapter describes the coverage of these two sets of tables.

8.2.2. *The Binomial Distribution.* The primary purpose in considering sets of binomial trials is to learn how the results of actual experimentation can be used to tell something about the true fraction of S's, p. Remembering that we are still acting as though we really knew the true value of

p, this ultimate purpose suggests several additional points that can be made with respect to the basic probability expression (8.1).

It appears intuitively reasonable that the sample fraction of S's, which will be designated as u/n, bears a more direct relation to the value of p than does the sample number of S's. This relation has been stressed in preceding chapters, especially in Sec. 3.1.3 where it was noted that the sample fraction of S's provides an estimate of the population fraction. Furthermore, since there is a one-to-one correspondence between the possible values of u and the possible values of u/n, i.e., for each value of u there corresponds one and only one value of u/n and vice versa, expression

TABLE 8.3. THE BINOMIAL DISTRIBUTION FOR ARBITRARY VALUES OF n AND p

Possible values of u	*Possible values of u/n*	Pr(u) or Pr(u/n)
0	$0/n$	$\frac{n!}{0!n!} p^0q^n$
1	$1/n$	$\frac{n!}{1!(n-1)!} p^1q^{n-1}$
2	$2/n$	$\frac{n!}{2!(n-2)!} p^2q^{n-2}$
.	.	.
.	.	.
.	.	.
u	u/n	$\frac{n!}{u!(n-u)!} p^uq^{n-u}$
.	.	.
.	.	.
.	.	.
$n-1$	$(n-1)/n$	$\frac{n!}{(n-1)!1!} p^{n-1}q^1$
n	n/n	$\frac{n!}{n!0!} p^nq^0$
Total........	....	1

(8.1) gives the probability of obtaining various sample fractions. For example, Table 8.2 shows the probability of obtaining a particular number of S's in a set of six binomial trials where $p = 1/6$. The sample number of S's can assume the values 0, 1, 2, . . . , 6 while u/n, or $u/6$ in this case, assumes the successive values $0/6 = 0$, $1/6 = .167$, $2/6 = .333$, $3/6 = .500$, . . . , $5/6 = .833$, and $6/6 = 1.000$. Thus $\Pr(u/6 = .833) = \Pr(u = 5) = .0015$; or $\Pr(u/6 = .333) = \Pr(u = 2) = .2458$.

For fixed values of n and p, substitution of successive values of u in (8.1) gives a probability for each possible value of u or u/n. The resulting assemblage of all possible values of u and their respective probabilities constitutes a *binomial distribution*. This distribution provides a complete description of what is expected in an indefinitely large number of sets of n binomial trials, each set being characterized with respect to the number or fraction of S's that occur in the set. Thus Table 8.2 gives the complete distribution of u for sets of six binomial trials where p is $1/5$. An evaluation of how well a sample fraction of S's can be expected to approximate the true value of p must be made with respect to such a distribution. The general binomial distribution, i.e., for arbitrary values of n and p, is given in Table 8.3. The term *binomial* comes from the fact

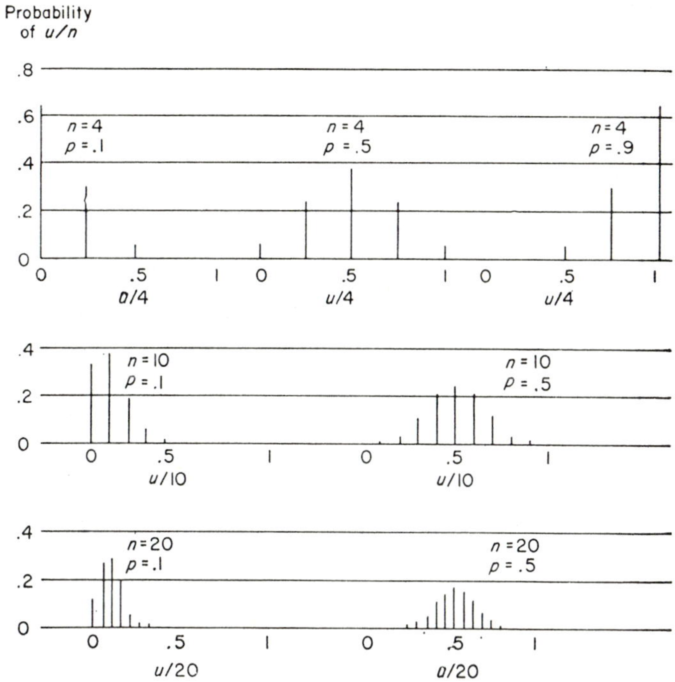

FIG. 8.1. Binomial distributions for various combinations of n and p.

that if the quantity $(p + q)$ is multiplied by itself n times, the individual terms entered in the $\Pr(u)$ column will be obtained.

The contents of Table 8.3 define the binomial distribution in a formal manner, but they do not provide any sort of picture of the distribution as a whole. Since extensive use will be made of this distribution, it is essential that we develop an appreciation, or "feel," for its general characteristics. The most direct way of facilitating this process is to present graphically the distributions for a number of specific values of n and p. Figure 8.1 shows seven binomial distributions arising from the following combinations of n and p: $(n = 4,\ p = .1)$, $(n = 4,\ p = .5)$, $(n = 4,\ p = .9)$, $(n = 10,\ p = .1)$, $(n = 10,\ p = .5)$, $(n = 20,\ p = .1)$, and $(n = 20,\ p = .5)$. The use of the variable u/n permits the presentation of these distributions on a comparable basis, the value of u/n always being between 0 and 1 no matter what the value of n. The height of the vertical line at a particular value of u/n is equal to the probability of obtaining that value of u/n.

A careful examination of the seven distributions of Fig. 8.1 leads to the following observations:

1. Whenever $p = .5$, the binomial distribution is symmetric about the point $u/n = .5$. Moreover, the probabilities of the individual u/n's become successively smaller as u/n moves away from .5 in either direction. This behavior is exhibited for ($n = 4$, $p = .5$), ($n = 10$, $p = .5$), and ($n = 20$, $p = .5$) and can be proved in general by using expression (8.1).

2. By comparing the distributions for ($n = 4$, $p = .1$), ($n = 4$, $p = .5$), and ($n = 4$, $p = .9$), we see that the binomial distribution is highly skewed for $p = .1$ and .9 whereas it is symmetric for $p = .5$. This result can also be demonstrated in a general manner. For *fixed* n, the distribution becomes more and more skewed to the right as p moves toward 0 and more and more skewed to the left as p moves toward 1.

3. Not only does the skewness of the distribution change as p moves away from .5, but the location of the distribution also moves. Thus a comparison of ($n = 20$, $p = .1$) and ($n = 20$, $p = .5$) shows the first distribution to be located some distance to the left of the second.

4. If we focus attention on a particular value of p, .1 or .5, and compare the distributions for increasing values of n, we see that the binomial distribution becomes less "spread out" as n increases. For example, ($n = 4$, $p = .5$) is spread over the entire range from 0 to 1 while ($n = 20$, $p = .5$) is primarily concentrated in the range from .30 to .70, inclusive.

5. Another important characteristic of the binomial distribution is obtained by comparing ($n = 4$, $p = .1$), ($n = 10$, $p = .1$), and ($n = 20$, $p = .1$). Each of these distributions is skewed to the right, but the last is much less skewed than the first. If we were to examine such distributions for still larger values of n, we would find that the skewness would become progressively less pronounced. This "approach to symmetry" will be basic to the considerations of Chap. 9.

6. Finally, we should like to call attention to the fact that the distributions of Fig. 8.1, with the exception of ($n = 4$, $p = .5$), appear to assign zero probability to some "possible" values of u/n; that is, no vertical line is drawn above these values. A "possible" value is one which can arise from an integer value of u. Thus .1 is a possible value for $u/20$ ($u = 2$) while .125 is an impossible value for $u/20$ (u would have to be 2.5). Actually, no possible value of u/n ever has zero probability. A probability may be so small that it cannot be represented on a graph, or it may be so small that it can be neglected for all practical purposes, but it will never be exactly zero. For example, the distribution ($n = 4$, $p = .1$) shows no line for $u/4 = .75$ or $u/4 = 1$. Nevertheless, direct computa-

tion shows that $\Pr(u/4 = .75 \text{ if } p = .1) = .0036$ and $\Pr(u/4 = 1 \text{ if } p = .1) = .0001$. Similarly, $\Pr(u/20 = .95 \text{ if } p = .5) = .000{,}019{,}0$. The fact that truly zero probabilities are impossible means that the only advance prediction about u/n which can be made with absolute certainty (i.e., with probability 1) is that its value will lie between 0 and 1, inclusive.

In spite of the fact that useful predictions cannot be made with probability 1, it is quite clear from an examination of the distributions of Fig. 8.1 that u/n will more frequently provide a "good" approximation to p in some situations than in others. Thus the distribution for ($n = 20$, $p = .5$) is less spread out than the distribution for ($n = 4$, $p = .5$). This means that the values of $u/20$ cannot usually "stray" as far away from the central point $u/20 = .5$ as can the values of $u/4$. Such statements as these are of course primarily subjective, and it is therefore essential that we obtain some method of stating them in a precise fashion.

These comparisons can be made in an objective manner by using selected probabilities. For example, the application of expression (7.4) gives

$$\begin{aligned}
&\Pr\left(\frac{u}{4} \text{ differs from .5 by more than .2} \mid p = .5\right)\\
&= \Pr\left(\frac{u}{4} \text{ is less than .3 or more than .7} \mid p = .5\right)\\
&= \Pr\left(\frac{u}{4} = 0 \mid p = .5\right) + \Pr\left(\frac{u}{4} = .25 \mid p = .5\right) + \Pr\left(\frac{u}{4} = .75 \mid p = .5\right)\\
&\qquad + \Pr\left(\frac{u}{4} = 1 \mid p = .5\right)\\
&= .0625 + .2500 + .2500 + .0625\\
&= .625
\end{aligned}$$

while

$$\begin{aligned}
&\Pr\left(\frac{u}{20} \text{ differs from .5 by more than .2} \mid p = .5\right)\\
&= \Pr\left(\frac{u}{20} \text{ is less than .3 or more than .7} \mid p = .5\right)\\
&= \Pr\left(\frac{u}{20} = 0 \mid p = .5\right) + \Pr\left(\frac{u}{20} = .05 \mid p = .5\right) + \cdots\\
&\qquad + \Pr\left(\frac{u}{20} = .25 \mid p = .5\right) + \Pr\left(\frac{u}{20} = .75 \mid p = .5\right) + \cdots\\
&\qquad + \Pr\left(\frac{u}{20} = 1 \mid p = .5\right)\\
&= .0000 + .0000 + .0002 + .0011 + .0046 + .0148 + .0148 + .0046\\
&\qquad + .0011 + .0002 + .0000 + .0000\\
&= .0414
\end{aligned}$$

These two probabilities reaffirm the visual impression gained from Fig. 8.1 that $u/4$ will more frequently "stray" farther from .5 than will $u/20$.

8.2.3. *The Mean of the Binomial Distribution.* The binomial distribution, for stated values of n and p, describes an infinite population of *sets* of n binomial trials (i.e., a single element of the population is a set of n binomial trials) with respect to the variable u/n, the fraction of S's occurring in a set. That is, we think of carrying out an indefinitely large number of sets of n binomial trials and $\Pr(u/n)$ is the relative frequency with which a stated value of u/n occurs in these sets. It is therefore possible to apply expression (4.6) in computing the mean of the binomial distribution. In using this expression we shall of course replace the possible values of x ($x_1, x_2, \ldots, x_k$) by the possible value of u/n ($0/n$, $1/n, \ldots, n/n$) and the respective probabilities $p_1, p_2, \ldots, p_k$ by $\Pr(0/n), \Pr(1/n), \ldots, \Pr(n/n)$.

The formula for the mean then becomes

$$E\left(\frac{u}{n}\right) = \sum_{u=0}^{n} \left(\frac{u}{n}\right)\Pr\left(\frac{u}{n}\right)$$

The special notation of Chap. 4, $E(u/n)$, has been used for the mean in order to emphasize the particular character of the variable that is being considered here.

This expression can be used to obtain the mean of any particular binomial distribution. That is, given numerical values for n and p, it is only necessary to substitute into the formula and compute the desired result. This process is illustrated in Table 8.4 for an n of 4 and a p of

TABLE 8.4. COMPUTATION OF THE MEAN OF THE BINOMIAL DISTRIBUTION FOR $n = 4$ AND $p = .2$

u	$u/4$	$\Pr(u/4)$	$(u/4)\Pr(u/4)$
0	0	.4096	0
1	.25	.4096	.1024
2	.50	.1536	.0768
3	.75	.0256	.0192
4	1.00	.0016	.0016
Total...		1.0000	.2000

$$E(u/4) = \sum_{u=0}^{4} (u/4)\Pr(u/4) = .2000$$

.2. It is found that $E(u/4) = .2$, which is the value of p, the probability of an S on a single binomial trial. Actually, this is only a special case of a general result that can be proved by performing appropriate manipulations on the previously given formal expression for $E(u/n)$. This result is:

> If an indefinitely large number of *sets* of n binomial trials are conducted, where p is the probability of an S on a single trial, then the average of all the sample fractions of S's is
>
> $$E\left(\frac{u}{n}\right) = p \tag{8.2}$$

This result holds for any combination of values of n and p and means that, even though the sample fraction of S's u/n will vary from set to set, its long-run average will always be equal to p. Thus each distribution pictured in Fig. 8.1 has the appropriate value of p as a measure of its location.

8.2.4. *The "Closeness" of u/n to p in Terms of Probability.* Since our ultimate goal is to use u/n to estimate p, it is of interest to know that u/n will, on the average, be equal to p. This long-run behavior of u/n is frequently summarized by saying that u/n is an *unbiased* estimate of p. However, simply saying that u/n is an unbiased estimate of p does not tell the entire story. The distributions of Fig. 8.1 make it quite apparent that u/n may vary to a considerable extent about this long-run average. It is therefore necessary to learn to control this variability.

The solution to this problem of controlling sample-to-sample variability depends upon our willingness to make a slight modification in approach. Roughly speaking, we must be willing to relax the natural desire for certain statements, which leads to the completely useless result that u/n will fall between 0 and 1, and to consider the possibility of employing a useful but uncertain statement. In a sense we are adopting a "nothing ventured, nothing gained" philosophy, and as we shall see, this leads to procedures which are under our control to the extent that we can arbitrarily choose the probability to be associated with the correctness of our statements.

The specific procedures to which the foregoing general approach leads can be easily illustrated in terms of particular binomial distributions. Table 8.5 gives three of these distributions, n being 20 in each case, for p equal to .1, .3, and .5. Let us first focus attention on the distribution for $p = .5$. That is, repeated samples of size 20 are being drawn from a population which has one-half S's and one-half F's. The entries $\Pr(u/20)$ give the relative frequencies with which the various values of $u/20$ would occur in an infinite number of such samples. Any value of $u/20$ (0, .05, . . . , .95, 1) could conceivably occur on any given set of 20 trials, and therefore the only statement that can be made with a probability of

TABLE 8.5. BINOMIAL DISTRIBUTIONS* FOR p EQUAL TO .1, .3, AND .5
(Number of trials n in a set is 20.)

$u/20$	$p = .1$		$p = .3$		$p = .5$	
	Pr(u/20)	Cumulated Pr(u/20)'s	Pr(u/20)	Cumulated Pr(u/20)'s	Pr(u/20)	Cumulated Pr(u/20)'s
0	.1216		.0008	.0008		
.05	.2702		.0068	.0076		
.10	.2852		.0278	.0354	.0002	.0002
.15	.1901		.0716		.0011	.0013
.20	.0898		.1304		.0046	.0059
.25	.0319	.0433	.1789		.0148	.0207
.30	.0089	.0114	.1916		.0370	.0577
.35	.0020	.0025	.1643		.0739	
.40	.0004	.0005	.1144		.1201	
.45	.0001	.0001	.0654		.1602	
.50			.0308	.0479	.1762	
.55			.0120	.0171	.1602	
.60			.0039	.0051	.1201	
.65			.0010	.0012	.0739	
.70			.0002	.0002	.0370	.0577
.75					.0148	.0207
.80					.0046	.0059
.85					.0011	.0013
.90					.0002	.0002
.95						
1.00						
Total.....	1.0002		.9999		1.0000	

* A blank for any value of $u/20$ means that its probability is less than .00005.

exactly 1 is that $u/20$ will lie between 0 and 1, inclusive. However, we can make much more useful statements about how well $u/20$ approximates p by merely allowing the probability of a correct statement to drop a little below 1. For example, we obtain from Table 8.5, using expression (7.4), such probability statements as the following:

$$\Pr\left(.15 < \frac{u}{20} < .85 \mid p = .5\right) = .9974$$

$$\Pr\left(.20 < \frac{u}{20} < .80 \mid p = .5\right) = .9882$$

$$\Pr\left(.25 < \frac{u}{20} < .75 \mid p = .5\right) = .9586$$

$$\Pr\left(.30 < \frac{u}{20} < .70 \mid p = .5\right) = .8846$$

The form in which the preceding statements have been presented was chosen with a definite purpose in mind. Thus the third expression says that in 95.86 per cent of *sets* of 20 binomial trials, where $p = .5$, the value of $u/20$ will be greater than .25 and less than .75 (i.e., it will have one of the values .30, .35, . . . , .50, . . . , .65, or .70). These are the very values which lie closest to the true value of p. In 2.07 per cent of the sets $u/20$ will have a value of .25 or less (i.e., deviate from $p = .5$ by being small) and in another 2.07 per cent of the sets it will have a value of .75 or more (i.e., deviate from $p = .5$ by being large). That is, we have defined a relatively narrow interval around .5, as compared to the interval from 0 to 1, within which we can reasonably expect $u/20$ to lie by merely giving up roughly 5 per cent of the outlying values of $u/20$. Of course if we feel that we have given up too much in this respect, we can still go back to the first expression, $\Pr(.15 < u/20 < .85 \mid p = .5) = .9974$, and we shall still have made a considerable gain over the 0 to 1 interval.

The reader may have observed that the phrase "probability of a statement" was used in the immediately preceding discussion. The ideas implied by this phrase will become increasingly important as we proceed into the details of statistical inference, and so a few words of clarification are in order. The relationship $\Pr(.25 < u/20 < .75 \mid p = .5) = .9586$ means that if an indefinitely large number of sets of 20 binomial trials are made, then $u/20$ will be greater than .25 and less than .75 for 95.86 per cent of the sets. Suppose now that before each set is obtained the unqualified statement is made that "$u/20$ for this coming set will be greater than .25 and less than .75." If this is done for each of an indefinitely large number of sets, then it follows that the statement will be "correct" for 95.86 per cent of the sets and "incorrect" for 4.14 per cent of the sets. Thus the original probability relationship has been merely reworded so that it refers to the probability that a given unqualified statement will be correct or incorrect. After a set of 20 trials has been conducted, we naturally find out whether the statement is correct or incorrect and there is no longer any question of probability for that particular set.

One of the aims in the present investigation is to obtain results which will permit the comparison of different binomial distributions (i.e., for different values of n and p). The questions that arise in this respect can be treated in the present context by considering the distributions for ($n = 20, p = .3$) and ($n = 20, p = .1$) in relation to the foregoing treatment of ($n = 20, p = .5$). There are many different ways in which such a comparison can be made, but the most generally useful one has proved to be that of using *equal probability* intervals which are in some sense "centered" at the known value of p. That is, one arbitrarily chooses some high and convenient probability level, say 95 per cent for the purposes of the present discussion, and always attempts to find an interval "centered" at p such that the probability is .95 that $u/20$ will fall in this interval. A comparison of these intervals, with respect to both length and location, will provide a comparison of distributions.

The word "centered" has been placed in quotation marks to indicate that it will have a special meaning here. That is, the intervals will be centered *in probability*. Not only will we attempt to choose intervals around p which have 95 per cent probability, but we shall attempt to choose them so that 2.5 per cent of the sets will have values of $u/20$ which fall outside the interval on the lower side and 2.5 per cent will fall outside on the upper side. There are some difficulties involved in fulfilling these conditions exactly for a binomial distribution with a relatively small value of n, and an n of 20 fits this category. (The difficulty arises because $u/20$ is a discrete variable, and the problems are similar to those raised in Sec. 4.3.2 for the median.) Therefore the following conditions are imposed on the search for 95 per cent centered probability intervals:

1. The centered interval will contain as close to 95 per cent probability as possible, but in no event will its probability be allowed to fall below 95 per cent.
2. The probability that $u/20$ will fall outside the interval on the lower side will be as close to 2.5 per cent as possible, but it will never be allowed to be greater than 2.5 per cent.
3. The probability that $u/20$ will fall outside the interval on the upper side will be as close to 2.5 per cent as possible, but it will never be allowed to be greater than 2.5 per cent.

These ideas on centered probability intervals can be applied to the three distributions of Table 8.5, and the results are shown graphically in Fig. 8.2. The shaded interval in each instance is the 95 per cent centered probability interval, or as close as it is possible to come under the restrictions stated above. The interval for p equal to .5 is the only one which is near the ideal, while the one for p equal to .1 is quite far away from the ideal. An examination of the actual probabilities in Table 8.5 shows why this is so. When p is equal to .1, the smallest possible value of $u/20$ has probability .1216 of occurring on repeated sets of 20 binomial trials. It is therefore impossible to take away any values of $u/20$ on the lower side of the desired interval without taking more than 2.5 per cent. Similarly on the right side of the distribution, if the cutoff is made between .25 and .30, 1.14 per cent of the values of $u/20$ will fall to the right of the desired interval. On the other hand, if the cut is made between .20 and .25, 4.33 per cent will fall to the right and one of the restrictions is violated. The possible cutting points are easily obtained by cumulating from each tail of the distribution. These cumulated probabilities are given in Table 8.5, and horizontal lines are drawn for each distribution to indicate how the intervals were defined for Fig. 8.2.

Although it is not necessary for us to go into the details at this point, it is important to recognize that as n increases, for a fixed value of p, one is able to come closer and closer to the ideal 95 per cent centered probability interval.

It is apparent that Table 8.5 and Fig. 8.2 are only special examples of what can be done for any value of p lying between 0 and 1. (Other values of n can also be used, but the treatment of these cases will be deferred to the next section.) If continued use can be made of such intervals, it seems only reasonable to determine these intervals for all

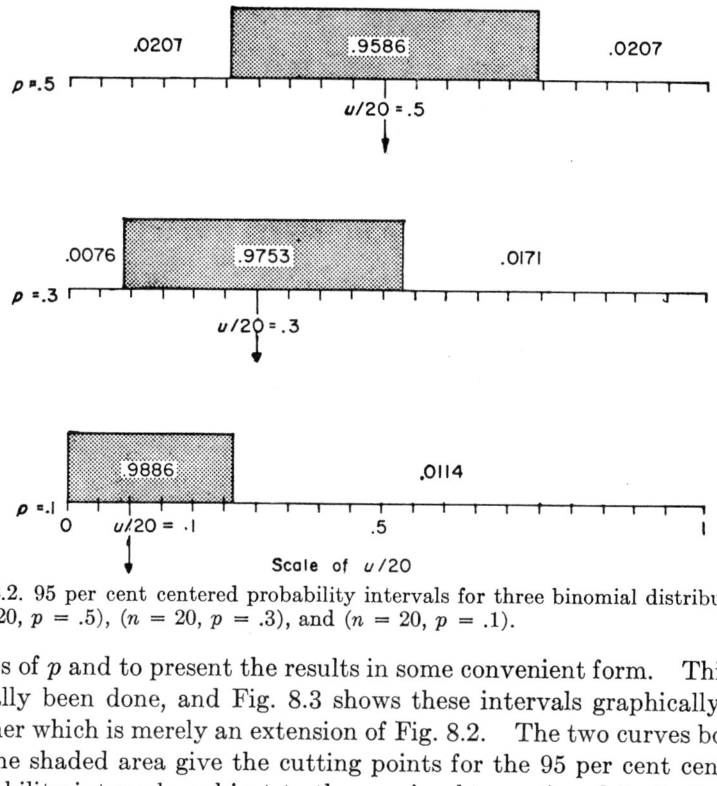

FIG. 8.2. 95 per cent centered probability intervals for three binomial distributions, (n = 20, p = .5), (n = 20, p = .3), and (n = 20, p = .1).

values of p and to present the results in some convenient form. This has actually been done, and Fig. 8.3 shows these intervals graphically in a manner which is merely an extension of Fig. 8.2. The two curves bounding the shaded area give the cutting points for the 95 per cent centered probability intervals, subject to the previously mentioned limitations.

The method of using this chart is illustrated by the dashed lines in Fig. 8.3. Assuming that we are interested in sets of 20 binomial trials, where p equals .80, we read horizontally from .80 on the p scale to the two points where this line intersects the boundaries of the shaded area. Reading down to the $u/20$ scale, the two values .58 and .98 are obtained. Thus we know that the probability is *at least* .95—and is as close to .95 as possible—that $u/20$ will lie between .58 and .98. Reference to the previously cited table of the binomial distribution shows that

$$\Pr\left(.58 < \frac{u}{20} < .98 \mid p = .80\right) = .9785$$

As a final point in this treatment of 95 per cent centered probability intervals, it is of interest to note that, if desired, these results (e.g., Fig.

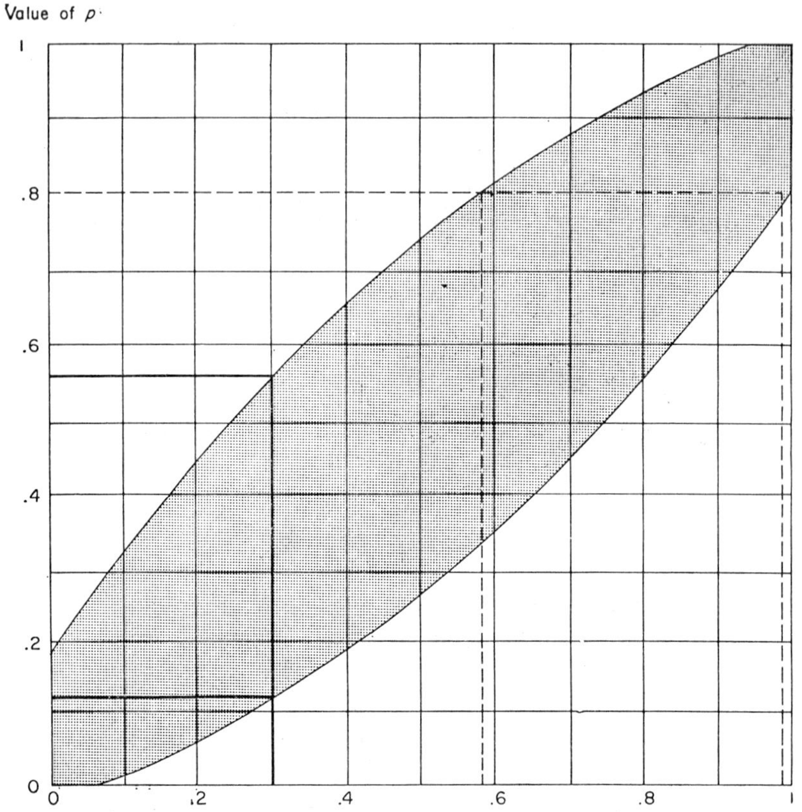

FIG. 8.3. 95 per cent centered probability intervals for the sample fraction of S's in sets of 20 binomial trials.

8.3) for $u/20$ can easily be transformed into results for u, the number of S's in a set of 20 binomial trials. Thus if

$$\Pr\left(.58 < \frac{u}{20} < .98 \mid p = .80\right) = .9785$$

then $u/20$ lying between .58 and .98 requires that u lie between

$$.58 \times 20 = 11.6 \qquad \text{and} \qquad .98 \times 20 = 19.6.$$

Therefore $\Pr(12 \leq u \leq 19) = .9785.$

8.3. Obtaining from a Sample an Approximate Location for an Unknown p

8.3.1. *The Procedure Introduced and Discussed for Sets of 20 Binomial Trials.* The entire development in the preceding section required knowing the value of p. In practice this information will not be available, and the entire purpose in conducting a stated number of binomial trials is to use the observed sample fraction of S's to tell something about this unknown value of p. In other words, it is now necessary to reverse the argument and go from an observed u/n to p rather than from p to u/n.

The broad outlines of a "reasonable" procedure can be discerned by referring back to the treatment of the problem of going from known p to u/n. We saw there that a knowledge of p did not permit us to predict with certainty anything about a coming value of u/n, other than the uninteresting fact that it would lie between 0 and 1. This problem was solved by relaxing, in a controlled manner, the natural desire for perfect prediction. This is a price that must be paid for the convenience and necessity of using samples.

It is apparent from these considerations that an observed value of u/n will not permit any useful judgment to be made with probability 1 about the value of p which is governing the binomial process under study. Therefore the goal must be to find some way of using an observed value of u/n to provide an approximate location for p, i.e., to assign p to an interval, and of measuring in some precise way how "well" the interval locates p.

Using the contents of Fig. 8.3, it is possible to devise a procedure which will give a solution to the stated problem for sets of 20 binomial trials. The actual mechanical details of the procedure will be given first, and this will be followed by an interpretation of the results obtained through application of the procedure.

Let us suppose that a set of 20 binomial trials is obtained under circumstances where the value of p is unknown (see Sec. 8.1.1 for illustrations of such situations) and that six S's are observed. That is,

$$\frac{u}{20} = \frac{6}{20} = .30$$

for this particular set. Enter the chart of Fig. 8.3 at the .30 point on the $u/20$ scale; follow up vertically from .30 to the two points where this vertical line intersects the boundaries of the shaded area; and read horizontally from these points to the p scale. This process is illustrated by the solid lines on Fig. 8.3, and the two resulting points on the p scale are approximately .12 and .55. These two values are called the 95 *per cent confidence limits for the unknown p*. The interval defined by the two limits is the 95 *per cent confidence interval.* Notice that a different pair of limits is obtained for each different value of $u/20$. Thus if u were observed to be 7, then $u/20 = 7/20 = .35$ and the corresponding limits are .15 and .59. Let us now see in what sense limits obtained in this manner do provide an approximate location for an unknown p.

Figure 8.4 reproduces the major features of the chart of Fig. 8.3, and with the aid of this chart we can examine the characteristics of 95 per cent confidence limits or intervals. Only one horizontal line is drawn through

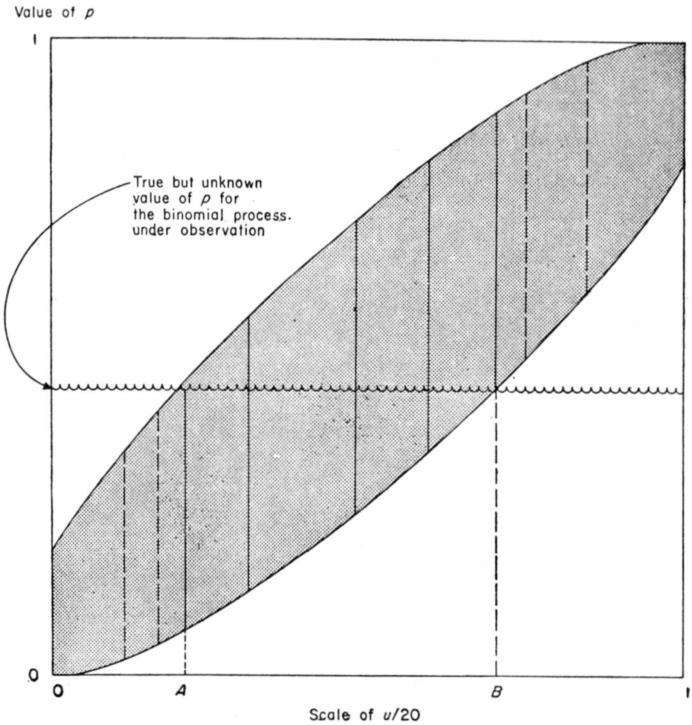

FIG. 8.4. Graphical interpretation of 95 per cent confidence limits for an unknown p. (Sets of 20 binomial trials.)

this chart, and it is drawn in a wavy fashion to indicate that it represents the true but *unknown* value of p. It could just as well have been drawn at any other point between 0 and 1, since the argument holds for any value of p. Now we know from the manner in which Fig. 8.3 was constructed that $u/20$ will fall between A and B for at least 95 per cent of an indefinitely large number of sets of 20 binomial trials, that it will fall below A for at most 2.5 per cent of the sets, and that it will fall above B for at most 2.5 per cent of the sets.

Confidence intervals obtained from various values of $u/20$ are indicated by the solid vertical lines in Fig. 8.4. Notice the following points:

1. The value of $u/20$ will vary from set to set of 20 binomial trials and since the confidence interval depends on the observed value of $u/20$, this confidence interval also changes from set to set.

2. Every time that $u/20$ falls between A and B, the resulting confidence interval *covers* or *includes* the true but unknown value of p. That is, this true value of p is contained between the limits of the interval.

3. Every time $u/20$ falls below A, the interval misses p by being too low.

4. Every time $u/20$ lies above B, the interval misses p by being too high.

Putting together these results, we have

> If an indefinitely large number of sets of 20 binomial trials are performed, where p may have any value between 0 and 1 (just so long as it remains the same for all trials), and if an interval is associated with each $u/20$ in the manner shown in Fig. 8.3, then
>
> $$\Pr(\text{interval covers the true but unknown value of } p) = .95$$

This result can also be related to the probability that a particular statement will be correct. Suppose that *in advance* of performing each set of 20 trials, we state "The interval I obtain will cover the true value of p." This statement will be "correct" for 95 per cent of all sets and "incorrect" for the remaining 5 per cent. Notice, however, that it will be impossible to distinguish the correct statements from the incorrect ones since the true value of p is unknown.

There now remains only a final step to take in providing a complete interpretation of 95 per cent confidence intervals. All the discussion has been built around what would happen if an indefinitely large number of sets of 20 binomial trials were performed (i.e., we have been talking in terms of probability), while in actual practice there will be *one* set, *one* observed value of $u/20$, and *one* confidence interval. Thus an observed u of six S's out of 20 trials gives an interval from .12 to .55. The unknown p either does or does not lie in this interval, and there can be no direct recourse to the ideas of probability. Nevertheless, we are reasonably confident that p does lie in the obtained interval in the sense that if we were to repeat the procedure for many samples the value of p would be contained within about 95 per cent of the resulting intervals. Therefore, we adopt the following mode of operation: We conduct a set of 20 binomial trials, determine the 95 per cent confidence interval, state that p lies in the interval, and attach 95 per cent confidence to the statement. Thus the term *confidence* is used to avoid the inappropriate application of the term *probability* to the result of a single set of 20 trials. Probabilities, as defined, refer only to the assemblage of results from an infinite number of sets of trials.

As an illustration of the foregoing results, the performing of repeated sets of 20 binomial trials has been simulated by drawing from a random-digit table such as Table 6.3. That is, 100 sets of 20 consecutive digits were taken from the table, and the number of *even* digits was counted in each set drawn. This is equivalent to performing 20 binomial trials where the probability p of an S (the occurrence of an even digit) on a single trial is .5. The 95 per cent confidence limits for p were obtained from Fig. 8.3 for each set, and the results are summarized in Table 8.6.

TABLE 8.6. 95 PER CENT CONFIDENCE LIMITS FOR p OBTAINED FROM 100 SETS OF 20 BINOMIAL TRIALS
(For illustrative purposes p was taken to be .5.)

Observed value of $u/20$	Confidence interval from Fig. 8.3	Number of times given interval occurred
.20	.05–.45	1
.25	.08–.49	2
.30	.12–.55	3
.35	.15–.59	8
.40	.19–.64	18
.45	.23–.69	20
.50	.27–.73	14
.55	.31–.77	16
.60	.36–.81	4
.65	.40–.85	7
.70	.45–.89	6
.75	.51–.92	0
.80	.56–.95	0
.85	.62–.97	1
Total..........	....	100

It will be noted that only four of the 100 intervals fail to cover the known value of p, and this is roughly what might be expected in 100 sets of trials. In practice, of course, only one of these intervals would be obtained and the true value of p would be unknown, but the odds are very much weighted in favor of coming up with an interval which does cover the true p.

It was observed in Sec. 8.1.3 that the binomial model, and therefore the confidence interval chart of Fig. 8.3 which was derived from this model, does not provide an exact description of what happens if the elements of a simple random sample are drawn one by one without replacement from a finite population. It would clearly be very convenient if we had some idea concerning the degree of approximation involved in using these results to describe the precision of simple-random-sampling estimates, especially if comparable charts are obtained for other sample sizes. (This extension to other sample sizes will be made in the following subsection.) There are two questions that can be asked in this respect: (1) What is the general character of the approximation? and (2) Under what circumstances can we expect the approximation to be "very good"? Partial answers to these questions will now be provided.

A very simple example will indicate the general character of the approximation. Consider a finite population containing five elements, three of which are S's and two of which are F's. Suppose samples of size 2 are to be drawn from this population for the purpose of estimating p, the population fraction of S's. Table 8.7 shows the distribution of

TABLE 8.7. DISTRIBUTIONS OF SAMPLE FRACTION OF S'S FOR SIMPLE RANDOM SAMPLES DRAWN WITH AND WITHOUT REPLACEMENT ($N = 5$, $n = 2$, $p = .6$)

Value of u	Value of $u/2$	Sampling with replacement $\Pr(u/2)$	Sampling without replacement $\Pr(u/2)$
0	0.00	.16	.10
1	0.50	.48	.60
2	1.00	.36	.30
Total....	....	1.00	1.00

$u/2$, the sample fraction of S's, under the condition that the sample is drawn with replacement and under the condition that the sample is drawn without replacement. The "sampling with replacement" probabilities are obtained by using expression (8.1) with $n = 2$ and $p = .6$, while the "sampling without replacement" probabilities are obtained by direct enumeration. That is, there are 10 possible simple random samples [see expression (6.1)], each of which is to have equal probability. One of these samples has two F's, six of them have one F and one S, and three of them have two S's. We thus obtain the "sampling without replacement" probabilities.

For present purposes, the important feature of Table 8.7 is that the "sampling without replacement" distribution is *less spread out* than is the binomial distribution. That is, there is more probability on the center value of $u/2$ and less probability on the two extreme values. This relationship will always hold no matter what the size of the population, what the size of the sample, or what the value of p. Therefore we make the following general observation: *If simple random samples are drawn without replacement and if confidence intervals are determined from a binomial chart like that of Fig.* 8.3, *the intervals will always be wider than they need to be to give* 95 *per cent confidence.* Methods will be developed in Chap. 9 that will permit us to make this general observation more precise.

The second question concerned the circumstances under which the binomial distribution would provide a "very good" approximation to sampling without replacement. Again the general result can be illustrated with a numerical example. Samples of size 3 are to be drawn without replacement from a population which has 40 per cent S's and 60 per cent F's, and attention will be focused on the probability that the sample drawn will contain exactly three S's. That is, we wish to find $\Pr(u/3 = 1.00 \mid p = .4)$. Now this probability will depend on the size of the population, and Table 8.8 shows the computation for various values

TABLE 8.8. THE PROBABILITY OF OBTAINING THREE S'S IN A SIMPLE RANDOM SAMPLE OF SIZE 3 DRAWN WITHOUT REPLACEMENT FROM POPULATIONS OF VARIOUS SIZES
($p = .4$ for each population)

Size of population	Computation of $\Pr(u = 3)$	Value of $\Pr(u = 3)$
10	$(4/10)(3/9)(2/8)$	.0333
20	$(8/20)(7/19)(6/18)$	.0491
50	$(20/50)(19/49)(18/48)$	.0582
100	$(40/100)(39/99)(38/98)$	.0611
1,000	$(400/1000)(399/999)(398/998)$	.0637
Infinite (binomial) . .	$(.4)(.4)(.4)$	.0640

of N. For example, suppose $N = 20$. This means there are 8 S's and 12 F's in the population since p is constant at .4. We wish to find Pr (first element drawn is an S *and* second is an S *and* third is an S). Applying expression (7.5),

$$\Pr(\text{first is an } S) = 8/20$$
$$\Pr(\text{second is an } S \mid \text{first is an } S) = 7/19$$
$$\Pr(\text{third is an } S \mid \text{first two are } S\text{'s}) = 6/18$$

and therefore the desired probability is $(8/20)(7/19)(6/18) = .0491$. The entry on the last line is for an infinite population, where the probability of an S does not change from trial to trial, and is the binomial model probability for this situation.

The principal feature to notice in this table is that the value of $\Pr(u = 3)$ draws closer and closer to the value of the binomial trial probability as the population size increases. That is, for large populations the probability of an S on a single trial changes very little from trial to trial. Actually, it is the size of the sample in relation to the size of the population that really counts, and a frequently used rough rule of thumb is: If a simple random sample, drawn without replacement, constitutes 5 per cent or less of a finite population, then the approximation obtained by using binomial model results are "very close" to the true situation. The motivation for the 5 per cent value will be given in Chap. 9.

8.3.2. *The Effects of Sample Size and Confidence Coefficient on Confidence Intervals.* The contents of Fig. 8.3, which have been used thus far in determining 95 per cent confidence intervals, refer to sets of 20

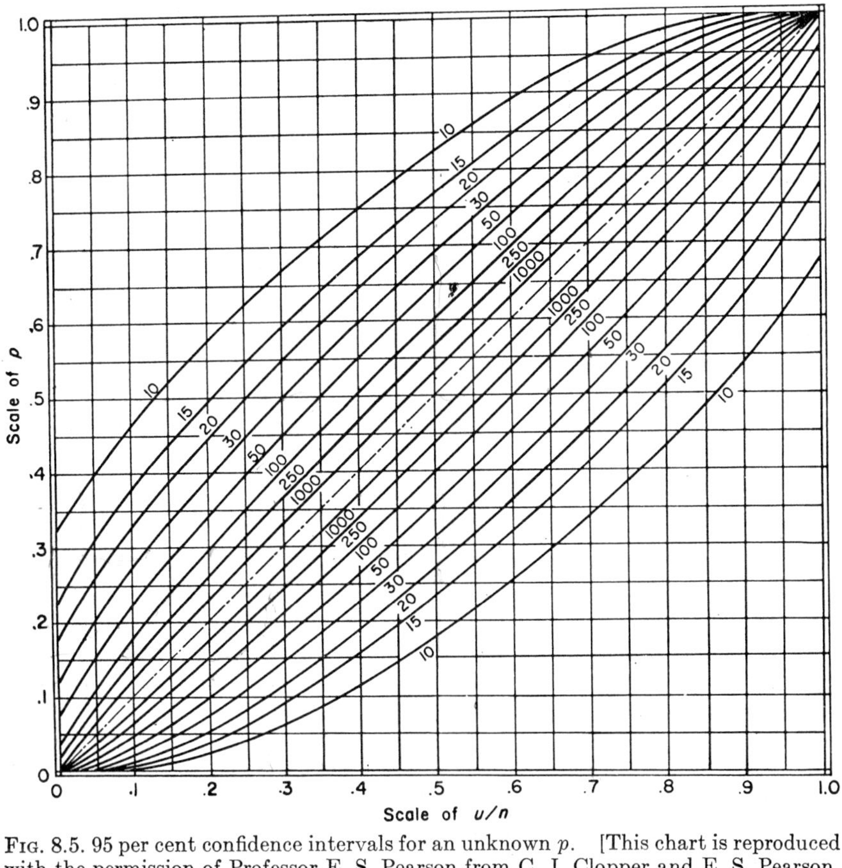

Fig. 8.5. 95 per cent confidence intervals for an unknown p. [This chart is reproduced with the permission of Professor E. S. Pearson from C. J. Clopper and E. S. Pearson, "The Use of Confidence or Fiducial Limits Illustrated in the Case of the Binomial," *Biometrika*, **26**:404 (1934).]

binomial trials. The same computational procedure can be used to develop similar materials for sets containing any other desired number of binomial trials, and one obtains a pair of curves for each sample size. Figure 8.5 presents such pairs of curves for n = 10, 15, 20, 30, 50, 100, 250, and 1,000. Notice that pairs of curves draw closer together—i.e., the length of the confidence interval becomes smaller—as the size of sample increases. This is due to a fact already noted in connection with

Fig. 8.1, namely, that the distribution of u/n for an indefinitely large number of sets of binomial trials becomes less and less spread out as n increases. Or, putting this in a more natural way, larger and larger samples provide more and more precise estimates for the value of the unknown parameter p. If one wishes to "pin down" the value of p very precisely, then it is necessary to pay a price by obtaining large samples.

The use of the chart of Fig. 8.5 is very straightforward once it has been ascertained that one's data have been obtained by a procedure which conforms reasonably well to the requirements of the binomial probability model. For example, Whiting and Child recently reported the results of a comparative study on child training and personality in primitive societies [4]. This study was based on data obtained from the Cross-cultural File at the Institute of Human Relations of Yale University. This file contains ethnographic reports about various primitive societies or cultures that have been observed by anthropologists. If a culture or society is regarded as a unit, if one thinks of the population of all primitive cultures that have existed or do exist at the present time, and if there is no reason to believe that anthropologists have been unduly "selective" in choosing such cultures for observation, then the cultures included in this file can be regarded as a simple random sample of human cultures in general.

One of the items of interest to Whiting and Child was the extent to which primitive cultures explained bodily illness by either soul loss (i.e., a person gets sick because his soul has left his body) or spirit possession (i.e., a person gets sick because an alien spirit has entered his body). Accounts of 75 cultures from the file were examined, and 38 of them showed evidence of using such explanation of bodily illness while 37 showed that such explanations were not used. What would be the 95 per cent confidence limits for the true fraction of primitive human cultures that did use this type of explanation? The value of u/n in this case is $38/75 = .507$. Entering the u/n scale of Fig. 8.5 at .507, we wish to read up to the curves for an n of 75. Actually there are no curves for this value of n, and we must therefore perform a rough interpolation between an n of 50 and an n of 100. Notice that these curves are not evenly spaced. Taking an n halfway between 50 and 100 requires going closer to the curve for 100 than to the curve for 50. Roughly then, we might take our 95 per cent confidence limits to be .39 and .62. Therefore we state that the true value of p lies between these two numbers, and we have 95 per cent confidence in this statement in the sense previously explained.

Although the preceding discussion of confidence intervals was phrased in terms of the 95 per cent confidence level, no difficulty would have been encountered in carrying through the computations and argument for any other stated level of confidence. For example, the tables of the binomial distribution can be used to obtain 90 per cent centered probability intervals for the sample fraction of S's just as was done for 95 per cent in Fig. 8.3. It is merely necessary to determine cutting points on the u/n scale (for fixed values of p and n) such that not more than 5 per cent of the

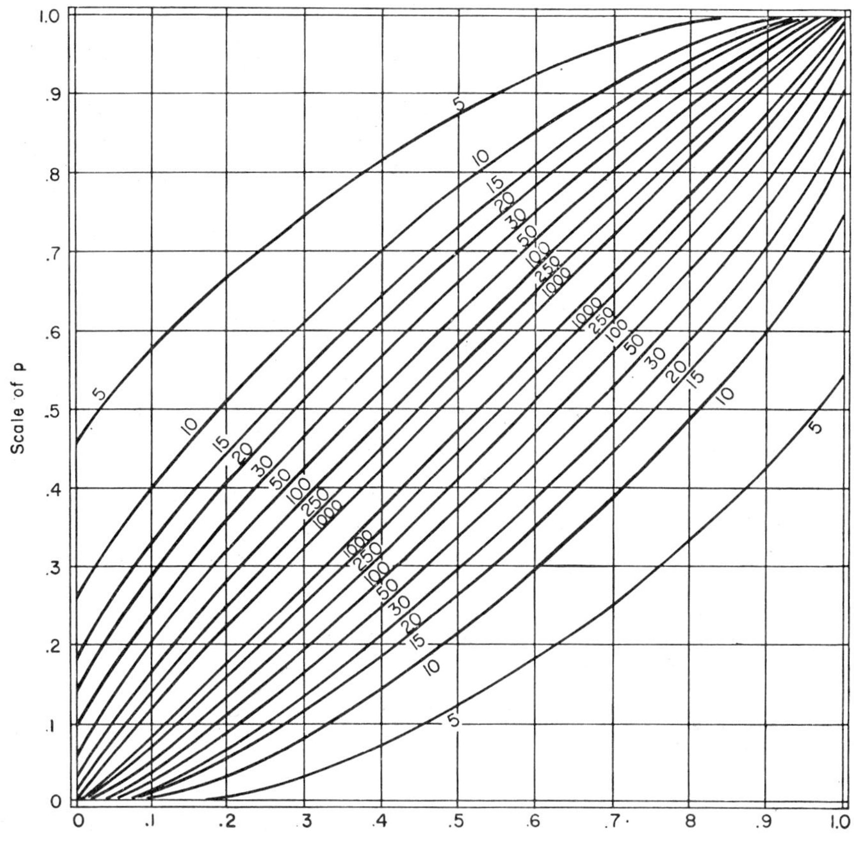

FIG. 8.6. 90 per cent confidence intervals for an unknown p. SOURCE: Reproduced with permission from Wilfrid J. Dixon and Frank J. Massey, Jr., *Introduction to Statistical Analysis*, 2d ed., p. 414, McGraw-Hill Book Company, Inc., New York, 1957.

values of u/n will fall above the upper cutting point and not more than 5 per cent will fall below the lower cutting point—subject to the restriction that as close to 5 per cent as is possible fall in each. The complete assemblage of the results of such computations can again be presented in the form of the chart of Fig. 8.5 and can be used to obtain 90 per cent confidence intervals for the value of an unknown p. A chart from which

90 per cent confidence intervals can be obtained is presented in Fig. 8.6. Finally, a chart giving 99 per cent confidence intervals is contained in Fig. 8.7. For the 99 per cent chart the cutting points are, in terms of probability, as close as possible to .5 per cent in the lower tail, 99 per cent in the centered probability interval, and .5 per cent in the upper tail.

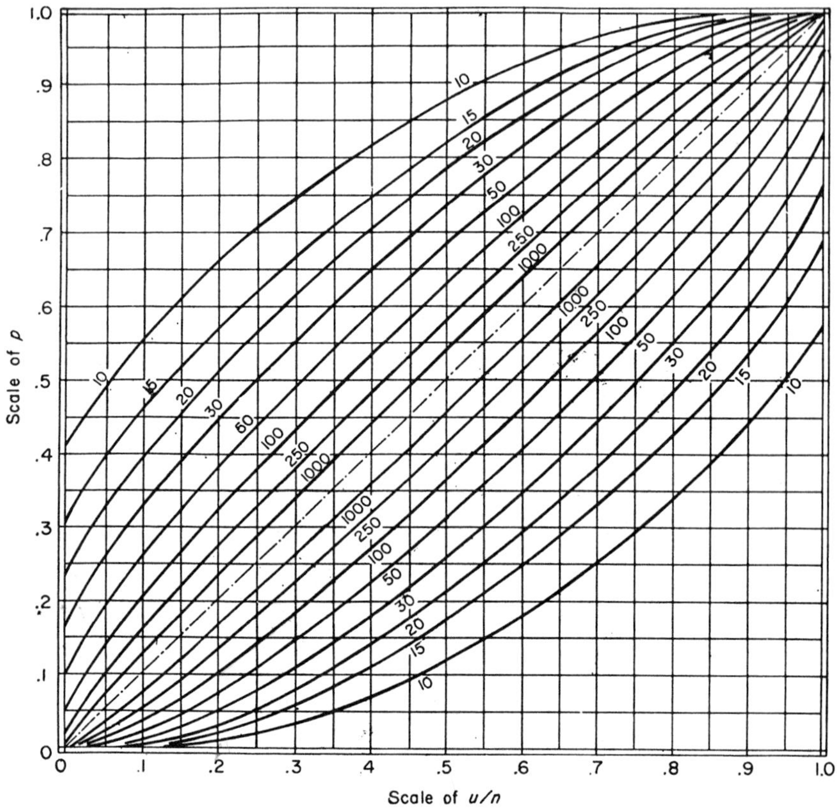

FIG. 8.7. 99 per cent confidence intervals for an unknown p. SOURCE: Reproduced with the permission of Professor E. S. Pearson from C. J. Clopper and E. S. Pearson, "The Use of Confidence or Fiducial Limits Illustrated in the Case of the Binomial," *Biometrika*, **26**:404 (1934).

These 90 and 99 per cent charts are used in the same manner as has already been illustrated for the 95 per cent charts.

One's first reaction at being given a choice between 90 per cent, 95 per cent, and 99 per cent levels of confidence, and any other levels which might be considered, could very well be, "Why give me a meaningless choice? Since it doesn't cost me anything I'll use the highest confidence level for which I have charts." Actually, a very cursory examination of the three charts will show that the increased confidence cannot be obtained without paying a price. For example, suppose a set of 50 binomial trials are carried out and $u/50$ is observed to be .30. Then the 90, 95, and 99 per cent confidence intervals for p are:

Confidence level, per cent	Confidence interval	Length of confidence interval
90	.21 to .40	.19
95	.18 to .45	.27
99	.15 to .49	.34

Thus for a given sample size, the interval within which we state that p lies becomes longer as the confidence level increases. This is the price that must be paid for having greater confidence.

8.3.3. *Sample Size Required to Obtain a Confidence Interval of Not More Than a Prescribed Length.* The precision with which a confidence interval estimates an unknown p is described by the confidence coefficient and the length of the interval. For a fixed confidence coefficient and a particular observed value of u/n, the length of the interval decreases as the size of the sample is increased. No direct use has yet been made of this fact since it has been taken for granted that the sample size would be given in any specific problem. A given sample size roughly corresponds to situations where we are making use of the results obtained by others and have no control over the size of sample or to situations where the resources allocated to a study determine the size of the sample that can be obtained. Another type of situation arises if the goals of a study state that an unknown p is to be estimated by a confidence interval with prescribed length and confidence coefficient. Under these circumstances the problem is to find the sample size required.

Actually, it is not possible to determine a sample size that will provide confidence intervals having *exactly* the prescribed length for the binomial distribution. Although the point has not been emphasized, the contents of Figs. 8.4, 8.5, 8.6, 8.7, and Table 8.6 show that the length of a confidence interval depends upon the particular value of u/n that is observed in the sample. This follows from the fact that the pair of curves giving the confidence intervals for a fixed sample size are not always the same distance apart. They draw closer together as the observed value of u/n approaches 0 or 1. Since the value of u/n will not be known until after the sample has been drawn, there is no way of guaranteeing that a confidence interval of the exact prescribed length will be obtained. We can, however, choose our sample size so that the length of the confidence interval will be *not greater* than the prescribed length, no matter what value of u/n is observed in the sample.

Our ability to determine the maximum length of confidence interval that can arise in advance of drawing the sample depends upon the following observable characteristic of the confidence interval curves. For any fixed sample size, the length of the interval—that is, the distance between the curves—is greatest when $u/n = .5$. Therefore, if we determine the sample size required to give the prescribed length as though u/n were going to be .5, the actual length of the interval obtained with this sample size cannot be greater than the prescribed length. This procedure is "conservative" in that the length of the obtained interval will usually be less than the prescribed length and will be equal to it only when u/n is .5.

Let us suppose that a company having a large number of professional employees wishes to determine the percentage of these employees who would give a "take it" response to the question "If a professional friend of yours were offered a job with your company, and you felt he had the proper qualifications for the job, would you advise him to take it or advise him to go somewhere else if he could?" It is decided that this percentage need only be known within plus or minus 10 percentage points, with 95 per cent confidence. The problem is to determine the size of the simple random sample that is required to satisfy this restriction. If we assume that u/n is going to be .5, for the reasons described in the preceding paragraphs, the following results are obtained from Fig. 8.5:

Size of sample n	Confidence interval	Length of confidence interval
50	.36–.64	.28
100	.40–.60	.20
250	.43–.57	.14

Thus a simple random sample of size 100 will give a 95 per cent confidence interval whose length cannot be greater than 20 percentage points.

After the simple random sample of 100 professional employees has been drawn and questioned, an observed value of $u/100$ will be computed and the actual length of the confidence interval obtained from Fig. 8.5. This length cannot be greater than 20 percentage points and may be considerably smaller. For example, if the observed value of $u/100$ were .80, the 95 per cent confidence interval would be from .71 to .87 with a length of 16 percentage points.

As noted in Sec. 8.3.1, these illustrative figures are obtained on the assumption either that the simple random sample is drawn with replacement or that the population from which the sample is drawn is very large. If the sample is drawn without replacement from a small population, relative to the sample size of 100, then the true length of the 95 per cent confidence interval will be smaller than that obtained from the charts.

8.4. Judging from a Sample whether p has a Stated Value

8.4.1. *A Procedure for Accepting or Rejecting a Stated Value of p.* The preceding section assumed that we were dealing with a population in which an unknown fraction p of the elements were of a certain kind and that the goal was to obtain from a sample an approximate location for p. This is not the only way to view the problem of obtaining information about an unknown p. It may happen that a specific value of p can be set forth for consideration in advance of any experimental data and that the mere acceptance or rejection of this value, on the basis of sample data, can aid considerably in formulating plans for further studies and courses of action.

As an illustration of this type of approach, let us consider the experiment on empathy described in Sec. 8.2.1. Each individual in a seven-member group is asked to place himself in one of *five* ordered categories on the trait "friendliness." Individual A, a member of the group, is asked to predict how each of the other six members rates himself. Individual A can score 0, 1, . . . , 6 "successes," where a low score indicates that he is not very sensitive in judging how others rate themselves and a high score indicates that he may possibly have some sensitivity in this respect. At the start of this experimental work, the important question would be, Do the results lead one to believe that any such phenomenon as social sensitivity exists? An affirmative answer to this question would lead to further studies which might be aimed at obtaining quantitative measures of the degree of social sensitivity to be expected under various conditions. A negative answer might mean that further experimental work would be suspended.

The six persons on whom A is asked to make predictions may be regarded as a simple random sample from an extremely large population, all of whose members might conceivably be presented to A for predictions. Let p be the fraction of this large population for which correct predictions would be made.

The magnitude of p is a rough measure of A's degree of social sensitivity for this particular experiment. If social sensitivity does not exist, A will be forced to guess on each of his judgments and the probability that he will be correct on a single judgment will be $1/5$ (since he must place an individual in one of five categories). As his degree of social sensitivity increases, either because there are certain types of people for whom he can always make a correct judgment or because he can do somewhat better than guessing for each person, the fraction of correct guesses will also increase over the guessing fraction of $1/5$.

Now recall that the primary purpose of this experiment is to answer the question, Do the results lead one to believe that any such phenomenon as social sensitivity exists? If social sensitivity does not exist, A's correct judgments will be obtained solely by chance and we expect that he will be correct on roughly 20 per cent of a large number of guesses. Therefore, we argue as follows. Assume, or *hypothesize*, that A is actually guessing. Compare the number of correct predictions made by A with the number of correct guesses that might reasonably be expected *on the basis of the hypothesis.* This comparison can be used to answer the original question. Thus rejecting the hypothesis of guessing means that the experimental evidence has led us to believe that social sensitivity exists. Notice that the hypothesis to be tested states that social sensitivity does not exist. This is the only way of obtaining a numerical value of p that can be used in the computations that will now be described. The existence of social sensitivity implies a value of p greater than $1/5$, but there is no rational means of focusing attention on any particular value between $1/5$ and 1. It is a historical convention to call the hypothesis to be tested the *null hypothesis*, the term *null* referring to the fact that one frequently wishes to test the hypothesis that some population characteristic has the value zero. The null hypothesis is to be accepted or rejected on the basis of observed data.

Not only has the preceding discussion set forth a hypothesis to be tested, but it has also implied that some member of a class of *alternative hypotheses* will be accepted in the event the null hypothesis is rejected. That is, if we reject the hypothesis of guessing, we wish to reject it in favor of the general alternative hypothesis that individual A is doing better than he would do by guessing. It does not seem reasonable to even consider situations in which A does worse than he would do by guessing, and one would almost certainly not wish to conduct further experimentation if this were the case. Thus we have the following analysis of this particular experiment:

Various ways of phrasing null hypothesis:

1. Individual A does not possess social sensitivity in the present sense.
2. Individual A guesses on each judgment he is asked to make.
3. The probability p that A makes a correct judgment is $1/5$.

Various ways of phrasing alternative hypothesis that will be accepted if the null hypothesis is rejected:

1. Individual A does, to some degree, possess social sensitivity.
2. Individual A does better in the long run than he would do by guessing alone.
3. The probability p that A makes a correct judgment is some value larger than $1/5$.

We shall now deal only with the third (and most explicit) forms of the null and alternative hypotheses.

Having decided upon a null hypothesis and upon an alternative hypothesis to be accepted in the event the null hypothesis is rejected, we can devise what appears to be a "reasonable" test procedure and can determine the characteristics of this procedure. A high number of correct judgments by A would seem to indicate that he is doing better than might be expected from guessing alone. But what is meant by a "high" number? Can three correct judgments out of six be regarded as high, or must we obtain six out of six in order to have a high number? Actually, there is no way to answer these questions without carrying the analysis a bit further, and so let us choose a particular test procedure and examine its consequences. A procedure that conforms with the preceding discussion is: (1) If A makes three or fewer correct judgments, the null hypothesis that $p = 1/5$ will be accepted. (2) If A makes four or more correct judgments, the null hypothesis will be rejected in favor of the alternative that $p > 1/5$.

If the null hypothesis is true, the probability that A makes any given number of correct judgments between 0 and 6 inclusive can be computed from (8.1). These probabilities were obtained earlier and are presented in Table 8.2. Combining these probabilities in accordance with the stated test procedure, we have: (1) *If the null hypothesis is true*, that is, $p = 1/5$, the probability that A will make three or fewer correct judgments and therefore that the null hypothesis will be accepted is $.2621 + .3932 + .2458 + .0819 = .9830$. (2) *If the null hypothesis is true*, the probability that it will be rejected is

$$.0154 + .0015 + .0001 = .0170$$

Notice that the probability given in 2 is the chance of making an error. That is, if an extremely large number of individuals were used in this experiment—each individual being placed in a group with six other members and being asked to make judgments on the other six—and if each of these was guessing on his six judgments, the stated test procedure would incorrectly say that 1.7 per cent were doing better than they would by guessing. The remaining 98.3 per cent are correctly classified as "guessers."

The error made by rejecting a true hypothesis is called a *Type I error*, and the probability of committing a Type I error is called the *level of significance* of the test procedure. It is customary to designate the level of significance by the Greek letter α (alpha), and this notational convention will henceforth be used. This definition of level of significance is completely general and applies whenever statistical tests of hypotheses are used. That is, no matter what hypothesis is being tested and no matter how complex may be the manipulations required to obtain a test procedure, the level of significance is always equal to the probability that the null hypothesis will be rejected when it is actually true.

The foregoing test procedure was obtained by first analyzing the empathy experiment so as to arrive at the general form of the procedure (that is, by saying that a "high" number of correct judgments would lead to the rejection of the null hypothesis) and by then arbitrarily stating that four or more correct judgments would be counted as a high number. This leads to a level of significance of 1.7 per cent. A more reasonable approach, now that we have seen some of the characteristics of the resulting test, is to determine the general form of the test, decide upon the actual value of the level of significance, and to determine the details of the test so that this level of significance will be attained. Thus one may say he wishes a test having a 1, 5, or 10 per cent level of significance. With a small number of binomial trials, one does not have too much freedom in this matter. (This is analogous to the difficulty encountered in obtaining exact 95 per cent centered probability intervals in Sec. 8.2.4.) However, in more general situations test procedures can be devised that will have any stated level of significance. For the empathy experiment, with a total of only six judgments, the following levels of significance are possible, as can be seen from Table 8.2:

Test procedure	*Level of significance* α
Reject null hypothesis if the number of correct judgments is:	
2 or more	.3447
3 or more	.0989
4 or more	.0170
5 or more	.0016
6	.0001

Since the level of significance of a test is the probability of making a Type I error and since its value is under the control of the person setting up the test procedure, one's natural inclination is to make it as small as possible. Why should an error rate of 1.7 per cent be tolerated when a simple change in procedure will reduce it to .01 per cent? The fallacy in this argument is immediately apparent if we adopt an extreme test procedure, namely, that of always accepting the null hypothesis no matter what the experimental evidence. The level of significance of this procedure is certainly zero since a null hypothesis is never rejected and therefore, in particular, a *true* null hypothesis is never rejected. However, individuals who actually do possess social sensitivity to some degree, and hence do better in the experiment than would be expected on a purely chance basis, will also be classified as "guessers." In other words, not only must we consider errors made by rejecting true null hypotheses, but we must also take into account the possibility that we may accept null hypotheses when they are actually false. *The error committed in accepting a false null hypothesis is called a Type II error.* Just so long as one is dealing with sample data, he must admit the possibility of committing these two types of error. It is possible, by appropriately choosing the test procedure, to control the probability of committing each type. However, the probability of making either error cannot be reduced to 0 without at the same time increasing the probability of the other error to 1. What is desired is a balance between the two.

In order to illustrate the foregoing notion of a Type II error, let us consider one of the specific procedures that was obtained for testing the null hypothesis that $p = 1/5$ against the alternative hypothesis that $p > 1/5$. Suppose that we (1) *accept* the null hypothesis if the experimental individual makes three or fewer correct judgments out of six, and (2) *reject* the null hypothesis if he makes four or more correct judgments. The level of significance of this test procedure has been shown to be 1.7 per cent. Now let us assume that, without the experimenter's knowledge, he is given a large number of individuals who can actually make 50 per cent correct judgments rather than the 20 per cent postulated by the null hypothesis. Each of these individuals makes six judgments and is accepted or rejected as a "guesser" in accordance with the stated procedure. We would like to know what fraction of them will be incorrectly classified as "guessers."

The probability that such an individual will be incorrectly classified as a "guesser" is equal to the probability that he obtains either 0, 1, 2, or 3 correct judgments on his six trials. From (8.1),

$$\begin{aligned}
\Pr(\text{zero correct judgments} \mid p = .5) &= .0156\\
\Pr(\text{one correct judgment} \mid p = .5) &= .0938\\
\Pr(\text{two correct judgments} \mid p = .5) &= .2344\\
\Pr(\text{three correct judgments} \mid p = .5) &= .3125
\end{aligned}$$

The desired probability is the sum of these four and is .6563. That is, 65.6 per cent of all individuals who are actually able to make 50 per cent correct judgments in the long run and who are processed through the experiment and test procedure will be incorrectly classified as "guessers." If an ability to make 50 per cent correct judgments in the experimental situation were regarded as sufficient evidence to make further study of social sensitivity worthwhile, this probability of a Type II error is so large that the experiment and test procedure cannot be regarded as adequate.

Computations similar to the foregoing have been performed for all the test procedures considered for the empathy experiment and for several different assumed true values of p, namely, .30, .50, and .80. These results are summarized in Table 8.9. For example, if we use the test

TABLE 8.9. PROBABILITIES OF COMMITTING TYPE I AND TYPE II ERRORS IN TESTING THE NULL HYPOTHESIS THAT $p = .20$ ON THE BASIS OF SIX BINOMIAL TRIALS

Test procedure	Level of significance	Probability of accepting the null hypothesis if the true value of p is:		
		.30	.50	.80
Reject the null hypothesis if the number of correct judgments is:				
2 or more	.3447	.4201	.1094	.0016
3 or more	.0989	.7442	.3438	.0170
4 or more	.0170	.9294	.6563	.0989
5 or more	.0016	.9889	.8907	.3447
6	.0001	.9991	.9845	.7379

procedure of rejecting the null hypothesis if the number of correct judgments is 5 or more, the level of significance is .16 per cent. However:

1. 98.89 per cent of individuals who can actually make 30 per cent correct judgments will be incorrectly classified as "guessers."
2. 89.07 per cent of individuals who can actually make 50 per cent correct judgments will be incorrectly classified as "guessers."
3. 34.47 per cent of individuals who can actually make 80 per cent correct judgments will be incorrectly classified as "guessers."

Notice that, for each assumed true value of p, the probability of making a Type II error increases as the level of significance decreases.

"Yes," you say, "but I want a test procedure that will have a 5 per cent level of significance and that will incorrectly classify as 'guessers' only 10 per cent of all individuals who can actually make 50 per cent correct judgments. How can I obtain such a procedure?" An examination of Table 8.9 shows that it is not possible to obtain such a test procedure with six binomial trials. However, the experimenter still has under his control the size of the group used, and therefore the number of judgments each individual will be asked to make. It can be shown that the following test procedure will have approximately the desired performance characteristics:

1. Use 20-member groups so that each individual makes judgments on 19 other individuals.

2. If 6 or fewer correct judgments are made out of 19, *accept* the null hypothesis that $p = \frac{1}{5}$.

3. If 7 or more correct judgments are made, *reject* the null hypothesis that $p = \frac{1}{5}$ in favor of the alternative that $p > \frac{1}{5}$.

The level of significance of this test is 6.76 per cent. Furthermore, only 8.35 per cent of individuals who can actually make 50 per cent correct judgments will be incorrectly classified as "guessers." The values of these Type I and Type II errors are obtained from the tables given in references [2] and [3].

A general summary of the argument and successive steps needed in devising and analyzing a procedure for accepting or rejecting a stated value of p is:

1. A specific value of p, say p_0, was set forth for consideration, and we wished to test, on the basis of data obtained from a number of binomial trials, the *null hypothesis* that this was the true value of p.

2. In the event that the null hypothesis were rejected, we wished to accept the general alternative hypothesis that $p > p_0$.

3. It appeared reasonable that a "high" value of u, the number of S's in n binomial trials, should lead to the rejection of the hypothesis and that a "low" number should lead to its acceptance.

4. We first recognized that any reasonable definition of "high" and "low" would sometimes lead to rejecting the null hypothesis when it was actually true, a *Type I error*. We then saw that we could almost arbitrarily set the probability of making this type of error, the *level of significance* of the test, at some low value. We thus translated the difficult distinction "high" and "low" into the more objective task of specifying a probability.

5. Not only will one sometimes reject a true null hypothesis, but he may also accept a null hypothesis when it is false. This latter is a *Type II error*. It is not possible to compute the probability of accepting a null hypothesis when it is false unless one specifies how "false" it is. Thus when the true p is close to p_0, the probability of a Type II error is larger than when the true p is far away from p_0. This is illustrated in Table 8.9.

6. Finally, we observed that the chance of accepting a false null hypothesis decreases as the size of the sample is increased. Furthermore, we observed that it is possible to devise a test that will have a stated level of significance and that will have a stated probability of making a Type II error for a specified value of p that is different from the null hypothesis value p_0.

8.4.2. *Obtaining Acceptance and Rejection Regions from Charts of Centered Probability Intervals.* The determination of a procedure for testing the hypothesis that $p = p_0$ against the alternative hypothesis that p is greater than p_0, as outlined at the end of the preceding subsection, depended fundamentally upon dividing the possible values of u, the observed number of successes in n binomial trials, into two groups. One group consisted of those values of u which appeared to be consistent with the hypothesis, and an observed value of u in this group led to accepting the null hypothesis. Let us call this group of u's the *acceptance region* of the test procedure. The other group contained those values of u which appeared to be most at variance with the null hypothesis, and an observed value of u in this group led to rejecting the null hypothesis. This group of u's defines the *rejection region* of the test. It was recognized that u might fall in the rejection region even though the null hypothesis were true, and the probability of this happening was called the level of significance of the test. One of the basic problems in devising a test procedure is that of finding a rejection region having the desired level of significance.

Although acceptance and rejection regions have thus far been defined in terms of the number of S's in a sample, they could also have been defined in terms of the sample fraction of S's, u/n. For example, when there are six binomial trials, the rejection region containing the u values of 4, 5, and 6 corresponds (and has the same probability) as the region containing the $u/6$ values of 2/3, 5/6, and 1. Thus in the empathy experiment, the rejection region contains those values of $u/6$ which are larger than and which differ most from the hypothesized long-run fraction of S's, $p_0 = 1/5$.

Now recall the discussion in Sec. 8.2.4 of 95 per cent centered probability intervals. For specified values of p and n, we attempted to find two cutting points on the scale of u/n such that (1) the probability of u/n falling below the smaller cutting point was as close to 2.5 per cent as possible, but was never allowed to be greater than 2.5 per cent; and (2) the probability of u/n falling above the upper cutting point was as close to 2.5 per cent as possible, but was never allowed to be greater than 2.5 per cent. For example, from Fig. 8.3, where $n = 20$, we obtain an upper cutting point of .42 when $p = .20$. But this is exactly the cutting point that is needed for defining acceptance and rejection regions for testing the hypothesis that $p = .20$ against the alternative hypothesis $p > .20$ on the basis of 20 binomial trials at the 2.5 per cent level of significance. The null hypothesis is accepted if $u/20$ is less than .42; it is rejected if $u/20$ is greater than .42. Furthermore, the probability that the hypothesis is rejected when it is actually true is approximately equal to, but never greater than, 2.5 per cent. This same argument shows that the curves in Figs. 8.5, 8.6, and 8.7 can be used to define acceptance and rejection regions for testing hypotheses about unknown values of p. From Fig. 8.5 we obtain test procedures having a 2.5 per cent level of significance; Fig. 8.6 gives tests having a 5 per cent level of significance; and Fig. 8.7 gives cutting points for a 0.5 per cent level of significance.

As an illustration of this use of the confidence-limit charts, let us consider an example whose general characteristics are quite representative of situations that frequently arise in opinion and attitude survey work. The civic leaders in a large community know from data gathered in previous studies that approximately 30 per cent of the adults living in the community are "well informed" on some matter of considerable local interest. Furthermore, these leaders feel that a larger fraction should be well informed, and so they sponsor a radio and newspaper campaign designed to make more people aware of the topic. At the conclusion of the campaign, the civic leaders wish to determine whether the campaign has been effective in increasing the fraction of well-informed adults. They decide to draw a simple random sample of 250 adults and judge the effectiveness of the campaign on the basis of the fraction of well-informed in the sample.

If the campaign had no effect, then a simple random sample is being drawn from a population in which 30 per cent of the members are well informed. On the other hand, an effective campaign would increase the population fraction above 30 per cent. There is no reason to suppose that the campaign might decrease the population fraction below 30 per cent. Assuming that the total adult population is so large in relation to the sample size of 250 that the binomial model can be used, we wish, on the basis of the results of 250 binomial trials, to test the

Null hypothesis: The campaign had no effect, i.e., $p = .30$

against the

Alternative hypothesis: The campaign did have an effect, i.e., $p > .30$

A sample fraction of well-informed "enough larger" than .30 should lead to rejection of the null hypothesis, where "enough larger" must be interpreted in terms of the level of significance of the test. Let us arbitrarily set this at 2.5 per cent. It is now necessary to determine a cutting point that will separate the possible values of $u/250$ into an acceptance region and a rejection region having the desired characteristics.

The desired cutting point can be obtained from Fig. 8.5. Entering the chart on the *vertical p scale at* $p = .30$, follow horizontally across to the *rightmost* curve labeled 250. Then follow down vertically to the u/n scale. The value of u/n is approximately .36. This is the desired cutting point. That is, if p is .30 and sets of 250 binomial trials are conducted, the probability is 2.5 per cent that the observed value of $u/250$ will be greater than .36. Therefore, the test procedure is:

1. Draw a simple random sample of 250 adults from the community and observe u, the number of well-informed in this sample.

2. If $u/250$ is greater than .36, *reject* the null hypothesis that the campaign has had no effect in favor of the alternative hypothesis that the fraction of well-informed is greater than 30 per cent.

3. If $u/250$ is less than .36, *accept* the null hypothesis that the campaign had no beneficial effect.

A cutting point determined in a similar manner from Fig. 8.6 would give a test procedure having a 5 per cent level of significance, while one obtained from Fig. 8.7 would give a 0.5 per cent level of significance.

Not only may the stated test procedure reject the null hypothesis when it is actually true, as measured by the 2.5 per cent level of significance, but it may also accept the null hypothesis when the campaign has really increased the fraction of well-informed over 30 per cent. It is of interest to cite a few figures to show how well this test procedure will do in detecting increases in the fraction of well-informed. These figures are:

True fraction of well-informed after the campaign	*Probability that the test procedure will accept the hypothesis that the campaign has had no effect*
.35	.629
.40	.099
.45	.002
.50	.000+

These values show that if the campaign has increased the fraction of well-informed from 30 to 35 per cent, the test procedure will, over half the time, say that there has been no increase; if the increase is to 40 per cent, the probability of making a Type II error is only 9.9 per cent; if the increase is to 45 or more, the test procedure will practically always reject the null hypothesis. Note that the .000 entry does not denote a zero probability. It does mean that the probability is less than .0005. If the civic leaders wished to be reasonably sure of detecting an increase from 30 to 35 per cent, they would have to take a larger sample. The level of significance could still be kept at 2.5 per cent, but the increase in sample size would decrease the chances of committing a Type II error.

8.4.3. *Another Form for an Alternative Hypothesis.* If a test procedure is determined so that the rejection of the null hypothesis implies the acceptance of the alternative hypothesis that the true value of p is greater than the hypothesized p_0, we say that we have a *one-sided* alternative and therefore a one-sided test. We also have a one-sided test if analysis of a particular problem leads to considering only the alternative that the true value of p is less than the hypothesized p_0. In this latter case, the rejection region of the test contains those values of u/n which are "enough smaller" than the hypothesized value to give the desired level of significance. The cutting point on the u/n scale which determines the rejection and acceptance regions can now be obtained by using the left-hand curve for the appropriate sample size on Figs. 8.5, 8.6, and 8.7.

Another form of alternative hypothesis arises when we wish to reject the null hypothesis p_0 when the true value of p merely differs from p_0, the distinction between "larger than p_0" and "smaller than p_0" being of no direct interest. In this case, we are testing the

Null hypothesis: $p = p_0$

against the

Alternative hypothesis: $p \neq p_0$

This is called a *two-sided* alternative, and the appropriate test procedure is a two-sided test. That is, the null hypothesis is rejected if the observed value of u/n is *either* "enough smaller" than p_0 *or* "enough larger" than p_0. The rejection region of this test contains both "small" and "large" values of u/n. The level of significance of the test is therefore equal to the probability that u/n falls in either the "small" or the "large" part of the region, assuming that the null hypothesis is true.

An illustration of a situation in which a two-sided alternative seems appropriate is the following. A manufacturing plant has a very large number of identical machines which produce a certain item. These machines require attention from a maintenance crew in more or less a random manner. Too small a crew will lead to lost machine time, while too large a crew will cause excessive maintenance costs.

A consulting firm makes recommendations concerning the optimum size of crew, based on the assumption that, on the average, 10 per cent of the machines will require maintenance on any given day. That is, the probability is .1 that a machine observed for a period of a day (a machine-day) will need attention. Management is willing to accept the conclusions drawn from this fact but questions whether the assumption is correct. It therefore decides to observe 1,000 machine-days. On the basis of the fraction of these 1,000 machine-days for which maintenance is required, a decision will be made on whether to follow the recommendations or to require further study by the consulting firm.

With the problem stated in this manner, management wishes to test on the basis of 1,000 binomial trials the

Null hypothesis: $p = .1$

against the

Alternative hypothesis: $p \neq .1$

That is, further study will be required if p is either smaller or larger than .1. An actual test procedure, at the 5 per cent level of significance, can be obtained from the 95 per cent centered probability intervals of Fig. 8.5. Entering the vertical p scale at a value of .1 and reading across to the *two* curves labeled 1,000, we obtain the two u/n values .08 and .12. We know from the manner in which these curves were constructed that, if the true value of p is .1, the probability is 2.5 per cent that $u/1{,}000$ is less than .08 and is also 2.5 per cent that $u/1{,}000$ is greater than .12. Therefore, the test procedure is:

1. Observe 1,000 machine-days and compute the fraction $u/1{,}000$ of these days on which maintenance was required.
2. If $u/1{,}000$ is less than .08 or more than .12, *reject* the null hypothesis.
3. If $u/1{,}000$ is between .08 and .12, *accept* the null hypothesis.

The probability that $u/1{,}000$ falls in the rejection region if the null hypothesis is true is $.025 + .025 = .05$, and this is the desired level of significance. Rejection regions leading to a 10 per cent level of significance for two-sided tests can be obtained from Fig. 8.6 and to a 1 per cent level of significance from Fig. 8.7.

8.5. Some General Problems Underlying the Applications of Statistical Inference

Starting with a population in which a fraction p of the elements are of a specified type S, the preceding sections have (1) developed the binomial distribution for describing the sample-to-sample behavior of the sample fraction of S's, (2) showed how to use the binomial distribution to infer an approximate location for p from an observed u/n, and (3) set forth a procedure for using an observed u/n to test whether p has some stated value.

Although the specific details of the derived techniques are only appropriate for the binomial model, the general characteristics of a confidence interval and of a test of a statistical hypothesis are the same no matter where they are used. Thus one can draw a simple random sample of families in the United States and ask for a 95 per cent confidence interval for the mean family income in the year preceding the interview; or one may wish to use the sample results to test the null hypothesis that the mean family income has not changed since the last decennial census against the alternative that it has increased. The techniques must of necessity differ from those used for the binomial model, but the interpretation of a confidence interval and of a statistical test will remain the same. Before developing techniques for treating some of these other problems we shall discuss briefly a number of implied but unstated questions that exist in the approaches of Secs. 8.3 and 8.4.

A person first introduced to the ideas of confidence intervals and statistical tests of hypotheses is usually most troubled by the fact that it is impossible to know *with certainty* whether a single observed confidence interval covers the true value of p or whether the results of a single test mean an hypothesis is true or false. He is apt to say, "You tell me that 95 per cent of a very large number of 95 per cent confidence intervals will cover the true value of p. But I want to know whether *this* interval does or does not cover the true p"; or "You tell me that, working with tests having a 5 per cent level of significance, I will reject true null hypotheses 5 per cent of the time. But I want to know whether *this* hypothesis I have just rejected is really false."

This desire for statements having probability one is natural, but it cannot be usefully fulfilled when results are based on samples. It is necessary to recognize that it is better to know that, over accumulated experience, one will be right a specified fraction of the time (without knowing whether any one decision is correct) than it is to give up all control over the process of inference. Notice that the term "accumulated experience" can include results obtained in different types of studies and results obtained by others. Measuring errors in terms of probability puts all of these on a common footing.

A second question concerns the generation of null hypotheses. If the reader will examine the examples discussed in Sec. 8.4, he will see that the null hypotheses were stated before any sample data had been obtained and came from one of two sources: (1) an analysis of the experimental situation (as in the experiment on social sensitivity) or (2) past experience (as in the test of effectiveness of an educational campaign). In no instance was the null hypothesis suggested by the data that were to be used to test the hypothesis. If hypotheses are obtained from the sample that is to be used for the test, there can be no control of the level of significance of the test. An extreme example will illustrate this fact. Suppose a simple random sample is drawn from a population and the sample fraction of S's is observed to be .35. The null hypothesis is then set up that the true value of p in this population is also .35. Since the sample result and the hypothesis agree, one must accept the hypothesis as being true and no argument can assign a level of significance to this test.

A third and much more difficult set of questions arises when one goes into the details of confidence-interval and hypothesis-testing procedures. Many aspects of these procedures are under the control of the individual who sets them up. For example, this individual can choose a sample size and a confidence coefficient when confidence intervals are to be obtained, or he can choose a sample size and a level of significance when an hypothesis is to be tested. (The contents of Figs. 8.5, 8.6, and 8.7 restrict one to certain sample sizes, confidence coefficients, and levels of significance. However, procedures can be devised that will frequently permit complete freedom in choosing values of these quantities. An example will be described in Chap. 9.) What criterion can be employed in determining the sample size, the confidence coefficient, or the level of significance?

It is not difficult to enumerate the items which should be considered in devising a criterion of the type just mentioned. These are (1) the cost of obtaining sample data, and (2) the consequences that follow from errors of various kinds. However, the combination of these items into a single criterion, which involves balancing the consequences of one type of error against those of another type, is a much more difficult task, and the solution will vary from situation to situation.

As an example of the foregoing, consider the exploratory experiments on social sensitivity described in Secs. 8.2.1 and 8.4.1. One might argue somewhat as follows. It is not possible to use groups having more than seven members because the individuals in larger groups would not get to know one another in the required intimate fashion. Furthermore, the level of significance of the test procedure should be set at a moderate level (say 10 per cent or even larger) rather than small (say 1 per cent or less) because (1) rejecting the null hypothesis when it is true will simply mean that more of these relatively cheap experiments will be conducted and the accumulation of experience will make it clear that the hypothesis is true, while (2) accepting the null hypothesis when it is false (and the chances of doing this are large if the level of significance is small) might mean the abandonment of further experimentation that would prove profitable.

These brief comments on the social-sensitivity experiments suggest that it may often prove useful to attempt to set down the consequences that will follow from various errors and to use these as guides in deciding upon sample size, confidence coefficient, and level of significance. Thus the test of the effectiveness of a community educational campaign, described in Sec. 8.4.2, might be analyzed as follows:

Null hypothesis (campaign ineffective)	Test procedure leads to:	
	Accepting hypothesis	Rejecting hypothesis
True	No error	The results may lead other communities to inaugurate such campaigns, and a large amount of time and money will be spent to no advantage.
False	An opportunity truly exists for raising the general level of information, and this opportunity is lost because of the test error.	No error

The entries in this table would seem to indicate that the level of significance should be kept low, to guard against wasteful expenditure of time and money. Also, the sample size should be made large enough so that the probability of a Type II error is small for whatever increase in the fraction of well-informed is regarded as worthwhile.

Analyses of the preceding type can be carried still further if it is possible to assign relative values to the losses that will arise when various types of errors are made. This approach recognizes that statistical investigations are undertaken as a basis for action and that decisions should be made in the light of their consequences. A precise formulation of the theory of statistical decisions was first given by Wald, and a popular account of the theory has been written by Bross (see section Other Reading, Chap. 2).

EXERCISES

8.1. If one examines the employment records of workers over a period of time, he will find that some of the workers remain continuously employed in the same industry while others move among different industry groups. One study [5] examined, for a period of 3 years, the records of 2,950 male workers who had the following characteristics: (1) in the first quarter of 1947 they were in employment covered by the United States social security system; and (2) they were between 20 and 24 years of age. It was found that 754 of these workers were continuously employed in the same broad industry classification for the entire 3-year period. Suppose that a simple random sample of five workers had been drawn in 1947 from among the 2,950 and that these five were then followed through the 3-year period.

a. What are the probabilities that exactly 0, 1, 2, 3, 4, and 5 of these would be continuously employed for the period?

b. What is the probability that two or fewer of the five would be continuously employed?

c. What is the probability that four or more would be continuously employed?

8.2. A consumer-testing organization wishes to rate a particular brand of electric toaster. Four toasters are selected at random from the stock of finished toasters, and the brand is said to be satisfactory if no defectives are found.

a. What is the probability that the brand will be rated satisfactory on the basis of this test if $1/10$ of the finished products are actually defective?

b. What is the probability that the brand will be rated satisfactory if $2/5$ of the finished product are defective?

8.3. Each of 393 men employed in the United States were asked the question "If by some chance you inherited enough money to live comfortably without working, do you think you would work anyway or not?" The individuals who conducted the study [6] report that 314 of these employed men gave a "would keep working" response. Assuming that this is a simple random sample of all employed men, what are the 95 per cent confidence limits for the true fraction of employed men who would give such a response?

8.4. From Fig. 8.5 determine the length of the 95 per cent confidence intervals for samples of size 30, 50, 100, 250, and 1,000 when the sample fraction of S's is .2. Plot the length of the interval against the corresponding sample size and sketch in a curve between the points. This curve shows how the precision of our estimate varies with n. For example, does doubling the sample size halve the length of the interval?

8.5. Suppose that you are an executive of a large department store which has one outstanding competitor. The two stores have maintained a price balance in the past in that your store *A* has sold any one item at a higher price than *B* one-half the time and at a lower price one-half the time. However, you suspect that store *B* is deliberately destroying this balance and is consistently underselling your merchandise. In order to investigate this possibility, you decide to send a "shopper" into store *B* to check the prices of 250 randomly selected articles. On the basis of the number of these articles on which *B* undersells *A*, you wish to decide whether there is evidence of a significant deviation from past policy.

a. Set this problem up as an example of hypothesis testing, stating the null hypothesis to be tested, the alternative hypothesis that will be accepted if the null hypothesis is rejected, and the level of significance that you would use in the test.

b. Determine the actual procedure (giving numbers) that you will use in accepting or rejecting the null hypothesis.

c. Suppose store *B* undersells on 155 out of the 250 randomly selected articles. Do you accept or reject the null hypothesis at your level of significance?

d. The level of significance is only one measure of the manner in which a statistical test performs. What is a second characteristic of the test procedure that should be considered?

e. What general advantage is gained by increasing the sample size, assuming that the level of significance remains unchanged?

f. Suppose the "shopper" checks the prices of 250 articles in only three departments (e.g., men's furnishings, large household appliances, and cosmetics) instead of randomly selecting them from among all articles sold by store *B*. Would it be appropriate to use the binomial model to describe the variability in sample results? Why or why not?

8.6. In a certain industry it was known that the median length of time that workers remained with any one company was 5 years. After the start of an industry-wide pension system, a particular company observed the records of a random sample of 100 workers who were hired during the first year of the program. After 5 full years, it is found that 62 of these are still with the company. Test the hypothesis that the median length of stay is the same as in the prepension period. State the alternative hypothesis, the level of significance, and whether the hypothesis is accepted or rejected. (*Hint:* What is the definition of the median?)

8.7. The personnel department of a company is charged with the responsibility for selecting individuals to fill certain types of positions. Ordinarily, some form of selection device will be used to determine whether an applicant will be "successful" or "unsuccessful" in a stated position. No such device can perform perfectly. Some individuals will be hired who will actually prove unsuccessful, and others will be rejected who would have turned out successfully.

This situation is related to the testing of statistical hypotheses. For each applicant, one wishes to test the hypothesis that he will be successful against the alternative that he will prove unsuccessful. The risk of rejecting a potentially successful applicant corresponds roughly to the level of significance of a test, while the risk of hiring an individual who will be unsuccessful corresponds roughly to the probability of committing a Type II error.

Suppose that it is possible to devise a selection criterion for a particular position that will have stated probabilities of committing these two types of error. What considerations would be relevant in setting the levels of these two probabilities for the following four cases:

a. Potential supply of qualified persons for the position is large—relatively inexpensive training period at the end of which it will be known whether a person hired is suitable or not.

b. Potential supply is large—relatively expensive training period.

c. Potential supply is very small—relatively inexpensive training period.

d. Potential supply is very small—relatively expensive training period.

REFERENCES CITED

1. Dymond, Rosalind F.: "A Scale for the Measurement of Empathic Ability," *Jour. Consult. Psych.*, **13**:127–133 (1949).
2. National Bureau of Standards: *Tables of the Binomial Probability Distribution*, Applied Mathematics Series, no. 6 (1950). These tables give the individual binomial probabilities and certain cumulated probabilities for p going from .01 to .50 in steps of .01, n going from 2 to 49 in steps of 1, and u going from 0 to $(n - 1)$ in steps of 1.
3. Romig, Harry G.: 50–100 *Binomial Tables*, John Wiley & Sons, Inc., New York, 1952. These tables give the individual binomial probabilities for p going from .01 to .99 in steps of .01 and n going from 50 to 100 in steps of 5.
4. Whiting, John W. M., and Irvin L. Child: *Child Training and Personality: A Cross-cultural Study*, Yale University Press, New Haven, Conn., 1953.
5. Blumen, Isadore, Marvin Kogan, and Philip J. McCarthy: *The Industrial Mobility of Labor as a Probability Process*, Cornell Studies in Industrial and Labor Relations, vol. VI, Ithaca, N.Y., 1955.
6. Morse, Nancy C., and Robert S. Weiss: "The Function and Meaning of Work and the Job," *Amer. Sociol. Rev.*, **20**:191–198 (1955).

OTHER READING

Ackoff, Russell L.: *The Design of Social Research*, chap. 2, The University of Chicago Press, Chicago, 1953. This chapter discusses the formulation of the problems for social research. Hypotheses, alternative hypotheses, and the consequences of various types of error are considered.

Wilks, S. S.: *Elementary Statistical Analysis*, chap. 6, Princeton University Press, Princeton, N.J., 1948. Obtains by formal proof some of the important properties of the binomial distribution.

CHAPTER 9

DRAWING STATISTICAL INFERENCES FROM THE ARITHMETIC MEAN OF A LARGE SAMPLE

9.1. Some Problems of Inference for Quantitative Variables

Problems of statistical inference that arise when samples are used to provide information on the population distribution of a dichotomous variable have been treated in detail. These problems also arise in studying quantitative variables. For example, the Survey of Consumer Finances of the Federal Reserve Board is described in Chap. 2. This survey selects a sample of spending units to represent the population of all spending units in the United States. When analysts compute and report arithmetic means, medians, variances, or interquartile ranges of sample distributions of quantitative variables such as annual money income before taxes, they do so not because the sample is the principal concern but because these measures of sample characteristics provide information concerning the values of the corresponding population measures. Since a sample value will not ordinarily be exactly the same as the corresponding population value, an objective determination of the amount of expected agreement is required.

Although one can describe the relation between a sample distribution and the corresponding population distribution with respect to any desired characteristic, the contents of this chapter will deal only with the location of the distribution as measured by the arithmetic mean. That is, a *simple random sample* will be drawn from the population under study.* The value of a particular quantitative variable, say x, will be obtained for each element in the sample. The arithmetic mean $\bar{x}$ of these observations will then be used either to estimate the unknown population mean μ or to test hypotheses concerning the value of μ. We wish to develop techniques for making these inferences, the principal requirement being that the techniques must provide a means for controlling the chances of committing various types of error. The fundamental ideas will be the same as in Chap. 8; only the specific techniques will differ.

* The Survey of Consumer Finances does not employ simple random samples. Modifications of simple random sampling will be described in Chap. 10.

9.2. The Long-run Behavior of Sample Means

9.2.1. *Empirical and Theoretical Sampling Distributions for $\bar{x}$.* The drawing of a simple random sample may be usefully viewed as a single trial of a conceptually repeatable experiment where the outcomes of present interest are the possible values of $\bar{x}$, the sample arithmetic mean. The value of $\bar{x}$ cannot be predicted with certainty for any single sample. It has been shown in Secs. 6.2 and 7.3 that the model for simple random sampling leads to probability statements concerning $\bar{x}$. The basic probabilities for one example are given in Table 6.1. These probabilities were obtained by direct enumeration of all samples that could be drawn from a small finite population.

Direct enumeration cannot be used in general since most populations are so large that it is not feasible (and is impossible for infinite populations) to prepare a listing of all possible samples of a given size. However, a formal approach related to that used in deriving the binomial distribution of Chap. 8 can be used to describe some aspects of what would be expected if all possible samples were enumerated. Although the actual details are beyond the scope of this book, some of the more useful results can be stated and applied.

The general features of the underlying theory can be easily illustrated by actually drawing a number of simple random samples from a known population, computing $\bar{x}$ for each sample, and observing the sample-to-sample variability in $\bar{x}$. Let us consider distribution A in Fig. 9.1 as the known population. This is the percentage distribution of the number of acres in all crops for a population of 556 full-time commercial farms [1]. The technique described in Sec. 6.3.3 was used to draw 100 simple random samples from distribution A, each sample containing 10 farms. The samples were drawn *with replacement.* The behavior of the resulting 100 sample means is summarized in distribution B of Fig. 9.1. This is called an *empirical* sampling distribution of $\bar{x}$ since it is based on samples that have actually been drawn.

Visual examination of distributions A and B brings forth the following points:

1. Distribution B shows the fluctuations in the values of 100 sample means, each mean being obtained from a simple random sample of 10 farms. It is this variability that must be described and controlled in a precise way.

2. Although quite different in shape, distributions A and B appear to have somewhat similar locations. This can be verified by direct computation. Thus the mean number of acres in crops for the population of 556 farms is approximately 98 acres, while the mean of the 100 sample means pictured in distribution B is 100 acres.

3. The spread of distribution B is much less than the spread of distribution A. Again this can be verified by direct computation. The standard deviation of distribution A is 68 acres, while the standard deviation of distribution B is 22 acres.

4. Distribution A is highly skewed, while distribution B is more nearly symmetric.

Basing the empirical sampling distribution of $\bar{x}$ on the results of 100 samples was completely arbitrary. Setting aside the practical problem of finding time to draw samples, 500 or 1,000 or 1 million simple

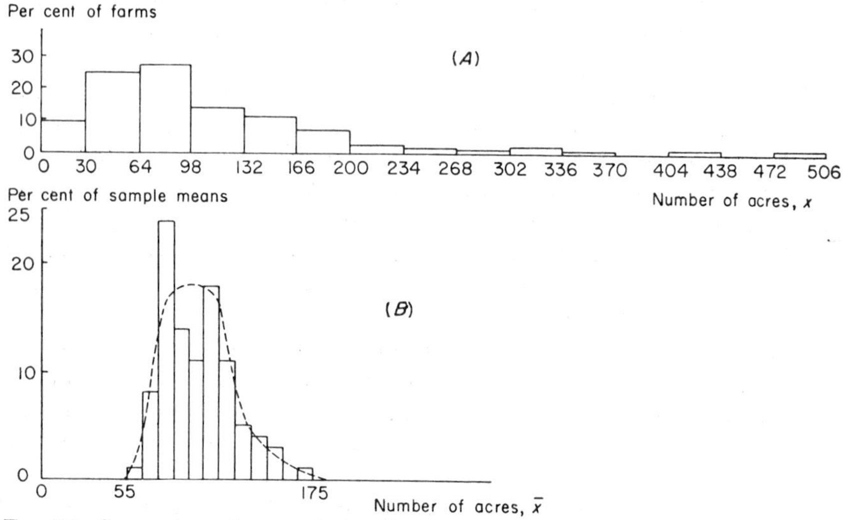

FIG. 9.1. Comparison of a population distribution with the distribution of the arithmetic means of 100 samples drawn from the population. Distribution A gives the distribution of the number of acres in crops for 556 farms. Distribution B is the distribution of the arithmetic means of 100 simple random samples, each sample containing 10 farms, drawn with replacement from the population of 556 farms.

random samples could have been drawn to determine an empirical distribution. However, with an increasing number of samples one would expect that the jumps between adjacent cells of the resulting histogram would become less and less marked. As a matter of fact, we can think of drawing an indefinitely large number of such samples. The distribution of sample means might then be expected to conform to some smooth curve such as is indicated by the dotted curve in distribution B of Fig. 9.1 (see Sec. 3.3.2). The distribution based on an indefinitely large number of samples describes, with respect to the variable $\bar{x}$, the infinite and hypothetical population of all possible simple random samples that one might draw. Such a distribution will be called the theoretical *sampling dis-*

tribution of $\bar{x}$. The binomial distribution of Chap. 8 was such a theoretical sampling distribution, appropriate for dealing with the sample fraction of S's.

We must now attempt to describe the circumstances under which we can specify the theoretical sampling distribution of $\bar{x}$, can give the characteristics of this sampling distribution, and can use this sampling distribution to solve problems of inference concerning unknown population means.

9.2.2. *The Mean and Variance of the Sampling Distribution of* $\bar{x}$. There are two characteristics of the theoretical sampling distribution of $\bar{x}$ which will be of paramount importance in succeeding considerations. The *location* of the distribution, as compared with the mean μ of the population from which the samples are drawn, tells whether sample means in general can be expected to cluster around the population mean or whether they may tend to cluster around some value that is removed from the population mean. Thus distribution B of Fig. 9.1 (taken as an approximation to the sampling distribution of $\bar{x}$) is located near the population mean of distribution A, and one obtains the general impression that $\bar{x}$ is "aimed at the correct target." If distribution B were displaced 50 or 60 units to the right, one would feel that the "aim" of $\bar{x}$ was poor and would hesitate to use it as an estimate of μ. The dispersion of the sampling distribution of $\bar{x}$ is also of importance. If the distribution has a large dispersion, then we cannot expect a large fraction of sample $\bar{x}$'s to be close to the population mean, even though $\bar{x}$ is properly "aimed." On the other hand, small dispersion (coupled with correct aim) indicates that $\bar{x}$ cannot stray too far away from the population mean. The first goal will therefore be to obtain measures of the location and dispersion of the sampling distribution of $\bar{x}$.

As a starting point, let us assume that the mean μ and the variance σ^2 of the population are known. With this knowledge at our disposal, the sampling distribution of $\bar{x}$ can be specified in sufficient detail to provide solutions for the inferential problems. This approach is identical with that presented in Chap. 8 for the binomial distribution. We first studied the behavior of the sample fraction of S's in terms of a known population fraction and then reversed the argument so as to learn something about the value of an unknown p from an observed sample fraction.

Taking the arithmetic mean of the sampling distribution of $\bar{x}$ as a measure of location and designating this arithmetic mean by $\mu_{\bar{x}}$ or $E(\bar{x})$, we have the following:

> If *simple random samples* of size n are drawn with or without replacement from a finite population having mean μ, or if

samples of n *independent* observations are obtained from an infinite population with mean μ, then

$$\mu_{\bar{x}} = E(\bar{x}) = \mu \tag{9.1}$$

This result states that if an indefinitely large number of simple random samples, each of size n, are drawn from a specified population, then the arithmetic average of all these sample means will be equal to the mean of the original population. Expression (9.1) is sometimes summarized in words by saying that $\bar{x}$ is an *unbiased* estimate of μ. Notice particularly that (9.1) holds subject only to the requirement that simple random samples are drawn from a finite population or that independent observations are obtained from an infinite population. There are no restrictions concerning population size, sample size, type of quantitative variable (i.e., discrete or continuous), or shape of the original population distribution. Thus the mean of a random sample has the correct aim for estimating μ.

A numerical verification of (9.1) has already been given in Sec. 6.2. Table 6.1 gives the sampling distribution of $\bar{x}$ for samples drawn from a small finite population. It was shown in the accompanying discussion that the mean of this sampling distribution was equal to the mean of the original population. Furthermore, empirical evidence concerning (9.1) was quoted in the discussion of Fig. 9.1. The mean of distribution B (describing the means of 100 rather than an infinite number of simple random samples) was found to be very close to the mean of distribution A.

As will be shown in Sec. 9.3, the variance (or its square root, the standard deviation) is the most useful measure of dispersion for the sampling distribution of $\bar{x}$. In order to give theoretical expressions for the variance of the sampling distribution of $\bar{x}$, we must distinguish several different situations that were handled simultaneously in the treatment of the mean of the distribution. Designating the variance of the sampling distribution by $\sigma_{\bar{x}}^2$ or $V(\bar{x})$, it can be proved that:

1. If simple random samples of size n are drawn *without replacement* from a finite population of N elements with variance σ^2,

$$\sigma_{\bar{x}}^2 = V(\bar{x}) = \left(1 - \frac{n}{N}\right)\frac{\sigma^2}{n} \tag{9.2}$$

2. If simple random samples of size n are drawn *with replacement** from a finite population having variance σ^2 or if n inde-

* The precise formula in this case is

$$\sigma_{\bar{x}}^2 = \frac{N-1}{N}\frac{\sigma^2}{n}$$

For most populations of practical interest, N is large enough so that the factor $(N-1)/N$ is extremely close to 1 and can be ignored.

pendent observations are obtained from an infinite population with variance σ^2,

$$\sigma_{\bar{x}}^2 = V(\bar{x}) = \frac{\sigma^2}{n} \tag{9.3}$$

These expressions hold for samples of any specified size n, for both discrete and continuous variables, and for population distributions of any shape. Remembering that the standard deviation is the square root of the variance, the standard deviation of the sampling distribution of $\bar{x}$ is given by $\sigma_{\bar{x}}$. It is customary to refer to $\sigma_{\bar{x}}$ as the *standard error* of $\bar{x}$.

Some empirical evidence concerning the validity of (9.3) has already been presented. The variance σ^2 of distribution A of Fig. 9.1 was found to be 4,623 (acres)^2. Distribution B is an empirical distribution of the means of simple random samples of size 10 drawn with replacement from the population represented in A. Substituting in (9.3),

$$\sigma_{\bar{x}}^2 = \frac{4{,}623}{10} = 462.3 \text{ (acres)}^2$$

$$\sigma_{\bar{x}} = \sqrt{462.3} = 21.5 \text{ acres}$$

The variance and standard deviation of distribution B should be close but not necessarily equal to these quantities. The value of the variance is 497.9 (acres)^2, and the standard deviation is 22.3 acres. If a larger number of samples were used in obtaining the empirical distribution, then even closer agreement with the theoretical values would be expected.

The difference between expressions (9.2) and (9.3) deserves some comment. The factor $[1 - (n/N)]$ which multiplies σ^2/n in (9.2) is called the *finite-population correction.* It goes from 1 to 0 as n increases from 0 to N and provides a quantitative measure of one's intuitive feeling that it is "better," for example, to have a sample of size 100 from a population of 200 than it is to have a sample of 100 from a population containing 1,000 elements. If two such populations have the same variance σ^2, then $\sigma_{\bar{x}}^2$ in the first instance is

$$\left(1 - \frac{100}{200}\right)\frac{\sigma^2}{100} = .005\sigma^2$$

while in the second it is

$$\left(1 - \frac{100}{1{,}000}\right)\frac{\sigma^2}{100} = .009\sigma^2$$

The reason for a smaller variance in the first instance is readily apparent. If sampling is without replacement, a population element can appear at most once in any given sample. However, a population element can appear more than once if sampling is done with replacement. The repeated appearance of the same element in one sample essentially "wastes" information, and hence there is a larger variance (i.e., the estimate is less precise) if the sampling is with replacement.

The effects of the sample size and the finite-population correction on the variance and standard deviation of the sampling distribution of $\bar{x}$ can be easily illustrated in terms of a numerical example. Table 9.1 presents some computations relating to the data of distribution *A* of Fig. 9.1. We shall speak specifically about the data of Table 9.1, but the general observations that can be made will apply to any situation

TABLE 9.1. EFFECTS OF SAMPLE SIZE AND FINITE-POPULATION CORRECTION ON THE VARIANCE AND STANDARD DEVIATION OF THE SAMPLING DISTRIBUTION OF $\bar{x}$
(Data of distribution *A* of Fig. 9.1; $N = 556$, $\sigma^2 = 4{,}623$)

Sample size n	Finite population correction $\left(1 - \frac{n}{N}\right)$	Variance of $\bar{x}$ for sampling		Standard error of $\bar{x}$ for sampling	
		With replacement $\frac{\sigma^2}{n}$	Without replacement $\left(1 - \frac{n}{N}\right)\frac{\sigma^2}{n}$	With replacement $\sqrt{\frac{\sigma^2}{n}}$	Without replacement $\sqrt{\left(1 - \frac{n}{N}\right)\frac{\sigma^2}{n}}$
5	.9910	924.60	916.28	30.4	30.3
10	.9820	462.30	453.98	21.5	21.3
20	.9640	231.15	222.83	15.2	14.9
40	.9281	115.58	107.27	10.8	10.4
100	.8201	46.23	37.91	6.8	6.2
250	.5504	18.49	10.18	4.3	3.2
556	0	8.31	0	2.9	0

where the mean of a simple random sample is being used to provide information concerning a population mean. In particular:

1. The variance and standard error of $\bar{x}$ decrease as the sample size increases—the larger the sample size the smaller the dispersion of the sample means about the population mean.

2. If random samples are drawn without replacement, the variance and standard error of $\bar{x}$ are zero when the sample size is equal to the population size and the sample mean is therefore equal to the population mean. This does not hold if the samples are drawn with replacement. There will be enough duplication of elements in the sample so that sample means can still fluctuate about the population mean.

3. The variances and standard errors for the two types of sampling are very close to one another when the sample size is small relative to the population size. The factor $[1 - (n/N)]$ and its square root are then near 1. If a sample comprises 5 per cent or less of a population, these factors have values larger than .95 and .975, respectively, and their effects can usually be ignored. This is the basis for the general working

practice stated in Sec. 8.3.1 of applying the binomial model when simple random samples constitute 5 per cent or less of a finite population.

Point 3 has one implication which is often difficult to grasp on first meeting. If a sample is 5 per cent or less of a population, the variance of $\bar{x}$ is extremely close to σ^2/n, and this expression does not depend on the population size. Thus $V(\bar{x})$ will be essentially the same for a sample of 10 drawn from a population of 1,000 elements as it will be for a sample of 10 drawn from a population of 1 million elements, provided of course that σ^2 is the same for both populations.

9.2.3. *The Sample Fraction of S's for a Dichotomy Can Be Treated as a Sample Mean.* A very simple artifice can be used to translate the dichotomy problem of Chap. 8 into a problem which involves the arithmetic mean of a quantitative variable. Suppose that a 1 is assigned to every S element in the population and that a 0 is assigned to every F element. This assignment defines a quantitative variable x, which can assume only the two values 0 and 1.

Now let us ask for the mean μ of this variable for a finite population. By definition the mean is the sum of values of the variable for all elements in the population divided by the number of elements in the population. The sum of values of the variable in this case is a sum of 0's and 1's, a 1 referring to an S and a 0 to an F. It is therefore equal to the total number of S's in the population. Hence

$$\begin{aligned}\mu &= \frac{\text{number of } S\text{'s in the population}}{N}\\ &= \text{fraction of } S\text{'s in the population}\\ &= p\end{aligned}$$

The same argument shows that the sample mean $\bar{x}$ is equal to the sample fraction of S's, u/n. Thus using the sample fraction of S's to estimate the population fraction of S's is a special case of the problem being considered in this chapter, and the results stated in the preceding sections can be applied directly.

From (9.1), the mean of a simple random sample is an unbiased estimate of the population mean. It therefore follows that the sample fraction of S's is an unbiased estimate of the population fraction [see expression (8.2)]. Stated in specific terms:

> If simple random samples of size n are chosen from a finite population having a fraction of S's equal to p, or if samples of n independent observations are obtained from an infinite population with a fraction of S's equal to p, then

$$E\left(\frac{u}{n}\right) = p \tag{9.4}$$

This result is more general than that given by (8.2) since the binomial distribution refers only to sampling from an infinite population.

Expressions for the variance of u/n can be obtained from (9.2) and (9.3), provided that an appropriate expression is used for σ^2. No difficulty is encountered in obtaining σ^2 if we remember that we are dealing with a quantitative variable for which a fraction p of the population has the value 1 and a fraction q has the value 0. For example, consider an infinite population and the definition of σ^2 given by (5.5). Taking $k = 2$, $x_1 = 1$, $x_2 = 0$, $p_1 = p$, $p_2 = q$, and $\mu = p$,

$$\begin{aligned}\sigma^2 &= (p \cdot 1^2 + q \cdot 0^2) - p^2\\ &= p - p^2\\ &= p(1 - p)\\ &= pq\end{aligned}$$

For finite populations, the use of the fourth form of (5.4) gives

$$\sigma^2 = \frac{N}{N - 1} pq$$

Substitution of these expressions for σ^2 into (9.2) and (9.3), where the precise form of (9.3) is used for sampling with replacement, leads to:

1. If simple random samples of size n are drawn without replacement from a finite population of N elements in which a fraction p are S's, the variance of the sample fraction of S's is

$$V\left(\frac{u}{n}\right) = \frac{N - n}{N - 1} \frac{pq}{n} \tag{9.5}$$

2. If simple random samples of size n are drawn with replacement from a finite population of N elements in which a fraction p are S's, or if n independent observations are obtained from an infinite population with the same characteristics, the variance of the sample fraction of S's is

$$V\left(\frac{u}{n}\right) \doteq \frac{pq}{n} \tag{9.6}$$

Recognizing that the binomial distribution describes the sample-to-sample behavior of u/n when random samples are drawn from an infinite population, we conclude that (9.6) gives the variance of the binomial distribution. This variance expression can be employed in confirming some of the visual impressions that were obtained in examining the binomial distributions of Fig. 8.1. The values of $V(u/n)$ for these distributions are:

Distribution	*Variance of u/n*
$n = 4,\ p = .1$	$(.1 \times .9)/4 = .0225$
$n = 4,\ p = .5$	$(.5 \times .5)/4 = .0625$
$n = 4,\ p = .9$	$(.9 \times .1)/4 = .0225$
$n = 10,\ p = .1$	$(.1 \times .9)/10 = .0090$
$n = 10,\ p = .5$	$(.5 \times .5)/10 = .0250$
$n = 20,\ p = .1$	$(.1 \times .9)/20 = .0045$
$n = 20,\ p = .5$	$(.5 \times .5)/20 = .0125$

Therefore:

1. *For a fixed value of n,* the binomial distribution becomes less spread out as p moves away in either direction from .50. Thus for ($n = 4$, $p = .5$), $V(u/4) = .0625$, while for ($n = 4$, $p = .1$), $V(u/4) = .0225$.

2. *For a fixed value of p,* the binomial distribution becomes less spread out as n increases. Thus for ($n = 4$, $p = .5$), $V(u/4) = .0625$, while for ($n = 20$, $p = .5$), $V(u/20) = .0125$.

9.3. The Normal Distribution as an Approximation to the Long-run Behavior of Means of Large Samples

9.3.1. *The Normal Distribution as an Approximation to the Sampling Distribution of $\bar{x}$.* The mean and variance of the sampling distribution of $\bar{x}$ give many insights into the behavior of $\bar{x}$ as an estimate of μ. They show that $\bar{x}$ has the correct aim for estimating μ, and they give some general impressions concerning the amount by which $\bar{x}$ can be expected to vary about the population mean under differing circumstances. This, however, is not enough if the goals are to obtain confidence limits for μ or to devise tests of hypotheses with known probabilities of Type I and Type II errors. The attainment of these goals requires probability statements about $\bar{x}$, and such statements can come only from a reasonably exact specification of the entire sampling distribution.

For *small* samples, there is little that can be accomplished in this direction unless (1) the exact sampling distribution of $\bar{x}$ can be derived from the population distribution of x, either by enumerating all possible simple random samples or by some other probability argument; and (2) the desired probabilities can be computed from the exact sampling distribution of $\bar{x}$. Both of the approaches mentioned in 1 have already been used in this book to obtain the probabilities required in 2. Table 6.1 illustrated the enumeration approach, while the exact sampling distribution of u/n was derived in Chap. 8. The enumeration approach is of little value since most populations of practical interest are too large. A formal probability approach for small samples requires a consideration of many different population distributions and the use of techniques which are beyond the scope of this book. Moreover, this approach must use information about the population distribution of x that is frequently not known. For these reasons attention will be restricted to *large* samples for which it is possible to state results that can be used in a wide variety of circumstances.

The empirical distribution of $\bar{x}$ for samples of size 10, pictured in Fig. 9.1, is more nearly symmetric than the distribution of x in the population from which the samples were drawn. This "approach to symmetry" would have been more pronounced if a larger sample size had been used. Such empirical behavior of $\bar{x}$'s suggests that it might be possible to approximate the sampling distribution of $\bar{x}$ for large n by some standard symmetric distribution. This can actually be done as a consequence of the following theorem:

> If repeated simple random samples are drawn (with or without replacement) from a finite population or if repeated sets of independent observations are obtained from an infinite population with finite variance σ^2 and if n is sufficiently large (for finite populations, N must be considerably larger than n), then the sampling distribution of $\bar{x}$ can be closely *approximated* by a *normal distribution* having the same mean and variance as the sampling distribution of $\bar{x}$. The quality of the approximation improves as n increases.

A *normal distribution* is a theoretical distribution which is unimodal and symmetric. A knowledge of the mean and variance of a normal distribution permits one to evaluate any desired probability. The properties of this distribution will be examined in Sec. 9.3.2.

Two particular normal distributions are given by the smooth curves in Fig. 9.2 where they are superimposed on two empirical sampling distributions that were generated by drawing random samples with replacement from distribution A of Fig. 9.1. The top histogram gives the distribution of 100 means of samples of size 5, while the lower one gives the distribution of 100 means of samples of size 10. (The lower histogram is the same as distribution B of Fig. 9.1.) The mean of each normal distribution is 98 acres, this being the value of μ for the population. The standard deviation of the top normal distribution is 30.4 acres, and for the lower distribution it is 21.5 acres. As is shown in Table 9.1, these are the values of the standard errors of the sampling distribution of $\bar{x}$ for samples of sizes 5 and 10, respectively. It is difficult to gain a precise impression of the "fit" of these normal distributions to the empirical distributions, and even more difficult to see how they would compare with the theoretical distributions of $\bar{x}$. However, one's imagination should indicate that the fit may not be too bad, especially for samples larger than 10. (The report [1] gives empirical distributions of $\bar{x}$ for samples containing 100 farms.) Of course, the previously stated theorem is based upon exact proof and not upon visual impressions. The approach to normality of the sampling distribution of $\bar{x}$ is one of the most useful and remarkable tools available to the statistician. It is a special case of a series of central limit theorems that have been proved during the past two hundred years by some of the world's most eminent mathematicians. Notice that it depends neither upon the shape of the distri-

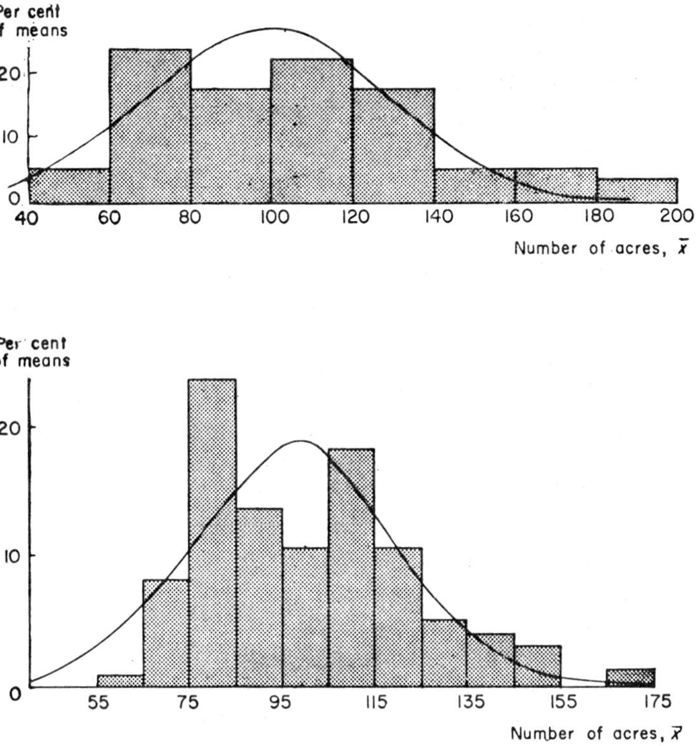

FIG. 9.2. Comparison of empirical distributions of the sample mean with normal distributions having the same mean and variance as the theoretical sampling distribution of $\bar{x}$. The upper histogram gives the distribution of $\bar{x}$ for 100 samples, each sample containing five farms, drawn with replacement from the population of Fig. 9.1. The lower histogram gives a similar distribution, where each sample contained 10 farms.

bution from which the samples are drawn nor upon the type of quantitative variable (i.e., discrete or continuous).

9.3.2. *The Properties of the Normal Distribution.* The two normal distributions given in Fig. 9.2 show several of the important properties of such distributions, namely, that they are unimodal and symmetric. As was observed in Chap. 4, the mean of a symmetric distribution is at the exact center of the distribution. Now let us see what additional information is required for making probability statements about the sample-to-sample variability of $\bar{x}$. For example, suppose samples of size 100 are to be drawn without replacement from the population ($\mu = 98$ acres) pictured in Fig. 9.1 and we wish to know the probability that the sample mean lies between 94 and 102 acres. The standard error of $\bar{x}$ for this sampling situation is obtained from Table 9.1 as 6.2 acres. Since the sample size is large, the normal distribution can be used to approximate the exact sampling distribution of $\bar{x}$. A normal distribution having a mean at 98 and a standard deviation of 6.2 is shown in Fig. 9.3. It follows from the development of Secs. 3.3.2 and 7.4 that the probability that a value of $\bar{x}$ will be between 94 and 102 is given by the fraction of the total area under the distribution curve between these two numbers. This area is shaded in Fig. 9.3, and it represents 48 per cent of the total area under the curve. If an indefinitely large number of samples of size 100 were drawn from the specified population, 48 per cent would have arithmetic means between 94 and 102 acres. We shall now describe how the mean and variance of a normal distribution can be used to obtain the values of this and similar probabilities.

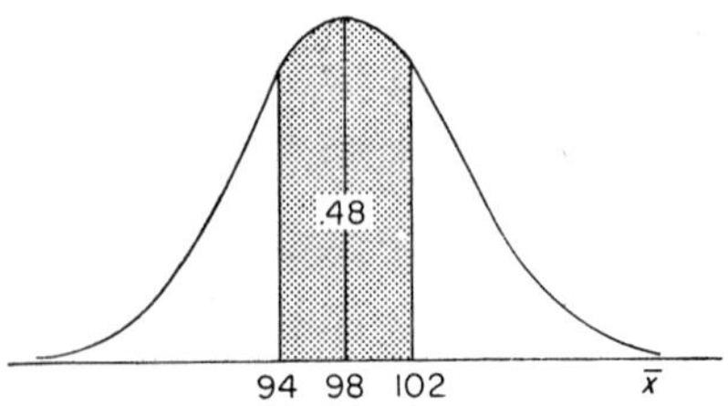

FIG. 9.3. Normal distribution with mean of 98 and standard deviation of 6.2. (Shaded area represents the probability that the mean of a sample of 100 farms drawn without replacement from population of Fig. 9.1 will lie between 94 and 102.)

All normal distributions have the general shape shown in Figs. 9.2 and 9.3. They can differ only with respect to their means and variances. Specifying μ and σ^2 completely determines a particular normal distribution.* In addition, the standard deviation σ is a "natural" measure of dispersion for a normal distribution. That is, if deviations from the mean are measured in units of σ, there exists a definite relation between these deviations and the area under the curve that is marked off by such deviations. This is illustrated in Fig. 9.4 which shows three normal distributions with different means and different standard deviations. The shaded area for each distribution is the area falling under the curve

* The equation of the normal distribution curve is given by

$$f(x) = \frac{1}{\sigma\sqrt{2\pi}} e^{-(x-\mu)^2/2\sigma^2}$$

where x = particular value of variable having normal distribution
μ = mean of distribution
σ^2 = variance of distribution
e = constant approximately equal to 2.7183
π = constant approximately equal to 3.1416

between the mean minus one standard-deviation unit and the mean plus one standard-deviation unit. In each instance this shaded area represents the same fraction of the total area under the curve, namely, 68.27 per cent. Moreover, this value holds for any normal distribution, no matter what the mean or standard deviation. Thus for any variable having a normal distribution the probability is .6827 that a random observation falls within one standard-deviation unit of the mean of the distribution.

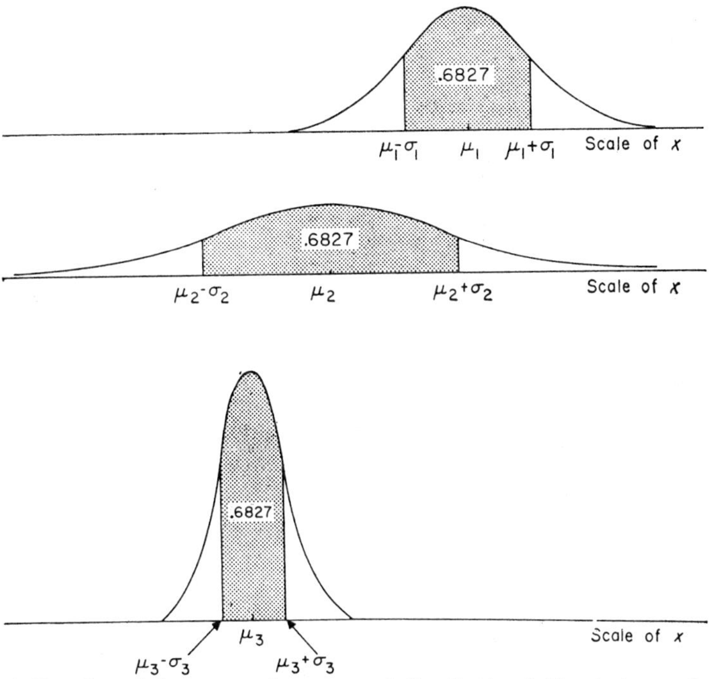

FIG. 9.4. Fraction of the area under a normal distribution falling between the mean minus one standard deviation and the mean plus one standard deviation.

Very extensive computations have been carried out to obtain the fraction of the area under a normal distribution that falls within any specified number, say k, of standard-deviation units of the mean. A small portion of these results are presented in Table 9.2. (One of the more detailed tabulations of the normal distribution is given in [2].) An interesting feature of the distribution, not heretofore mentioned, is illustrated by the contents of Table 9.2. Theoretically, the curve of the normal distribution extends indefinitely far on each side of the mean without ever coming in contact with the x axis. Actually, Table 9.2 shows that all but .17 per cent of the area is within three standard deviations of the mean and all but .00006 per cent is within five standard deviations of the mean. In practice we never deal with variables having a truly infinite range, as is implied by the foregoing property. However, such a high fraction of the normal distribution does fall within a finite range that difficulty is not ordinarily encountered in this respect.

Although Table 9.2 gives only the probabilities associated with intervals centered at the mean of a normal distribution, these values can be used to obtain probabilities for intervals not centered at the mean.

TABLE 9.2. FRACTION OF THE AREA OF A NORMAL DISTRIBUTION FALLING WITHIN A SPECIFIED NUMBER OF STANDARD-DEVIATION UNITS OF THE MEAN

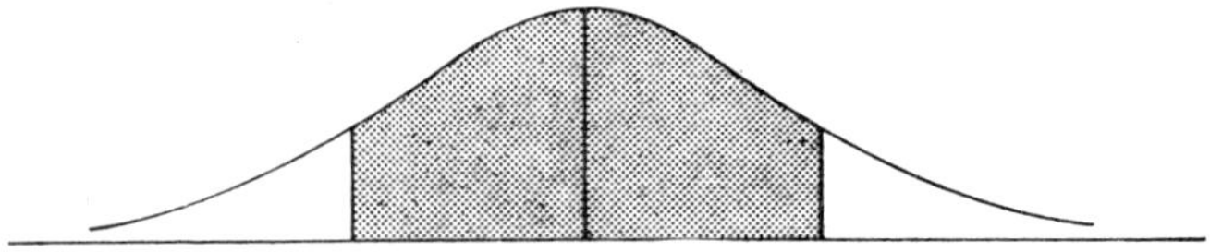

Interval	Fraction of total area	Interval	Fraction of total area
$\mu \pm 0\sigma$	.0000	$\mu \pm 1.7\sigma$	.9109
$\mu \pm .1\sigma$	.0797	$\mu \pm 1.8\sigma$	.9281
$\mu \pm .2\sigma$	.1585	$\mu \pm 1.9\sigma$	.9426
$\mu \pm .3\sigma$	.2358	$\mu \pm 2.0\sigma$	.9545
$\mu \pm .4\sigma$	.3108	$\mu \pm 2.1\sigma$	.9643
$\mu \pm .5\sigma$	.3829	$\mu \pm 2.2\sigma$	.9722
$\mu \pm .6\sigma$	.4515	$\mu \pm 2.3\sigma$	.9786
$\mu \pm .7\sigma$	.5161	$\mu \pm 2.4\sigma$	.9836
$\mu \pm .8\sigma$	.5763	$\mu \pm 2.5\sigma$	.9876
$\mu \pm .9\sigma$	.6319	$\mu \pm 2.6\sigma$	.9907
$\mu \pm 1.0\sigma$	.6827	$\mu \pm 2.7\sigma$	.9931
$\mu \pm 1.1\sigma$	.7287	$\mu \pm 2.8\sigma$	.9949
$\mu \pm 1.2\sigma$	.7699	$\mu \pm 2.9\sigma$	.9963
$\mu \pm 1.3\sigma$	.8064	$\mu \pm 3.0\sigma$	.9973
$\mu \pm 1.4\sigma$	.8385	$\mu \pm 4.0\sigma$	.99994
$\mu \pm 1.5\sigma$	.8664	$\mu \pm 5.0\sigma$	.9999994
$\mu \pm 1.6\sigma$	.8904		

These extensions depend only upon the fact that the distribution is symmetric about the mean. Some of the more useful possibilities are illustrated in Fig. 9.5. The first diagram merely points out that one-half the total area lies on each side of the mean, this being an immediate consequence of symmetry. In the second diagram, the shaded area represents the probability of an observation falling between $\mu - 1.5\sigma$ and μ. From Table 9.2 the probability of falling between $\mu - 1.5\sigma$ and $\mu + 1.5\sigma$ is .8664. Because of symmetry, one-half of this lies on each side of the mean. Therefore the required probability is .4332. The shaded area in the third diagram is the probability of falling below

$\mu - 1.5\sigma$. This is equal to the area to the left of the mean (.5000) minus the area between $\mu - 1.5\sigma$ and μ (.4332), namely, .0668. Because of symmetry, the fraction of the area above $\mu + 1.5\sigma$ is also .0668. Thus the shaded area of the fourth diagram is the probability that an observation is less than $\mu - 1.5\sigma$ or more than $\mu + 1.5\sigma$. This is equal to

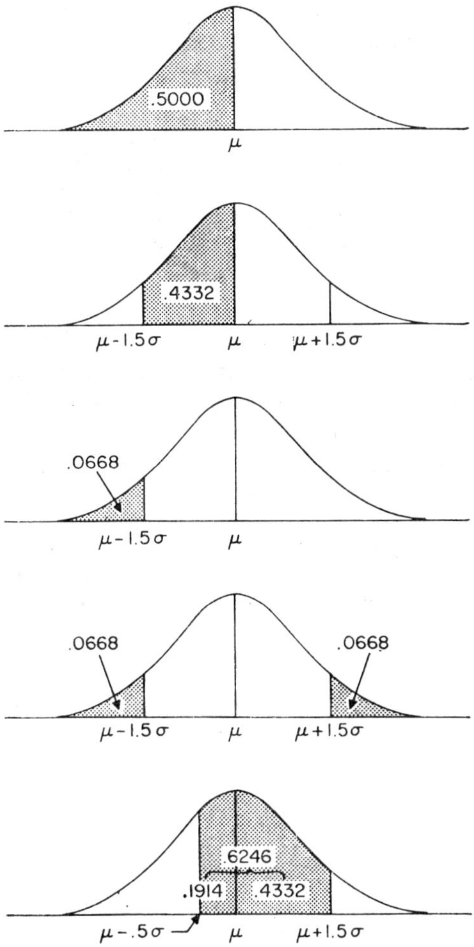

FIG. 9.5. Areas under a normal distribution.

.0668 + .0668, or .1336. This area can also be obtained by subtracting the area between $\mu - 1.5\sigma$ and $\mu + 1.5\sigma$, .8664, from the total area, 1.0000. The fifth and final diagram shows the probability of having an observation between $\mu - .5\sigma$ and $\mu + 1.5\sigma$. It is equal to one-half the area between $\mu - .5\sigma$ and $\mu + .5\sigma$, $(\frac{1}{2})(.3829)$, plus one-half the area between $\mu - 1.5\sigma$ and $\mu + 1.5\sigma$, $(\frac{1}{2})(.8664)$. The actual value is .6246.

It will frequently be convenient to know the number of standard-deviation units required on each side of the mean to define a centered interval having a specified fraction of the total area under the normal curve. Table 9.3 presents a few of the more commonly used values.

TABLE 9.3. CENTERED INTERVALS ABOUT THE MEAN CONTAINING A SPECIFIED FRACTION OF THE TOTAL AREA UNDER A NORMAL DISTRIBUTION

Fraction of area over the centered interval	*Centered interval*
.50	$\mu \pm .6745\sigma$
.80	$\mu \pm 1.282\sigma$
.90	$\mu \pm 1.645\sigma$
.95	$\mu \pm 1.960\sigma$
.98	$\mu \pm 2.326\sigma$
.99	$\mu \pm 2.576\sigma$
.999	$\mu \pm 3.291\sigma$

Thus the interval $\mu \pm .6745\sigma$ accounts for the middle 50 per cent of the area; the interval $\mu \pm 1.960\sigma$ accounts for the middle 95 per cent of the area; and the interval $\mu \pm 3.291\sigma$ accounts for the middle 99.9 per cent of the area.

9.3.3. *The Application of the Normal Distribution When Population Characteristics Are Known.* The approximate normality of the sampling

TABLE 9.4. DURATION OF PHYSICAL THERAPY FOLLOWING HOSPITAL DISCHARGE OF PARALYZED POLIOMYELITIS PATIENTS
(New York City: disease contracted in 1949)

Duration of physical therapy, months	*Number of patients*
0 up to 3	21
3 up to 6	30
6 up to 9	49
9 up to 12	47
12 up to 15	43
15 up to 18	30
18 up to 21	32
21 up to 24	25
24 up to 27	22
27 up to 30	5
30 up to 33	4
33 up to 36	0
36 up to 39	1
Total	309

SOURCE: Helen M. Wallace, Patricia Heely, Margaret A. Losty, Herbert Rich, "A Study of Follow-up of Poliomyelitis Patients," *Amer. Jour. Pub. Health,* **43**: 542–553 (1953).

distribution of $\bar{x}$, coupled with a knowledge of the mean and variance of this distribution, permits us to use the results of Sec. 9.3.2 for making probability statements concerning the behavior of the means of large samples. As an illustration, consider the data of Table 9.4. This distribution was obtained in a follow-up study made in New York City from public health nursing records and from records in the Bureau for Handicapped Children. The 309 patients represented in this distribution contracted the disease in 1949, were paralyzed 3 months after the onset of the disease, and received some physical therapy following hospital discharge. One of the reasons for making this study was to see how actual facts "correlated with the recommendation of a medical advisory committee in the city that, except for unusual circumstances, physical therapy is usually not necessary and not productive after a period of one year. . . . " Using the known characteristics of this distribution, it is possible to determine how well the mean of a simple random sample of these patients would have approximated the true mean.

The mean μ of the distribution of Table 9.4 is 13.5 months, while the variance σ^2 is 56.01 (months)^2. From (9.1) and (9.2),

$$\mu_{\bar{x}} = \text{mean of sampling distribution of } \bar{x} = 13.5 \text{ months}$$
$$\sigma_{\bar{x}} = \text{standard error of sampling distribution of } \bar{x}$$

$$= \sqrt{\left(1 - \frac{n}{N}\right)\frac{\sigma^2}{n}} \qquad \text{months}$$
$$= \sqrt{\left(1 - \frac{n}{N}\right)}\frac{\sigma}{\sqrt{n}} \qquad \text{months}$$
$$= \sqrt{\left(1 - \frac{n}{N}\right)}\frac{7.48}{\sqrt{n}} \qquad \text{months}$$

These results hold no matter what the size of sample. If, in addition, n is "large," the sampling distribution can be approximated by a normal distribution and the values given in Tables 9.2 and 9.3 furnish the information needed in making probability statements.

Consider samples of size 75. Then

$$\sigma_{\bar{x}} = \sqrt{\left(1 - \frac{75}{309}\right)}\left(\frac{7.48}{\sqrt{75}}\right)$$
$$= (.87)\left(\frac{7.48}{8.66}\right) = .75 \text{ months}$$

Using Table 9.3, we can now make such approximate probability statements as the following:

1. $\Pr(\mu_{\bar{x}} - 1.645\sigma_{\bar{x}} \leq \bar{x} \leq \mu_{\bar{x}} + 1.645\sigma_{\bar{x}})$
$$= \Pr[13.5 - 1.645(.75) \leq \bar{x} \leq 13.5 + 1.645(.75)]$$
$$= \Pr(12.3 \leq \bar{x} \leq 14.7) = .90$$
2. $\Pr(\bar{x} \geq 14.7) = .05$
3. $\Pr(\mu_{\bar{x}} - 2.576\sigma_{\bar{x}} \leq \bar{x} \leq \mu_{\bar{x}} + 2.576\sigma_{\bar{x}})$
$$= \Pr(11.6 \leq \bar{x} \leq 15.4) = .99$$
4. $\Pr(\bar{x} \geq 15.4) = .005$
5. $\Pr(\bar{x} \leq 11.6) = .005$

Notice that it is also possible to find the approximate probability that $\bar{x}$ lies in any stated interval. For example consider $\Pr(17 \leq \bar{x} \leq 18)$. Expressing 17 and 18 in standard-error units from the mean, we have

$$13 = 13.5 - \frac{.5}{.75}(.75) = 13.5 - .67(.75)$$
$$= \mu_{\bar{x}} - .67\sigma_{\bar{x}}$$

and

$$14 = \mu_{\bar{x}} + .67\sigma_{\bar{x}}$$

From Table 9.2

$$\Pr(\mu_{\bar{x}} - .67\sigma_{\bar{x}} \leq \bar{x} \leq \mu_{\bar{x}} + .67\sigma_{\bar{x}}) = .50$$

If the sample size were different from 75, it would only be necessary to determine the appropriate standard error from (9.2) and use this value in determining probability statements similar in form to the foregoing.

The approximate normality of the sampling distribution of $\bar{x}$, together with the analysis of Sec. 9.2.3, also provides probability statements for the sampling behavior of u/n, the sample fraction of S's. The case of a variable x which can assume only the two values 0 and 1 is included in the general statement of the approach to normality of the sampling distribution of $\bar{x}$. Consider drawing simple random samples of size n from a finite population in which the fraction of S's is p. Then if n is sufficiently large, the use of (9.4), (9.5), and Table 9.3 leads to approximate probability statements of the following type:

$$\Pr(\mu_{\bar{x}} - 1.96\sigma_{\bar{x}} \leq \bar{x} \leq \mu_{\bar{x}} + 1.96\sigma_{\bar{x}}) = .95$$
$$\Pr\left(p - 1.96\sqrt{\frac{N-n}{N-1}\frac{pq}{n}} \leq \frac{u}{n} \leq p + 1.96\sqrt{\frac{N-n}{N-1}\frac{pq}{n}}\right) = .95$$

As an illustration, assume that random samples of size 100 are drawn from the distribution of Table 9.4 and that an S is defined to be "duration of physical therapy is 12 months or more." Thus $n = 100$ and $p = 162/309 = .52$. The standard error of $u/100$ is

$$\sqrt{\left(\frac{309-100}{308}\right)\left(\frac{.52 \times .48}{100}\right)} = \sqrt{(.6786) \times (.002496)}$$
$$= \sqrt{.001694} = .041$$

Substituting,

$$\Pr\left[.52 - 1.96(.041) \leq \frac{u}{100} \leq .52 + 1.96(.041)\right] = .95$$

$$\Pr\left(.44 \leq \frac{u}{100} \leq .60\right) = .95$$

If the sampled population is indefinitely large, the binomial distribution of Chap. 8 gives the exact sampling distribution of u/n for any value of n. This distribution was used in determining the centered probability intervals of Figs. 8.5, 8.6, and 8.7. For sufficiently large n, the use of the normal approximation should give very nearly the same results as are obtained from the curves of these figures. For example, suppose $p = .40$ and $n = 50$. Then from Fig. 8.7,

$$\Pr\left(.22 \leq \frac{u}{50} \leq .59\right) = .99$$

The standard error of $u/50$, using (9.6), is $\sqrt{(.4 \times .6)/50}$, or .069. Therefore from Table 9.2,

$$\Pr\left[.40 - 2.576(.069) \leq \frac{u}{50} \leq .40 + 2.576(.069)\right] = .99$$

$$\Pr\left(.22 \leq \frac{u}{50} \leq .59\right) = .99$$

This result agrees with the one obtained from Fig. 8.7. Thus the normal approximation gives additional freedom in choosing a sample size and in determining centered intervals with stated probability for the binomial distribution.

At no point in the preceding discussion have we attempted to give a precise meaning to the phrase "sufficiently large n." The approximation becomes continually better as n increases, but for practical purposes one must have some idea of how good it is in a particular situation. The various factors which influence this matter are:

1. *The shape of the distribution from which the samples are drawn.* If the distribution is highly skewed, then n must be large enough to "pull" the sampling distribution into the symmetric shape of the normal distribution. On the other hand, if the population distribution is already symmetric, the sampling distribution of $\bar{x}$ will be symmetric for any value of n and not such a large sample size will be required. It can also be shown that *if the variable x has a normal distribution, the sampling distribution of $\bar{x}$ will be exactly normal for any size of sample* ($n = 1, 2, 3, \ldots$).

2. *The portion of the sampling distribution that is to be approximated by the normal distribution.* A smaller value of n will generally be required

to obtain a "good" approximation to the "center" of the sampling distribution of $\bar{x}$ than will be needed to approximate the "tails" of the distribution.

3. *The accuracy required of the approximation.* One may desire 95 per cent centered probability intervals for $\bar{x}$ but may be satisfied if the actual probability is any place between 93 and 97 per cent; or, one may wish the actual probability to be between 94.5 and 95.5 per cent. A larger sample size will be required in the second case.

Except under special circumstances, it is very difficult to translate the foregoing factors into any exact rule as to what is meant by "sufficiently large n." However, these factors can be kept in mind when using the normal approximation. Some specific recommendations concerning sample size when determining confidence limits for p will be given in the next section.

9.4. Confidence Limits and Tests of Hypotheses for the Mean of a Population

9.4.1. *Confidence Limits for an Unknown Population Mean.* Large random samples are to be drawn from a population having an unknown mean μ. The sample mean $\bar{x}$ will be used as an estimate of μ. A measure of precision is to be associated with an observed value of $\bar{x}$, that is, a measure of how "close" we can expect this $\bar{x}$ to be to the unknown μ. This goal can be realized by using the approximate normality of the sampling distribution of $\bar{x}$ to determine confidence intervals for μ. The interpretation of these confidence intervals will be the same as that set forth in Chap. 8 for the binomial distribution.

Suppose one wishes to obtain 95 per cent confidence intervals for the unknown mean of a population. (For the moment, it will be assumed that σ^2 is known. This restriction will soon be removed.) The procedure is as follows:

1. Draw a large random sample of size n from the population under study.
2. Compute the arithmetic mean $\bar{x}$ of the sample.
3. Determine the two numbers $\bar{x} - 1.96\sigma_{\bar{x}}$ and $\bar{x} + 1.96\sigma_{\bar{x}}$.
4. State that the true value of μ lies between the two numbers computed in step 3.

The value of $\sigma_{\bar{x}}$ is obtained from the known σ^2 and the appropriate expression as between (9.2) and (9.3). As will now be shown, the statement of step 4 can be made with 95 per cent confidence.

Since $\mu_{\bar{x}} = \mu$, the approximate normality of the sampling distribution of $\bar{x}$ leads to the probability statement

$$\Pr(\mu - 1.96\sigma_{\bar{x}} \leq \bar{x} \leq \mu + 1.96\sigma_{\bar{x}}) = .95$$

This probability relationship is illustrated in Fig. 9.6 where wavy lines have been drawn at $\mu - 1.96\sigma_{\bar{x}}$, at μ, and at $\mu + 1.96\sigma_{\bar{x}}$ to emphasize that these are fixed but unknown numbers. Every time that $\bar{x}$ falls between $\mu - 1.96\sigma_{\bar{x}}$ and $\mu + 1.96\sigma_{\bar{x}}$, and it will do this for 95 per cent of an indefinitely large number of samples, the random interval $\bar{x} - 1.96\sigma_{\bar{x}}$ to $\bar{x} + 1.96\sigma_{\bar{x}}$ will include the true value of μ. ($\bar{x}$ differs from μ by not

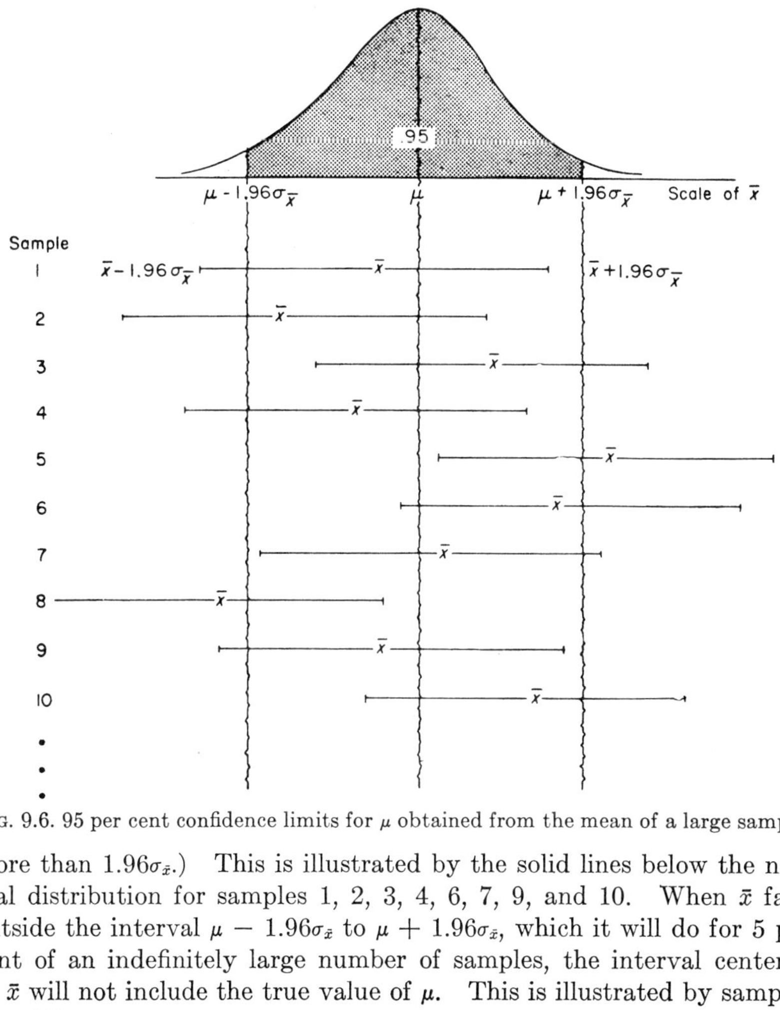

FIG. 9.6. 95 per cent confidence limits for μ obtained from the mean of a large sample.

more than $1.96\sigma_{\bar{x}}$.) This is illustrated by the solid lines below the normal distribution for samples 1, 2, 3, 4, 6, 7, 9, and 10. When $\bar{x}$ falls outside the interval $\mu - 1.96\sigma_{\bar{x}}$ to $\mu + 1.96\sigma_{\bar{x}}$, which it will do for 5 per cent of an indefinitely large number of samples, the interval centered at $\bar{x}$ will not include the true value of μ. This is illustrated by samples 5 and 8.

These results mean that if we were to draw an indefinitely large number of samples from the population, compute the limits $\bar{x} - 1.96\sigma_{\bar{x}}$ and $\bar{x} + 1.96\sigma_{\bar{x}}$ for each sample, and state each time that μ lies between the computed limits, then 95 per cent of the statements would be correct.

Of course any particular interval does or does not include μ, and it is not possible to speak in terms of probability. However, because 95 per cent of a large number of these intervals will include μ, we say that we have 95 per cent confidence in the particular interval that has been obtained. If some confidence coefficient other than .95 is desired, it is only necessary to choose the appropriate multiplier (to be used in place of 1.96) from Table 9.3.

The value of σ^2 will be unknown in actual practice. However, the sample variance s^2 can be used as an approximation to σ^2 when the sample size is large. For example, if a simple random sample of size n is drawn from a finite population of N elements, the 95 per cent confidence limits for the unknown mean are

$$\bar{x} - 1.96\sqrt{\left(1 - \frac{n}{N}\right)\frac{s^2}{n}}$$

and

$$\bar{x} + 1.96\sqrt{\left(1 - \frac{n}{N}\right)\frac{s^2}{n}}$$

Thus the sample provides its own measure of the precision of $\bar{x}$ as an estimate of μ. All these results flow from the model for simple random sampling, and this again stresses the necessity for drawing samples in accordance with known models. Similar results can be obtained for models other than simple random sampling, and some of these will be treated in Chap. 10.

When the sample size is small, it is not only necessary to consider how well the sampling distribution of $\bar{x}$ is approximated by the normal distribution but one must also take into account the fact that s^2 itself may be subject to considerable sample-to-sample variability. Although this book will not directly treat *small-sample theory*, as opposed to the *large-sample theory* of this chapter, some remarks on this topic will be found in Sec. 9.6.

As an application of this confidence-limit procedure, consider the data of Table 9.5. These were obtained from a national sample of approximately 1,700 tool-and-die makers who were interviewed in the spring of 1951 by the U.S. Bureau of Labor Statistics. Assuming that the 151 workers of this table constitute a random sample of the large number of tool-and-die makers employed in aircraft and aircraft-parts industries, the 95 per cent confidence limits for the true average age of all such workers are

$$\bar{x} - 1.96\left(\frac{s}{\sqrt{n}}\right) = 43.1 - 1.96\left(\frac{10.03}{\sqrt{151}}\right) = 41.5$$

and

$$\bar{x} + 1.96\left(\frac{s}{\sqrt{n}}\right) = 43.1 + 1.96\left(\frac{10.03}{\sqrt{151}}\right) = 44.7$$

The true average age may or may not be between 41.5 and 44.7 years, but we have 95 per cent confidence in the statement that it is between them in the sense that we know that this statement would be correct for approximately 95 per cent of a large number of pairs of limits. The 99 per cent confidence limits can be obtained by merely changing the multiplier from 1.96 to 2.58. These limits are 41.0 and 45.3 years. Wider limits are required to obtain increased confidence.

TABLE 9.5. DISTRIBUTION OF AGE FOR TOOL-AND-DIE MAKERS EMPLOYED IN AIRCRAFT AND AIRCRAFT-PARTS INDUSTRIES IN FEBRUARY AND MARCH, 1951
(National sample of tool-and-die makers)

Age, years	*Number of workers*
20 up to 30	8
30 up to 40	59
40 up to 50	46
50 up to 60	29
60 up to 70	9
Total	151

$$\bar{x} = \frac{\Sigma n_i x_i}{n} = 43.1 \text{ years}$$

$$s^2 = \frac{1}{n-1}\left[\sum n_i x_i^2 - \frac{(\Sigma n_i x_i)^2}{n}\right] = 100.54 \text{ (years)}^2$$

$$s = 10.03 \text{ years}$$

SOURCE: U.S. Department of Labor, *The Mobility of Tool and Die Makers*, Bulletin 1120, 1952.

It rarely happens, except for empirical sampling studies like those described in Sec. 9.2.1, that one can determine whether actual confidence limits do include the true population mean. However, Hansen and Hurwitz [3] report on a situation where this was possible. In 1945 the U.S. Bureau of the Census conducted a census of agriculture in the United States. In order to obtain quick estimates for some of the more important items the Bureau selected a sample of the returns and published, in 1946, national estimates for 61 agricultural items. Each estimate was accompanied by a statement of its precision (i.e., its standard error). Corresponding figures from the complete census became available about a year later, and Hansen and Hurwitz compare the sample estimates with the true population values. These comparisons demonstrate, for example, that about 5 per cent of the differences exceeded two standard errors, that is, approximately 95 per cent of the 95 per cent confidence intervals did include the true population value, and that none of the differences exceeded three standard errors.

These procedures also provide confidence limits for p, the true but unknown fraction of elements in a population having a stated characteristic. If simple random samples of size n are drawn from a finite population of N elements, the standard error of u/n is given by (9.5). This expression depends upon the unknown value of p, but the observed u/n can be used as an estimate of p. An approximation to the standard error of u/n is thereby obtained that will not be too far off if n is large. (This corresponds to using s^2 as an estimate of σ^2 in determining the standard error of $\bar{x}$.) Thus 95 per cent confidence limits for p may be determined as follows:

1. Draw a large random sample of size n from the population under study.
2. Compute the sample fraction of S's, u/n.
3. Determine the two numbers

$$\frac{u}{n} - 1.96\sqrt{\frac{N-n}{N-1}\frac{\frac{u}{n}\left(1-\frac{u}{n}\right)}{n}} \quad \text{and} \quad \frac{u}{n} + 1.96\sqrt{\frac{N-n}{N-1}\frac{\frac{u}{n}\left(1-\frac{u}{n}\right)}{n}}$$

4. State, with 95 per cent confidence, that the true value of p lies between the limits obtained in step 3.

The factor $(N - n)/(N - 1)$ will be close to 1 and can be ignored when the sample size is small in comparison with the population size. This will be true automatically if the population is infinite, as is required by the binomial model of Chap. 8.

Suppose we wish to determine from the data of Table 9.5 the 99 per cent confidence limits for the true fraction of tool-and-die makers employed in aircraft and aircraft-parts industries (February and March, 1951) who were under 40 years of age at that time. From the table we see that 67 out of 151 workers, or 44 per cent, fell in this category. Therefore if the factor $(N - n)/(N - 1)$ is ignored, the 99 per cent confidence limits are

$$.44 - 2.58\sqrt{\frac{(.44)(.56)}{151}} = .34$$

and

$$.44 + 2.58\sqrt{\frac{(.44)(.56)}{151}} = .54$$

If the population size is very large in comparison with the sample size, it is possible to make general recommendations concerning circumstances under which the normal distribution will provide a "close" approximation to the sampling distribution of u/n. Cochran [4, p. 41] gives a set of working rules for determining when the normal approximation may be used if the following restrictions are imposed: (1) the "95 per cent limits" will enclose the true value of p in at least 94.5 per cent of all cases; (2) the probability that the upper limit is below p is between 2.5 and 3.5 per cent; and (3) the probability that the lower limit exceeds p is between 2.5 and 1.5 per cent. These working rules are reproduced in Table 9.6. Notice that the required sample size becomes larger as u/n departs from 0.5. This arises from the fact observed in connection with Fig. 8.1 that for a given value of n the binomial distribution becomes more highly skewed as p approaches 0 or 1 and that a larger sample size is therefore required to "force symmetry."

9.4.2. *Testing Hypotheses about an Unknown Population Mean.* One often is interested in testing, on the basis of evidence supplied by a sample, the statistical hypothesis that the unknown mean of a population has some stated value, say μ_0. Since the rejection of this hypothesis implies acceptance of some alternative hypothesis, the appropriate alternative hypothesis must be stated explicitly for each problem. Three such alternatives will be treated in the following discussion. These are (1) μ has some value which is different from $\mu_0(\mu \neq \mu_0)$; (2) μ has some value which is less than $\mu_0(\mu < \mu_0)$; and (3) μ has some value which is greater than $\mu_0(\mu > \mu_0)$.

Since a decision to accept or reject the hypothesized value of the mean will be based upon the characteristics of a sample, it is necessary to admit

TABLE 9.6. SMALLEST VALUES OF n FOR USE OF THE NORMAL APPROXIMATION*

Observed u/n	n
0.5	30
0.4 or 0.6	50
0.3 or 0.7	80
0.2 or 0.8	200
0.1 or 0.9	600
0.05 or 0.95	1400

* Restrictions 2 and 3 apply if p is smaller than 0.5. If p is greater than 0.5, then the wording of 2 and 3 must be reversed. Thus 2 would then read "the probability that the lower limit is above p is between 2.5 and 3.5 per cent," and 3 would read "the probability that the upper limit is less than p is between 2.5 and 1.5 per cent."

SOURCE: Cochran [4, p. 41].

the possibility of making errors. As in the development of Sec. 8.4, two possibilities are distinguished: (1) the hypothesis may be rejected when it is actually true (Type I error); and (2) the hypothesis may be accepted when it is actually false (Type II error). The present goal is to obtain test procedures for which the probabilities of committing these two types of errors can be computed.

The general outlines of appropriate test procedures are easy to discern. The mean of a simple random sample is an estimate of the population mean μ. When the observed value of $\bar{x}$ is "close" to the hypothesized μ_0, it would appear that the hypothesis should be accepted. On the other hand, it seems reasonable to reject the hypothesis when the value of $\bar{x}$ is "far away" from μ_0. Since the value of $\bar{x}$ varies from sample to sample, as described by its sampling distribution, there can be no sharp division between "close" and "far away" and the distinction must be made in terms of probability. In other words, for fixed sample size, the appropriate test procedure will be determined by specifying the *level of significance of the test.* It will be recalled that the level of significance of a test is the probability of rejecting the hypothesis when it is actually true.

Now let us translate these considerations into an actual test procedure. Suppose we wish to test the hypothesis $\mu = \mu_0$ against the two-sided alternative $\mu \neq \mu_0$. If n is sufficiently large and if the hypothesis is true, the sampling distribution of $\bar{x}$ will be approximately normal with $\mu_{\bar{x}} = \mu_0$ and with $\sigma_{\bar{x}}$ given by the appropriate one of (9.2) and (9.3). It therefore follows, provided that a 5 per cent level of significance is desired, that

$$\Pr(\mu_0 - 1.96\sigma_{\bar{x}} \leq \bar{x} \leq \mu_0 + 1.96\sigma_{\bar{x}}) = .95$$

and $$\Pr(\bar{x} < \mu_0 - 1.96\sigma_{\bar{x}} \quad \text{or} \quad \bar{x} > \mu_0 + 1.96\sigma_{\bar{x}}) = .05$$

The test procedure is:

1. Draw a random sample from the population under study.
2. Compute the arithmetic mean $\bar{x}$ of the sample.
3. If $\bar{x}$ is between $\mu_0 - 1.96\sigma_{\bar{x}}$ and $\mu_0 + 1.96\sigma_{\bar{x}}$, *accept* the hypothesis $\mu = \mu_0$.
4. If $\bar{x}$ is less than $\mu_0 - 1.96\sigma_{\bar{x}}$ or greater than $\mu_0 + 1.96\sigma_{\bar{x}}$, *reject* the hypothesis in favor of the alternative $\mu \neq \mu_0$.

Since $\bar{x}$ will fall in the region of rejection 5 per cent of the time when the hypothesis is actually true, the level of significance of this test procedure is the desired 5 per cent. Of course $\bar{x}$ may fall in the acceptance region even though the hypothesis is false, and this leads to a Type II error. Some computations of the probabilities of committing Type II errors will be presented shortly.

In the event that the alternative hypothesis is of the form $\mu < \mu_0$ or $\mu > \mu_0$, the appropriate test procedures are:

$\mu < \mu_0$:

1. If $\bar{x}$ is equal to or greater than $\mu_0 - 1.96\sigma_{\bar{x}}$, *accept* the hypothesis $\mu = \mu_0$.
2. If $\bar{x}$ is less than $\mu_0 - 1.96\sigma_{\bar{x}}$, *reject* the hypothesis in favor of the alternative $\mu < \mu_0$.

$\mu > \mu_0$:

1. If $\bar{x}$ is equal to or less than $\mu_0 + 1.96\sigma_{\bar{x}}$, *accept* the hypothesis $\mu = \mu_0$.
2. If $\bar{x}$ is greater than $\mu_0 + 1.96\sigma_{\bar{x}}$, *reject* the hypothesis in favor of the alternative $\mu > \mu_0$.

The level of significance of both these tests is 2.5 per cent since the rejection regions contain only one tail of the normal distribution. If a 5 per cent level of significance for a one-sided alternative is required, reference to Table 9.3 shows that a multiplier of 1.645 should be used in place of 1.96. These three test procedures are pictured in Fig. 9.7. Other levels of significance for either two- or one-sided alternatives can be obtained by merely choosing the appropriate multiplier from Tables 9.2 and 9.3.

As an illustration of these procedures, consider the distribution given in Table 9.7 of errors made by a sample of 153 husbands in reporting

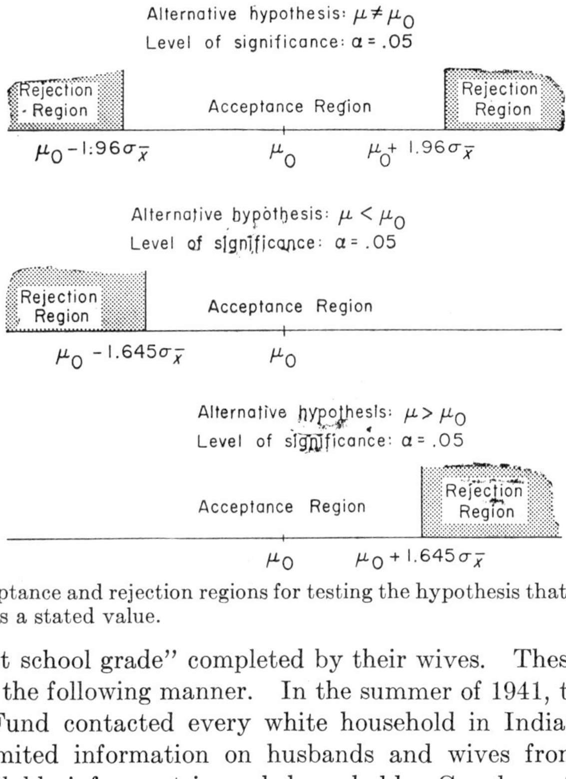

FIG. 9.7. Acceptance and rejection regions for testing the hypothesis that the mean of a population has a stated value.

the "highest school grade" completed by their wives. These data were obtained in the following manner. In the summer of 1941, the Milbank Memorial Fund contacted every white household in Indianapolis and obtained limited information on husbands and wives from the most readily available informant in each household. Couples satisfying certain stated requirements were thereby identified for further study (see Sec. 3.2.1). At a subsequent time, very detailed interviews were conducted with the wives. For the 153 couples represented in Table 9.7, the original information was obtained from the husband. The error is the difference between "highest school grade" as reported by the husband at the initial contact and "highest school grade" as reported by the wife in the detailed interview. (Both husband and wife in each of these couples were supposed to have completed at least an elementary school education.) A plus error indicates that the husband overreported his wife's education. The average error for this distribution is +.32 years, and this indicates a tendency for husbands to exaggerate (although possibly not intentionally) their wives' educational attainments.

TABLE 9.7. DISTRIBUTION OF ERRORS MADE BY HUSBANDS IN REPORTING THE HIGHEST SCHOOL GRADE COMPLETED BY THEIR WIVES

($n = 153$)

Error, years	*Per cent for 153 couples*
−3*	1.3
−2	1.3
−1	7.8
0	63.4
+1	11.8
+2	8.5
+3*	5.9
Total	100.0

* The cited reference reports these as "3 or more." In order to perform the computations, we assume all these cases were in error by exactly 3 years.

$$\bar{x} = \sum \frac{n_i}{n} x_i = +.32 \text{ years}$$

$$s^2 = \frac{n}{n-1}\left(\sum \frac{n_i}{n} x_i^2 - \bar{x}^2\right) = 1.14 \text{ (years)}^2$$

$$s = 1.07 \text{ years}$$

SOURCE: P. K. Whelpton and Clyde V. Kiser, *Social and Psychological Factors Affecting Fertility*, vol. I, p. 109, Milbank Memorial Fund, New York, 1946.

Viewing these 153 couples as a sample from a very large population of couples with similar characteristics, we might argue that the positive average error could have arisen by chance, even though underreporting and overreporting balance one another in the entire population. This can be phrased in hypothesis testing terms by stating that we wish to test the

$$\textit{Null hypothesis: } \mu = \mu_0 = 0$$

against the

$$\textit{Alternative hypothesis: } \mu > 0$$

The one-sided alternative is chosen in this instance because experience would suggest exaggeration on a prestige item such as education.

With a 1 per cent level of significance, the general treatment says that the hypothesis should be *accepted* if

$$\bar{x} < \mu_0 + 2.326\sigma_{\bar{x}} = 0 + 2.326\sigma_{\bar{x}}$$

and *rejected* if

$$\bar{x} > \mu_0 + 2.326\sigma_{\bar{x}} = 0 + 2.326\sigma_{\bar{x}}$$

Since $\sigma_{\bar{x}} = \sqrt{\sigma^2/n}$ and σ^2 is unknown, we follow the procedure used in Sec. 9.4.1 and use the sample s^2 as an estimate of σ^2. Thus from Table 9.7, $\sigma_{\bar{x}}$ is approximated by $\sqrt{s^2/n} = \sqrt{(1.14)/153} = .09$ year.

The boundary of the critical region, $0 + 2.326\sigma_{\bar{x}}$, therefore becomes $0 + (2.326)(.09) = .21$ year. Since $\bar{x}$ is greater than this critical value, the hypothesis is rejected, at the 1 per cent level of significance, in favor of the alternative that the true mean error is greater than zero.

The foregoing test procedure was determined so that the probability of rejecting a true null hypothesis is .01. The null hypothesis may also be accepted when μ is actually different from zero (a Type II error). The probability of committing a Type II error is equal to the probability that $\bar{x}$ falls in the acceptance region of the test, and this probability clearly depends upon the exact value of the true mean error. For example, we might ask, What is the probability that the test procedure will lead us to accept the hypothesis $\mu = 0$ when as a matter of fact the true mean error in the population of all similar couples is equal to .30 year? That is, we wish the value of

$$\begin{aligned}\Pr(\bar{x} \text{ falls in acceptance region of test when } \mu = .30)\\ = \Pr(\bar{x} < .21 \text{ when } \mu = .30)\\ = \Pr(\bar{x} < .21 \mid \mu = .30)\end{aligned}$$

Notice particularly that the acceptance and rejection regions of the test were determined by the stated null hypothesis, alternative hypotheses, sample size, and level of significance. These regions are now regarded as *fixed*, and we are simply asking for the behavior of the test when the null hypothesis is not true.

The desired probability can be obtained very easily from tables of the normal distribution if the assumption is made that $\sigma_{\bar{x}}$ remains unchanged no matter what the true value of μ may be. The estimated value of $\sigma_{\bar{x}}$ is obtained from the data of Table 9.7 as .09 year. All further computations will be done in terms of this value. Hence $\Pr(\bar{x} < .21 \mid \mu = .30)$ is approximately equal to the fraction of area which falls to the left of .21 for a normal curve with a mean of .30 and a standard deviation of .09. This area is shaded in the top normal curve of Fig. 9.8. The actual amount of this shaded area follows immediately from Table 9.2. Since .21 differs from .30 by one standard-deviation unit (.09), Table 9.2 shows that 68.27 per cent of the total area falls between .21 and .39. One-half of this (34.13 per cent) lies between .21 and .30. Since 50 per cent of the area lies to the left of the mean, 15.87 per cent (50 − 34.13) falls to the left of .21. Thus the probability is .1587 of accepting the hypothesis $\mu = 0$ when the true mean error is in fact .30 year.

The choice of $\mu = .30$ as a point at which to evaluate the probability of committing a Type II error was quite arbitrary. Any other value of μ,

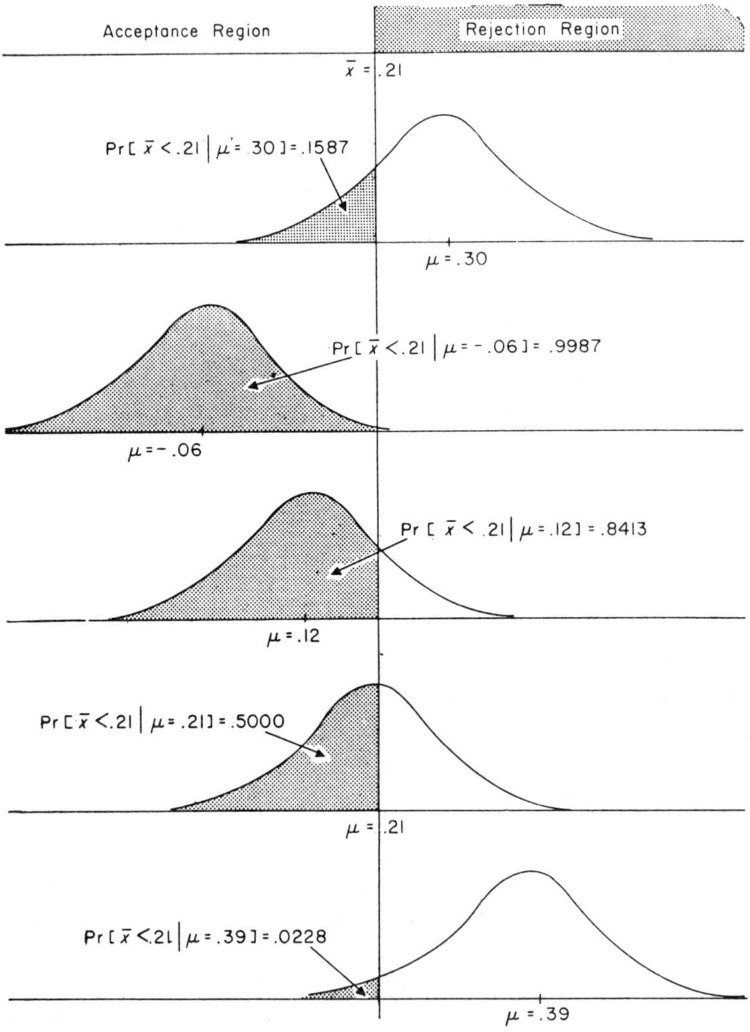

FIG. 9.8. Probability of accepting the hypothesis $\mu = 0$ for various true values of μ. ($n = 153$, $\sigma_{\bar{x}} = .09$, and level of significance $= .01$.)

different from zero, could have been used for the computation (a different probability of course being obtained for each), and the computations for a few further values are pictured in Fig. 9.8. These values demonstrate that the probability of accepting the hypothesis becomes progressively smaller as the true mean error departs from zero in the positive direction. Thus if the true mean error is .39 year, the probability is only .0228 that the null hypothesis will be accepted. If the true mean is less than zero, the probability is very large (greater than .99) of accepting the null hypothesis. This characteristic of the Type II error was built into the test procedure when it was stated that the null hypothesis was to be rejected only if it could be rejected in favor of the alternatives that the

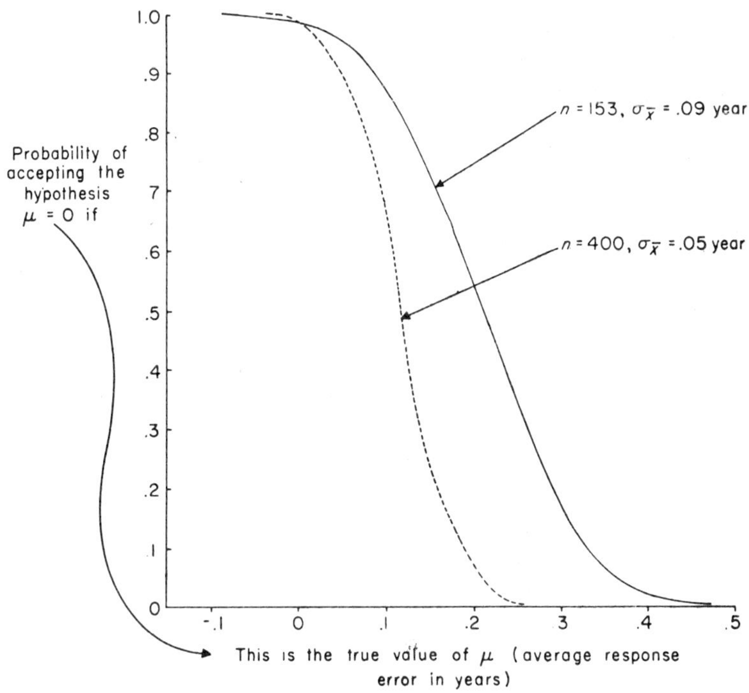

FIG. 9.9. Operating characteristic curves for one-sided tests of the hypothesis that $\mu = 0$. ($\sigma = 1.07$ years, level of significance = .01.)

true mean is larger than zero. In effect this says that no error is committed in accepting the hypothesis $\mu = 0$ when the true mean error is negative.

The preceding discussion has not singled out any particular value of μ, alternative to $\mu = 0$, upon which to focus attention when computing the probability of a Type II error. This suggests presenting the expected behavior of the test for all possible alternatives. This can be accomplished by plotting $\Pr(\bar{x} < .21 \mid \mu)$ as a function of μ. The resulting smooth curve is known as the *operating characteristic* of the test procedure and is given for the present example by the solid curve in Fig. 9.9. It will be observed that the particular values given in Fig. 9.8 fall on this curve and that the height of the curve at $\mu = 0$ is .99. If the hypothesis is true, the probability of rejecting it is .01, and the probability that it will be accepted is therefore .99. The operating characteristic summarizes the complete behavior of a test procedure and tells what can be expected for all possible values of μ.

The fact was noted in Sec. 8.4.1 that for a fixed level of significance an increase in the sample size decreases the chances of making Type II errors. This can be illustrated in terms of the operating characteristic curve. Let us assume that the data of Table 9.8 were obtained from a sample of 400 couples rather than 153 and that the standard deviation of errors is still 1.07 years. Then the appropriate test procedure, at the 1 per cent level of significance, is:

1. Accept the null hypothesis $\mu = 0$ if $\bar{x}$ is less than

$$0 + 2.326\sigma_{\bar{x}} = 2.326(1.07/\sqrt{400}) = .12$$

2. Reject the null hypothesis $\mu = 0$ if $\bar{x}$ is greater than .12.

The operating characteristic of this test procedure is given by the dashed curve in Fig. 9.9. It lies below the solid curve ($n = 153$) for all values of μ to the right of zero. For example, if the true value of μ is .20 year, the probability of a Type II error is approximately .05 for an n of 400 while it is approximately .53 for an n of 153. Thus the procedure based on the larger sample size is more sensitive or more powerful* in detecting situations where the null hypothesis is false.

The foregoing test procedures can also be employed in testing the hypothesis that the true fraction of elements in a population having a stated characteristic is some specified value, say p_0. Assuming that n is sufficiently large for the sampling distribution of u/n to be approximated by the normal distribution, where u is the number of S's in a sample, it is only necessary to compute the variance of u/n from the appropriate one of (9.5) and (9.6) and to use this variance in setting up the test procedure. For example, suppose the population size N is large enough in relation to the sample size so that (9.6) can be used. Two- and one-sided test procedures at the 5 per cent level of significance are:

Hypothesis: $p = p_0$
Alternative: $p \neq p_0$

1. If the observed u/n is between $p_0 - 1.96\sqrt{p_0q_0/n}$ and $p_0 + 1.96\sqrt{p_0q_0/n}$, accept the hypothesis.

* Statistical test procedures are often evaluated in terms of *power*. Let Pr(Type II error $|$ μ) be the probability of a Type II error for a fixed value of μ different from μ_0. The power of the test for the fixed value μ is defined to be

$$1 - \Pr(\text{Type II error} \mid \mu)$$

Hence the power is the probability of rejecting the hypothesis $\mu = \mu_0$ when in fact it should be rejected.

2. If u/n is less than $p_0 - 1.96 \sqrt{p_0 q_0/n}$ or greater than $p_0 + 1.96 \sqrt{p_0 q_0/n}$, reject the hypothesis.

Alternative: $p < p_0$

1. If u/n is equal to or greater than $p_0 - 1.645 \sqrt{p_0 q_0/n}$, accept the hypothesis.
2. If u/n is less than $p_0 - 1.645 \sqrt{p_0 q_0/n}$, reject the hypothesis.

Alternative: $p > p_0$

1. If u/n is equal to or less than $p_0 + 1.645 \sqrt{p_0 q_0/n}$, accept the hypothesis.
2. If u/n is greater than $p_0 + 1.645 \sqrt{p_0 q_0/n}$, reject the hypothesis.

Observe that in this instance all details of the test procedure can be determined before the sample has been drawn. This is possible because the variance of u/n depends only upon the hypothesized value p_0. In the general case discussed previously, it was necessary to estimate σ^2 by the sample s^2 in order to obtain an estimate of $\sigma_{\bar{x}}$.

In concluding the present treatment of statistical test procedures based on the means of large samples, it may be well to remind the reader that the points raised for consideration in Sec. 8.5 apply equally well here. In particular, the determination of the sample size and level of significance—which in turn fix the probabilities of committing Type II errors—must depend upon the cost of obtaining observations and upon the relative importance associated with the various errors which may arise as a result of applying the test procedure. If it is possible to measure the relative importance of these errors in quantitative terms, these values may sometimes be used in arriving at a procedure which is in some sense reasonable or best.

9.5. Confidence Limits and Tests of Hypotheses for the Difference between the Means of Two Populations

A frequent statistical problem is that of determining, on the basis of random samples, how much of a difference, if any, exists between the means of two distinct populations. Consider, for example, the distributions of straight-time hourly earnings for machine-tool operators and tool-and-die makers given in Table 5.1. These distributions are based on samples of machine-tool operators and tool-and-die makers selected from the populations of all such workers employed in St. Louis, Mo., in January, 1954. The average hourly earnings for machine-tool operators, say $\bar{x}_1$, was found to be \$2.13, while the average hourly earnings of tool-and-die makers, say $\bar{x}_2$, was \$2.49. The difference between these two averages, $(\bar{x}_1 - \bar{x}_2)$, is $-\$0.36$. Clearly this difference tells something about the difference in average hourly rates for all machine-tool operators and all tool-and-die makers in St. Louis at that time. And yet we cannot expect it to be exactly equal to the true difference between the population means. It is necessary to take account of the fact that the values of $\bar{x}_1$ and $\bar{x}_2$ were obtained from samples.

The foregoing example is only a special case of the following general situation. There are two populations, and the mean value of a variable x in the first population is to be compared with the mean value of the same variable in the second population. Let μ_1 and σ_1^2 be the mean and variance of x in the first population and μ_2 and σ_2^2 the mean and variance of x in the second. Suppose that a random sample of size n_1 is drawn from the first population and that the sample mean is $\bar{x}_1$. The corresponding quantities for a sample from the second population are n_2 and $\bar{x}_2$. Then the observed difference in sample means $(\bar{x}_1 - \bar{x}_2)$ is to be used for determining confidence limits for the difference of the population means $(\mu_1 - \mu_2)$ or to test hypotheses about this difference. This section will provide procedures for accomplishing this when the sample sizes n_1 and n_2 are both large.

A pair of random samples—one sample from each of the two populations—leads to an observed difference in sample means $(\bar{x}_1 - \bar{x}_2)$. This observed difference is the outcome of a single trial of a conceptual experiment that can be repeated under similar conditions an indefinitely large number of times. Moreover, the outcome of any single trial is uncertain in that the exact value of the difference cannot be predicted in advance of actually drawing a pair of samples. As in preceding considerations, the outcome of a single trial of such an experiment can be evaluated only in terms of what would be expected if an indefinitely large number of trials were performed. In brief, we need to know the sampling distribution of $(\bar{x}_1 - \bar{x}_2)$.

For present purposes, the most useful results that can be proved concerning the sampling distribution of $(\bar{x}_1 - \bar{x}_2)$ are the following:

> Consider a pair of random samples. Let the first member of the pair, of size n_1, be drawn from a population with mean μ_1 and variance σ_1^2. Let the second member, of size n_2, be drawn *independently* of the first from a population with mean μ_2 and variance σ_2^2. Then
>
> 1. The arithmetic mean of the sampling distribution of $(\bar{x}_1 - \bar{x}_2)$ will be
>
> $$E(\bar{x}_1 - \bar{x}_2) = \mu_{(\bar{x}_1 - \bar{x}_2)} = \mu_1 - \mu_2 \tag{9.7}$$
>
> 2. The variance of the sampling distribution of $(\bar{x}_1 - \bar{x}_2)$ will be
>
> $$V(\bar{x}_1 - \bar{x}_2) = \sigma^2_{(\bar{x}_1 - \bar{x}_2)} = \sigma^2_{\bar{x}_1} + \sigma^2_{\bar{x}_2} \tag{9.8}$$

3. If, in addition, n_1 and n_2 are large, the sampling distribution of $(\bar{x}_1 - \bar{x}_2)$ can be approximated by a normal distribution having mean and variance as given by (9.7) and (9.8).

Expressions (9.7) and (9.8) are valid no matter what the two sample sizes may be. Large samples are required only to obtain the approximate normality of the sampling distribution of $(\bar{x}_1 - \bar{x}_2)$. Also, observe that these results hold for random samples drawn with or without replacement from finite populations or for sets of independent observations obtained from an infinite population. It is only necessary to use the appropriate formulas for $\sigma^2_{\bar{x}_1}$ and $\sigma^2_{\bar{x}_2}$ in expression (9.8).

There is one feature of the foregoing that requires special comment. The variance expression (9.8) is valid only if the two sample means are independent of one another. That is, the individual elements (or the values of the variable assigned to them) obtained in the first sample have no influence on the choice of elements for the second sample (or on their associated values of the variable), and vice versa. If some form of dependence exists in the sampling procedure or in the measurement process, then there will be a tendency for large or small values of $\bar{x}_1$ to be accompanied by correspondingly large or small values of $\bar{x}_2$, or possibly with small or large values of $\bar{x}_2$, and the variance of $(\bar{x}_1 - \bar{x}_2)$ will be either larger or smaller than the value given by (9.8).

A simple example of dependence between $\bar{x}_1$ and $\bar{x}_2$ is the following. Suppose that the arithmetic mean of the distribution of a variable x in a population of elements at time 1 is to be compared with the arithmetic mean at time 2 of the same variable for the same population of elements. A random sample of n elements is chosen at time 1, and the sample mean is determined. At time 2 measurements are obtained for exactly the same n elements. The values of $\bar{x}_1$ and $\bar{x}_2$ are almost certain to be dependent on one another in such a situation, and (9.8) could not be used. However, (9.7) would still be appropriate. In studies of attitudes, opinions, consumer wants, and like variables, a research design of this type is called a *panel* study. For example, panels have been used extensively in studying changes in attitude that occur during the course of a Presidential election campaign [5]. One of the principal advantages of panel techniques is that they permit one not only to study changes that occur in the distribution of the variable but also to trace through time the manner in which individual population elements change with respect to the variable.

Now that the mean, variance, and approximate normality of the sampling distribution of $(\bar{x}_1 - \bar{x}_2)$ have been determined, confidence limits and tests of hypotheses for the true and unknown difference $(\mu_1 - \mu_2)$ between the two population means can be obtained directly from the treatment of Secs. 9.4.1 and 9.4.2. It is only necessary to set up the following correspondence between the quantities of that section and the quantities of this section:

Single population mean (Sec. 9.4)	*Difference between two population means*
μ	$(\mu_1 - \mu_2)$
$\bar{x}$	$(\bar{x}_1 - \bar{x}_2)$
$\sigma_{\bar{x}}^2$	$\sigma_{\bar{x}_1}^2 + \sigma_{\bar{x}_2}^2$
$\sigma_{\bar{x}}$	$\sqrt{\sigma_{\bar{x}_1}^2 + \sigma_{\bar{x}_2}^2}$

Thus 99 per cent confidence limits for $(\mu_1 - \mu_2)$ are given by

$$(\bar{x}_1 - \bar{x}_2) \pm 2.58 \sqrt{\sigma_{\bar{x}_1}^2 + \sigma_{\bar{x}_2}^2}$$

As an illustration, consider the data of Table 5.1 as discussed in the opening paragraphs of this section. We assume that the samples of

TABLE 9.8. COMPUTATION OF 99 PER CENT CONFIDENCE LIMITS FOR THE DIFFERENCE IN AVERAGE HOURLY EARNINGS OF MACHINE-TOOL OPERATORS AND TOOL-AND-DIE MAKERS
(Data from Table 5.1)

Machine-tool operators	Tool-and-die makers
$n_1 = 1{,}012$	$n_2 = 1{,}033$
$\bar{x}_1 = \$2.13$	$\bar{x}_2 = \$2.49$

$$\bar{x}_1 - \bar{x}_2 = -\$0.36$$

Machine-tool operators	Tool-and-die makers
$s_1^2 = .0409$	$s_2^2 = .0126$
$\sigma_{\bar{x}_1}^2 = \dfrac{\sigma_1^2}{n_1} \doteq \dfrac{s_1^2}{n_1} = \dfrac{.0409}{1{,}012}$	$\sigma_{\bar{x}_2}^2 = \dfrac{\sigma_2^2}{n_2} \doteq \dfrac{s_2^2}{n_2} = \dfrac{.0126}{1{,}033}$
$= .000{,}040{,}4$*	$= .000{,}012{,}2$*

$$\sigma^2_{(\bar{x}_1-\bar{x}_2)} = \sigma_{\bar{x}_1}^2 + \sigma_{\bar{x}_2}^2 = .000{,}052{,}6$$

$$\sigma_{(\bar{x}_1-\bar{x}_2)} = \sqrt{\sigma_{\bar{x}_1}^2 + \sigma_{\bar{x}_2}^2} = .0073$$

$$(\bar{x}_1 - \bar{x}_2) \pm 2.58 \sqrt{\sigma_{\bar{x}_1}^2 + \sigma_{\bar{x}_2}^2} = -\$0.36 \pm \$0.019$$

99 per cent confidence limits for $(\mu_1 - \mu_2)$ are $-\$0.341$ and $-\$0.379$.

* The symbol $\doteq$ means "approximately equal to."

machine-tool operators and tool-and-die makers were randomly and independently drawn from their respective populations, that the finite-population correction can be ignored so that $\sigma_{\bar{x}_1}^2$ and $\sigma_{\bar{x}_2}^2$ are given by (9.3), and that the population variances σ_1^2 and σ_2^2 are estimated by the sample variances s_1^2 and s_2^2. The computations are summarized in Table 9.8. The resulting limits, $-\$0.341$ and $-\$0.379$, are close together, and this of course occurs because the two samples sizes are quite large. Smaller sample sizes would lead to wider confidence intervals.

The hypothesis-testing procedures of Sec. 9.4.2 also carry over directly to the present situation. An observed difference in the sample means is to be used in testing the hypothesis that the difference in population means has some specified value, say d_0. That is, we wish to test the

Null hypothesis: $\mu_1 - \mu_2 = d_0$

against the appropriate one of the following three classes of possible

Alternative hypotheses:

1. $\mu_1 - \mu_2 \neq d_0$
2. $\mu_1 - \mu_2 < d_0$
3. $\mu_1 - \mu_2 > d_0$

If the observed value of $(\bar{x}_1 - \bar{x}_2)$ is "close" to the hypothesized difference d_0 the hypothesis should be accepted, while if it is "far away" the hypothesis should be rejected. Again "close" and "far away" must be interpreted in terms of the probability of rejecting the hypothesis when it is actually true, i.e., the level of significance of the test.

In actual practice, the most common null hypothesis is that the two populations have identical means. That is, $(\mu_1 - \mu_2) = d_0 = 0$. Actual test procedures follow from those of Sec. 9.4.2. For example, suppose we wish to test, at the 1 per cent level of significance, the hypothesis that $(\mu_1 - \mu_2) = 0$ against the two-sided alternative $(\mu_1 - \mu_2) \neq 0$. The test procedure would be:

1. Draw independent random samples from the two populations.
2. Compute the difference in sample means, $(\bar{x}_1 - \bar{x}_2)$.
3. If $(\bar{x}_1 - \bar{x}_2)$ is between

$$0 - 2.58\sqrt{\sigma_{\bar{x}_1}^2 + \sigma_{\bar{x}_2}^2}$$

and

$$0 + 2.58\sqrt{\sigma_{\bar{x}_1}^2 + \sigma_{\bar{x}_2}^2}$$

accept the hypothesis that there is no difference between the population means.

4. If $(\bar{x}_1 - \bar{x}_2)$ is less than $-2.58\sqrt{\sigma_{\bar{x}_1}^2 + \sigma_{\bar{x}_2}^2}$ or greater than $+2.58\sqrt{\sigma_{\bar{x}_1}^2 + \sigma_{\bar{x}_2}^2}$, *reject* the hypothesis in favor of the alternative $(\mu_1 - \mu_2) \neq 0$.

These test procedures may also be used in comparing two populations with respect to the fraction of elements having some defined characteristic. The sample fraction from the first population u_1/n_1 corresponds to $\bar{x}_1$, and the sample fraction from the second population to $\bar{x}_2$. As an illustration, consider the data of Table 9.9. These data were obtained in a study conducted in Pittsburgh in 1941 "to discover the influence of the social status of the interviewer upon the findings he reports." The study design was reported by the author as follows:

> To test the effects of the social status of the interviewer two separate interviewing staffs were employed. The *white-collar*, or *control* staff, consisted of five

Gallup interviewers, regularly employed by the American Institute of Public Opinion, and four new interviewers eligible for the American Institute, some of whom have been subsequently used by that organization. The *working-class*, or *experimental* staff was made up of eleven wage workers, some of them mill hands, and most of them with no college training. The two groups of interviewers were sent into the same low-rental areas of Pittsburgh with the same instructions and the same ballot or questionnaire during the two weeks beginning March 1, 1941. Pittsburgh was selected as an industrial town with a fair proportion of labor union people so that the results of the experiment are most applicable to industrialized, unionized sections of the United States.

The question to which the data of Table 9.9 refer was only one of 10 or more that were used in the study.

TABLE 9.9. COMPARISON OF RESPONSES OBTAINED BY WHITE-COLLAR AND WORKING-CLASS INTERVIEWERS TO THE QUESTION "Should there be a law against strikes in war industries?"

Response	White-collar interviewers		Working-class interviewers		Total	
	Number	Per cent	Number	Per cent	Number	Per cent
Yes	390	65	326	54	716	60
No	156	26	217	36	373	31
No opinion	54	9	60	10	114	9
Total	600	100	603	100	1,203	100

SOURCE: Daniel Katz, "Do Interviewers Bias Poll Results?" *Pub. Op. Quart.*, **6**:248–268 (1942).

In the sample of 600 individuals interviewed by the white-collar interviewers, 65 per cent gave a "yes" response to the question while in the sample of 603 interviewed by the working-class interviewers only 54 per cent gave a "yes" response. Is this 11 per cent sample difference "large enough" to indicate that there is some tendency for individuals to respond differently to this type of question, depending upon the social class of the interviewer?

A formal statement of this problem can be phrased as follows:

Consider the population of adults residing in the low-rental areas of Pittsburgh at the time of the study. Let p_1 be the fraction of adults in this population who, if approached by a white-collar interviewer, would give a "yes" answer to the question under discussion. Let p_2 be the fraction in the population who, if approached by a working-class interviewer, would give a "yes" answer. Let u_1/n_1 be the fraction of "yes" answers obtained by white-collar interviewers in a random sample of n_1 individuals, and let u_2/n_2 be the fraction obtained by working-class interviewers in an independent random sample of size n_2. Then we desire to test, on the basis of the difference in sample fractions, the hypothesis that $p_1 - p_2 = 0$. When the hypothesis is true, p_1 and p_2 are equal and their common value can be designated by p. The appropriate alternative in this instance would appear to be that $p_1 - p_2 > 0$. That is, a "yes" answer is the "patriotic" or "conservative" answer that would be expected in response to the question "Should there be a law against strikes in war industries?" If there were any tendency for respondents to attempt to conform or adapt to the social status of the interviewer, working-class respondents would be expected to give a higher fraction of "yes" answers to white-collar interviewers than to working-class interviewers.

Since the variance of $(\bar{x}_1 - \bar{x}_2)$ is $\sigma^2_{\bar{x}_1} + \sigma^2_{\bar{x}_2}$, the correspondence set up in Sec. 9.2.3 shows that the variance of $(u_1/n_1) - (u_2/n_2)$ is equal to $\sigma^2_{u_1/n_1} + \sigma^2_{u_2/n_2}$. If the finite population correction can be ignored, we have, from (9.6), that

$$\sigma^2_{u_1/n_1} + \sigma^2_{u_2/n_2} = \frac{p_1q_1}{n_1} + \frac{p_2q_2}{n_2}$$

Furthermore, the hypothesis to be tested states that $p_1 = p_2 = p$, and this variance then becomes

$$\frac{pq}{n_1} + \frac{pq}{n_2} = pq\left(\frac{1}{n_1} + \frac{1}{n_2}\right)$$

Assuming that a 1 per cent level of significance is desired for a one-sided test, previous considerations lead to the following procedure:

1. If $(u_1/n_1) - (u_2/n_2)$ is less than or equal to

$$0 + 2.326\sqrt{pq\left(\frac{1}{n_1} + \frac{1}{n_2}\right)}$$

accept the null hypothesis that $p_1 - p_2 = 0$.

2. If $(u_1/n_1) - (u_2/n_2)$ is greater than

$$2.326\sqrt{pq\left(\frac{1}{n_1} + \frac{1}{n_2}\right)}$$

reject the null hypothesis in favor of the alternative that $p_1 - p_2 > 0$.

Notice that this test procedure depends upon the common value p that p_1 and p_2 assume when the hypothesis is true. This common value must be estimated from the sample before the boundary between the acceptance region and the rejection region can be computed. Since the entire test procedure is determined on the assumption that the null hypothesis is true, it seems only reasonable to continue in this vein and to estimate p from the *combined* samples. That is, p will be estimated by $(u_1 + u_2)/(n_1 + n_2)$. Table 9.9 gives this value under the *total* column as $^{716}/_{1203}$, or .60. Therefore, the final test procedure becomes:

1. If $(u_1/n_1) - (u_2/n_2)$ is less than or equal to

$$2.326 \sqrt{(.6 \times .4)\left(\frac{1}{600} + \frac{1}{603}\right)} = .066$$

accept the null hypothesis that $p_1 - p_2 = 0$.

2. If $(u_1/n_1) - (u_2/n_2)$ is greater than .066, *reject* the null hypothesis in favor of the alternative that $p_1 - p_2 > 0$.

Since the observed difference is .11, the hypothesis is rejected.

The source cited for the data for Table 9.9 did not of course confine its analyses to the single question used in this illustration. Many other comparisons of a similar nature were made. Two of the conclusions drawn from these analyses were:

> Middle-class or white-collar interviewers, such as the public opinion polls employ, find a greater incidence of conservative attitudes among lower income groups than do interviewers recruited from the working class.
>
> It would be a mistake to assume from the results of this experiment that the most desirable situation for all studies is one in which the interviewer has membership-character in the group he is interviewing. For some types of information the subject will respond more completely and accurately to an outsider with the proper prestige. Thus in studies of family problems and sex matters the housewife may talk more freely to the nurse or psychiatrist than to someone from her own social group personally unknown to her.

9.6. Some Comments on Making Statistical Inferences from Small Samples

The means and variances of large random samples have been used in the preceding sections for developing procedures that provide confidence limits and tests of hypotheses for unknown population means. The large-sample requirement is important in two respects: (1) it guarantees the approximate normality of the sampling distribution of $\bar{x}$, which in turn permits control of the confidence coefficient and level of significance; and (2) it makes it possible to substitute the sample variance s^2 for the population variance σ^2 without having this substitution change to any great extent the preassigned confidence coefficients or levels of significance. If "small" samples must be used, then it is not necessarily true that the sampling distribution of $\bar{x}$ is even approximately normal and it becomes imperative that the sampling variability of s^2 play an important role in the procedure. These matters cannot be considered in detail, but some brief comments can be made concerning forms of solution.

Under certain circumstances, the exact sampling distribution of $\bar{x}$ for any stated sample size, no matter how small, can be specified in such a form that confidence limit and test procedures may be obtained directly from this distribution. The binomial distribution of Chap. 8 provides a case in point. If the sampling and measurement conform to the requirements of the binomial model as described in Sec. 8.1.2, the curves in Figs. 8.5, 8.6, and 8.7 can be used in cases where the normal approximation would not be appropriate for describing the sampling distribution of u/n.

Exact small-sample theory is also available when one can argue or demonstrate that the variable x itself has a normal distribution. It can then be shown that the sampling distribution of $\bar{x}$ will be truly normal for any value of n, with mean and variance as given by (9.1) and (9.3). If the population variance of x is known, the confidence limit and test procedures of Sec. 9.4 need no modification for small samples. They are exact. However, if σ^2 must be estimated by the sample s^2, these procedures must be changed slightly. The change merely consists of obtaining the appropriate multiplier from a table of *Student's t distribution** rather than from a table of the normal distribution.

Since the sample-to-sample variability of s^2 decreases as n increases, the value of the multiplier depends not only upon the desired confidence coefficient or level of significance but also upon the size of the sample. For example, suppose 95 per cent confidence limits are desired. For an n of 2, the multiplier should be 12.706; for an n of 3, it is 4.303; for $n = 10$, 2.262; for $n = 31$, 2.042; for $n = 121$, 1.980; and for n indefinitely large, 1.960. The value of the multiplier becomes progressively smaller as the size of the sample increases, and for n indefinitely large it is equal to 1.960. This we recognize as the value given by the normal distribution for obtaining 95 per cent confidence limits.

Student's t distribution can also be used when the means of two normally distributed variables are to be compared on the basis of such samples, but some additional problems arise in this connection. A detailed discussion of all these points will be found in Dixon and Massey [7, chap. 9].

Although much of current small-sample theory is based on the assumption that the variable x has a normal distribution, a new and fruitful line of approach has been developed in recent years. Many test procedures, and a few estimation procedures, have been devised that do not depend on the character of the population distribution of x. These are

* Student's t distribution is a theoretical distribution that describes the sample-to-sample variability of the quantity $(\bar{x} - \mu)/\sqrt{s^2/n}$ and of other related quantities that are encountered in statistical work. Tables of this distribution are given in terms of *degrees of freedom* and the probability that the above quantity will fall between or will exceed stated values of t. If s^2 is computed from a single sample of size n, as is the case here, the number of degrees of freedom is set equal to $(n - 1)$. Each particular application of the t distribution requires a determination of the appropriate number of degrees of freedom. A table of this distribution is given in Pearson and Hartley [6].

called *nonparametric*, or *distribution-free*, procedures. They should be particularly helpful in social-science research where it is frequently known that variables do not have normal distributions, even though the exact form of the distribution is unknown.

Exercise 8.6 provides a very simple example of the distribution-free approach. In that example, the hypothesis to be tested was that the median of a distribution had a stated value. The median is defined as that value of the variable such that half the elements in the population have a smaller value and half have a larger value. Therefore, if one hypothesizes the true value of the median, the probability is .5 that a random observation will have a value smaller than the true median and .5 that it will have a value larger than the median. A sample will of course not give exactly a 50-50 split, but the techniques of Chap. 8 can be used to test the hypothesis that the true probability of obtaining an observation less than the hypothesized median is equal to .5 against the alternative that this probability is different from .5. Rejecting this hypothesis about p is equivalent to rejecting the hypothesis that we have stated the true value of the median. This test procedure depends in no way on the character of the distribution of x and is therefore distribution-free.

This example indicates the general character of distribution-free procedures. They depend on counting, ordering, or ranking operations and do not use the actual numerical values of the observations for any other purpose. A general account of the application of these methods in social-science research has been provided by Smith in a book edited by Festinger and Katz [8], and by Siegel [15].

EXERCISES

9.1. A detailed analysis of the concept of accident proneness has been reported by Adelstein [9]. The data were obtained from the records of the South African Railways and concern the work experience of shunters (switchmen). The following table shows the distribution of number of accidents, arising out of and in the course of a workman's employment and resulting in personal injury, incurred during the first year of experience as shunters of 227 men between 26 and 30 years of age.

Number of accidents x	*Number of men*
0	121
1	85
2	19
3	1
4	0
5	0
6	1
Total	227

Determine the 90 per cent confidence limits for the true mean number of accidents per man for the distribution that would be obtained if one had the experience of an extremely large number of shunters.

9.2. Between November, 1951, and February, 1952, members of the field staff of the Bureau of Old-Age and Survivors Insurance visited a very large number of insurance beneficiaries to collect information on how they had been getting along during the preceding year [10]. The beneficiaries visited were selected randomly from the nationwide rolls of the Bureau. Included in the sample were 2,533 aged-widows, constituting 1 per cent of the total population of such beneficiaries. The average total money income (cash receipts from all sources) reported by these 2,533 aged-widows for the preceding year was $804 and the standard deviation s was $589. Compute the 99.9 per cent confidence limits for the average total money income for all aged-widow beneficiaries on the rolls at that time. 15.3 per cent of these aged-widows reported total money income of $1,200 or more. What are the 99.9 per cent confidence limits for the true fraction of such beneficiaries receiving $1,200 or more?

9.3. A random sample of 392 members of the Teamsters Union of the American Federation of Labor, Local 688, in St. Louis, Mo., was interviewed in 1949 [11]. Of these respondents 24.7 per cent gave a "no, not at all" or a "don't know" response to the question "Do union problems which do not concern your shop interest you?" Ignore the finite-population correction and use the normal approximation to the sampling distribution of u/n to determine 99 per cent confidence limits for the fraction of such responses that would have been obtained had all members of Local 688 been interviewed.

9.4. As explained in Sec. 9.4.1, 95 per cent confidence limits for an unknown population mean are given by $\bar{x} - 1.96\sigma_{\bar{x}}$ and $\bar{x} + 1.96\sigma_{\bar{x}}$. The length of the confidence interval, the upper limit minus the lower limit, is equal to $2(1.96)\sigma_{\bar{x}}$ and is a measure of the precision with which the unknown mean is estimated. If the sample size is small relative to the population size, $\sigma_{\bar{x}} = \sigma/\sqrt{n}$. Assume that $\sigma = 100$ and compute the length of the confidence interval for $n = 100$, 400, 900, and 1,600. Plot these points on a graph and sketch in a curve showing the relationship between sample size and length of confidence interval. Observe that it is necessary to quadruple the sample size in order to halve the length of the 95 per cent confidence intervals. This result holds no matter what the value of σ, provided of course that the finite-population correction can always be ignored.

9.5. Suppose that large samples are used to test the null hypothesis that the mean of a population has some specified value μ_0 against the alternative hypothesis that the mean is different from μ_0 ($\mu \neq \mu_0$). Draw a rough sketch (without performing any computations) of the operating-characteristic curve of the two-sided test procedure outlined in Sec. 9.4.2. Compare the shape of this curve with the shape of the operating characteristic given in Fig. 9.9.

9.6. Choose a pair of distributions contained in the illustrations or exercises of the preceding chapters and describe how you would use random-number tables to obtain an empirical approximation to the sampling distribution of $(\bar{x}_1 - \bar{x}_2)$.

9.7. In a study of the prestige ranking of occupations [12] each of a random sample of adults in Spokane, Wash., was requested to "Please rank the occupations listed below according to your opinion of their prestige or social standing, by making a check in the appropriate box—Very High, High, Average, Low, Very Low." In analyzing data that result from a question of this type, the procedure is sometimes followed of assigning numerical scores to the ordered categories, e.g., 1 to "very high," 2 to "high," 3 to "average," 4 to "low," and 5 to "very low," and then performing numerical operations on these scores. (This implies that the "distance" between any pair of consecutive categories is the same and that the categories have the same meanings to all respondents. Although this is a tenuous assumption, this difficulty will be ignored here.) The reported distributions of rankings for each of two occupations are as follows:

Category of ranking	Assigned score	Occupation	
		Civil servant, no.	Electrician, no.
Very High	1	38	18
High	2	151	95
Average	3	113	194
Low	4	7	5
Very Low	5	2	0
Total		311	312

Assuming that each respondent *independently* assigned a ranking to "civil servant" and to "electrician," determine 95 per cent confidence limits for the true difference in mean scores that would have been obtained if an extremely large number of adults had performed the ranking.

9.8. In the summer of 1950 a survey was conducted in Louisville, Ky., to ascertain the attitudes of individuals 18 years of age and older on mental illness [13]. Among other things, respondents were requested to classify as "true" or "false" the statement "Most mental illness is inherited." One of the points of interest in the analysis was the comparison of the attitudes of younger people with those of older people. The following table shows the distribution of responses to the statement for males between the ages of 18 and 24 and for males 65 years of age and over:

	Males 18–24, per cent	Males 65 and over, per cent
Response:		
False	75	52
True	18	29
Don't know	7	19
Total	100	100
Number interviewed	260	188

Assuming that random and independent samples were drawn from each of the two age groups, test the hypothesis that the fraction of "false" responses in the total populations of the two age groups is the same. In setting up the test procedure, state and justify the alternative hypothesis that you feel is appropriate.

9.9. Guetzkow and Simon [14] report the results of an experiment designed to compare the effectiveness of different communication networks that can be used in small groups. Five-man groups were given a certain task to perform, where task completion depended upon the passage of written messages among the members of the group. Among the different communication networks tested were the following: (1) the "All-Channel" network for which the members of the group could communicate with (or send messages to) any other member of the group, and (2) the "Wheel" network for which *one* man could send messages to any other man but the other four men could only send to the "hub" man. The efficiency of a network was measured by the time required for task completion. The following results were obtained:

Network	Number of groups	Average time for task completion, min	Standard deviation of times for task completion
All-Channel................	20	24.38	4.82
Wheel....................	15	19.12	3.09

Use large-sample theory (although the number of groups is somewhat small in this case) to test the hypothesis that the average time for task completion is the same for the two different networks.

REFERENCES CITED

1. West, Quentin M.: *The Results of Applying a Simple Random Sampling Process to Farm Management Data*, A. E. 743, Cornell University, Department of Agricultural Economics, Ithaca, N.Y., 1951.
2. National Bureau of Standards: *Tables of Probability Functions*, vol. II, New York, 1942.
3. Hansen, Morris H., and William N. Hurwitz: "Dependable Samples for Market Surveys," *Jour. Marketing*, **14**:363–372 (1949).
4. Cochran, William G.: *Sampling Techniques*, John Wiley & Sons, Inc., New York, 1953.
5. Berelson, Bernard R., Paul F. Lazarsfeld, and William N. McPhee: *Voting*, University of Chicago Press, Chicago, 1954.
6. Pearson, E. S., and H. O. Hartley: *Biometrika Tables for Statisticians*, vol. 1, Cambridge University Press, New York, 1954.
7. Dixon, Wilfrid J., and Frank J. Massey, Jr.: *Introduction to Statistical Analysis*, 2d ed., McGraw-Hill Book Company, Inc., New York, 1957.
8. Festinger, Leon, and Daniel Katz: *Research Methods in the Behavioral Sciences*, The Dryden Press, Inc., New York, 1953.
9. Adelstein, A. M.: "Accident Proneness: A Criticism of the Concept Based upon an Analysis of Shunters' Accidents," *Jour. Roy. Stat. Soc.*, **A**, **115**:354–410 (1952).
10. U.S. Bureau of Old-Age and Survivors Insurance: "More Selected Findings of the National Survey of Old-Age and Survivors Insurance Beneficiaries, 1951," mimeographed, 1954.
11. Rose, Arnold M.: *Union Solidarity*, University of Minnesota Press, Minneapolis, 1952.
12. Montague, Joel B., Jr., and Bernard Pustilnik: "Prestige Ranking of Occupations in an American City with Reference to Hall's and Jones' Study," *Br. J. Soc.*, **5**:154–160 (1954).
13. Woodward, Julian: "Changing Ideas on Mental Illness and Its Treatment," *Amer. Sociol. Rev.*, **16**:443–454 (1951).
14. Guetzkow, Harold, and Herbert A. Simon: "The Impact of Certain Communication Nets upon Organization and Performance in Task-oriented Groups," *Management Science*, **1**:233–250 (1955).
15. Siegel, Sidney: *Nonparametric Statistics for the Behavioral Sciences*, McGraw Hill Book Company, Inc., New York, 1956.

CHAPTER 10

ELEMENTS OF SAMPLE DESIGN

10.1. The Goals of Sample Design

Practical as well as theoretical considerations must condition the choice of procedures for selecting samples from finite populations. Simple random sampling serves admirably on the theoretical side, as has been demonstrated in Chaps. 8 and 9, but it is not adequate for meeting many situations that arise in practice. For example, a comprehensive list of population elements from which to draw the simple random sample may not be available; if the population is spread over a wide geographic area, the elements in a simple random sample will tend also to be widely spread and it may be extremely costly in travel time to obtain a personal interview with each sample element; or, previous investigations relating to the population may have provided information which should permit the design of sampling procedures that are "better" in some sense than simple random sampling. In this chapter we shall describe briefly how it is possible to design sampling procedures that retain the theoretical advantages of simple random sampling and that are also able to meet practical requirements.

Simple random sampling is defined as that procedure which assigns the same selection probability to each possible sample of a stated size. Since the total number of possible samples of size n that can be drawn from a finite population of N elements is ${}_NC_n$ (see Sec. 6.3.2), each sample has probability $1/{}_NC_n$. This assignment of a known probability to each possible sample, *in advance of the actual drawing of a sample,* defines the *probability model* for simple random sampling. The results stated in Chap. 9 concerning the mean, variance, and approximate normality of the sampling distribution of $\bar{x}$ are derived from this model. These results lead in turn to confidence limits and tests of hypotheses. Simple random sampling stands in sharp contrast to the selection procedures based solely on convenience and judgment that were described in Sec. 6.1. No probability models exist for these procedures, and hence there is no possibility of obtaining objective measures of the "representativeness" of the resulting samples.

These considerations suggest that in facing practical sampling situations one should attempt to retain the probability-model character of simple random sampling. Procedures which do have this character will be called *probability-model* sampling procedures. They are formally determined in the following manner:

1. A list is prepared (at least conceptually) which shows each possible sample of size n that can be drawn from a finite population of N elements.
2. In advance of actually drawing a sample, a known probability is assigned to each sample. These probabilities give the fraction of times that each possible sample would be expected to turn up if the sampling procedure were repeated an indefinitely large number of times. (These probabilities are all equal for simple random sampling, but they need not be equal for probability-model sampling procedures in general.)
3. The expected behavior of the sampling procedure is then deduced from the model. (In particular, expressions for the mean and the variance of the sampling distribution of a sample estimate are required.)
4. A mechanical operation is devised to draw samples in accordance with the model.

Although this definition of a probability-model sampling procedure is phrased as though one would have to actually enumerate all possible samples of a given size, this is never necessary. Various devices can always be used to avoid direct enumeration, both in developing theory and in selecting a sample for a study. Nevertheless, this form of definition brings out most clearly the fundamental nature of these procedures. The discussion of simple random sampling in Chap. 6 illustrates this fact for one probability-model procedure.

The decision to use a probability-model sampling procedure for a contemplated statistical investigation does not automatically provide a cut-and-dried sampling scheme. It merely states the ground rules which must be obeyed in designing a selection procedure. Sample design consists of working within these ground rules to obtain procedures which (1) can be used to draw samples from the defined population without undue expenditure of time and effort, (2) will provide a sample of elements from which data can be obtained within the limits imposed by the available resources of time, money, and personnel, (3) make use of known facts about the population to increase the precision of the sample, (4) can be analyzed so as to provide the characteristics of the sampling distributions of quantities like $\bar{x}$ and s^2, and (5) are in some sense "best" when compared with all probability-model procedures that satisfy the foregoing conditions.

It is clear that sample design can and often will be a highly complex and specialized undertaking. This is evidenced by the recent appearance of a number of books devoted to the subject, for example, Cochran [1] and Hansen, Hurwitz, and Madow [2]. Although it is not appropriate here to attempt any detailed coverage of the techniques available for sample design, it will be possible to set forth a few of the basic elements that permit one to meet practical problems with probability-model sampling procedures and to indicate how these elements influence the precision of sample estimates.

10.2. Cluster Sampling

10.2.1. *Groups of Population Elements as Sampling Units.* When a comprehensive list of population elements cannot be obtained or when it can be prepared only at prohibitive expense, alternatives to simple random sampling must be devised. Experience has shown that it is usually advantageous in these instances to carry out the sampling in several stages. That is, the population is first separated into a set of mutually exclusive and exhaustive groups of elements where the groups can be identified even though the elements within a group are unknown. A *sample of groups* is then selected, and a list of population elements is obtained for each of the chosen groups. The final sample may contain all elements in the chosen groups, or it may be obtained by subsampling within each of these groups. The groups of individual population elements which are to be used as sampling units are often called "clusters" since they are defined on the basis of some common characteristic which their elements possess.

As an illustration of cluster sampling, suppose that one wishes to go into a very large industrial community and obtain a sample of workers employed in some skilled occupation like hand molding and coremaking. (Hand molders and coremakers are employed in foundries to make the sand forms into which various types of molten metal are poured to produce castings.) No listing of individuals engaged in this occupation would ordinarily exist, and yet it would be relatively easy to obtain a complete list of foundries which employ hand molders and coremakers. Each employer therefore defines a cluster of hand molders and coremakers. The general outlines of a feasible sampling procedure would be the following: (1) select a sample of employers; (2) visit the chosen employers and obtain from each a list of hand molders and coremakers who work for them; and (3) select a sample of names from each of these lists for interview, or else interview all individuals on the lists.

It might be argued that lists could just as well be obtained from all employers and a simple random sample drawn from the combined lists. This procedure would lead to two difficulties. These difficulties would become insurmountable if one were thinking of extending the sampling to the entire nation. First, it would be very costly and time-consuming to contact each employer for a list, and second, the "spread-out" nature of a simple random sample would lead to large travel costs. Assuming that workers were to be interviewed in their place of employment, interviewers would have to visit many plants and would probably interview at most one or two workers in each plant. Much of their time would be spent in traveling from plant to plant. With the proposed procedure, an interviewer would visit a sample plant and without additional travel would interview some reasonable number of individuals. A modification of this procedure has been used to select a national sample of hand molders and coremakers [3].

Thus far we have made no attempt to specify a probability model for cluster sampling. In order to show that this can be accomplished, let us consider a greatly oversimplified situation. Suppose there are three

TABLE 10.1. COMPARISON OF PROBABILITY MODELS FOR SIMPLE RANDOM SAMPLING AND FOR ONE FORM OF CLUSTER SAMPLING
[$N = 6$, $n = 4$; clusters are (AB), (CD), and (EF).]

All possible samples of size 4	Probabilities under simple random sampling	Probabilities under cluster sampling
$(AB)(CD)$	1/15	1/3
$ABCE$	1/15	0
$ABCF$	1/15	0
$ABDE$	1/15	0
$ABDF$	1/15	0
$(AB)(EF)$	1/15	1/3
$ACDE$	1/15	0
$ACDF$	1/15	0
$ACEF$	1/15	0
$ADEF$	1/15	0
$BCDE$	1/15	0
$BCDF$	1/15	0
$BCEF$	1/15	0
$BDEF$	1/15	0
$(CD)(EF)$	1/15	1/3
Total.	1	1

plants, each containing two workers, and that a sample of four workers is desired. We shall designate the six workers as A, B, C, D, E, and F and consider that the grouping into plants is (AB), (CD), and (EF). A probability-model cluster sampling procedure can now be defined as (1) draw a simple random sample of two plants; (2) interview all workers in the selected plants. Notice that it is not necessary to know which workers are in which plants until after the sample of plants has actually been drawn. Table 10.1 gives the probability model for both simple random sampling and for this form of cluster sampling. All the possible samples that do not contain two complete clusters are assigned a probability of zero, and those samples that do have the required grouping are assigned equal probabilities.

It is quite clear that the foregoing probability-model sampling procedure can be applied to situations where the clusters are defined on any convenient basis and where there can be any number of clusters. (For purposes of the present discussion we shall continue to require that each cluster have the same number of elements, although comments will be made in Sec. 10.2.3 as to how this restriction can be removed.) Furthermore, it is not necessary that every element in the selected clusters be drawn into the sample. The probability model can be extended by requiring that *simple random samples of the same size* be drawn from each sample cluster.

Now the mere fact that we have been able to find a probability-model sampling procedure which can be used to meet certain practical considerations is no guarantee that it will have any other desirable properties. Further progress can be made only by attempting to deduce from the model some of the characteristics of sampling distributions generated by repeated application of the procedure. The present model also behaves well in this respect. Assuming that the population mean μ of a variable x is to be estimated, the following results can be proved:

1. One should use the sample mean $\bar{x}_{cl}$, computed for the sample as a whole without regard to clustering, as an estimate of the unknown population mean μ.

2. The sample mean is an unbiased estimate of the population mean. That is, $E(\bar{x}_{cl}) = \mu$.

3. It is possible to obtain from the model an expression for the variance of $\bar{x}_{cl}$, that is, $\sigma^2_{\bar{x}_{cl}}$ or $V(\bar{x}_{cl})$, and to estimate this variance from a sample. See Cochran [1, chap. 10].

4. If, in addition, the number of clusters in the population is large, the sampling distribution of $\bar{x}_{cl}$ can be approximated by a normal distribution and one can therefore compute confidence limits for μ. Thus the cluster sampling model not only helps to solve a frequently occurring practical problem, but it also retains the theoretical advantages of simple random sampling.

Although cluster sampling was introduced primarily as a device to get around the absence of a list, this form of sampling must often be used even though a complete listing of the individual elements in a population is available. An excellent illustration of this is provided by the experience of the English Social Survey, described in a paper by Gray and Corlett [4]. The Social Survey did have population registers from which to select simple random samples, but the cost of sending enumerators to such a sample was regarded as prohibitive. The Social Survey therefore defined clusters on the basis of geographic and administrative considerations, chose a sample of clusters, and used the registers to select simple random samples from within each of the sample clusters.

There are no hard and fast rules which can always be used for defining clusters of population elements. Each situation must be examined in its own right. However, several criteria of definition deserve special mention. These are *physical proximity* and *organizational attachment.* When a population is dispersed over a wide geographic area and each individual element can be associated either with a point in the area or with a certain portion of the area, it is always natural to consider the possibility of forming groups within which the members are in some sense physically "close" to one another. Such groups are usually defined on the basis of political and natural physical boundaries that are commonly used in reporting census figures or in drawing maps. Thus it is possible to consider in the United States units like counties, cities, townships, wards, blocks, and the like. Sometimes slightly different but related approaches are possible. For example, organizations have selected samples of business establishments (scattered over the United States) by first choosing a sample from a list and then associating with each chosen establishment those three or four additional establishments which are geographically closest. Magazine subscription lists have also been used in a similar manner.

When clusters of population elements are defined on an area basis and when a sample of these areas is chosen in accordance with some probability-model procedure, we speak of *area sampling.* Area sampling has been used extensively in the United States where population registers are not available to aid in sampling and in regions of the world where periodic census data do not even exist. Experience with the Current Population Survey of the U.S. Bureau of the Census is described in [2, vol. I, chap. 12B], and Shaul [5] has given an account of the use of sampling surveys in Central Africa.

The second general criterion that can often be used in defining clusters is that of organizational attachment. Most individuals are members of one or more identifiable and formally organized groups and may possibly be contacted through these groups. Thus we can consider members of different households or spending units, groups of workers having different employers, groups of workers belonging to different labor unions (or the locals thereof), doctors in different county medical societies, lawyers in different firms, farmers belonging to different units of the Grange, businessmen who are members of different service clubs, college students belonging to different fraternities or sororities, and so on. Special segments of a population may frequently be economically contacted only through means such as these. The major difficulty encountered in using groups of this type is that of securing assurance that all members of the desired population are contained in the defined system of groups.

10.2.2. *Comparison of the Precision of Cluster Sampling with That of Simple Random Sampling.* Assuming that either simple random sampling or cluster sampling can be used for a particular study, the previous discussion seems heavily weighted in favor of cluster sampling. Both procedures can be described in terms of probability models, but the cost per sample element is ordinarily going to be smaller for cluster sampling. However, this leaves entirely out of the picture the fact that samples are drawn to provide estimates of population characteristics. A comparison of the costs of two different sampling procedures must be made on the basis of the precision of resulting estimates. For example, it might be possible to obtain a cluster sample of 500 cases or a random sample of 200 cases for the same total cost, and yet the random sample might give a *more precise estimate* of the population mean, that is, *an estimate with smaller sampling variability.* The rational decision would clearly be to take the random sample. Although exact comparisons of the precision provided by simple random sampling and cluster sampling can only be made under completely specified circumstances, it is possible to develop some rough guides concerning the expected behavior of the two and to supply some data that illustrate this behavior.

As a starting point, let us examine a very simple situation and draw upon intuition to the fullest possible extent. Suppose we have a population which can be divided into subgroups or clusters, each cluster having the same number of population elements. If one wishes to talk in terms of a concrete example, it is possible to think of the adult population in a city as living on blocks, each block having the same number of adults. A two-stage, or cluster, sample would be obtained in this situation by first selecting a random sample of blocks (where not all blocks are included in the sample) and then taking all or a sample of individuals from each of the chosen blocks.

In order to illustrate the effects of cluster sampling, let us assume that we can arbitrarily assign individual population elements to clusters, and let us distinguish between the following three methods of assignment:

Method of assignment	*Description*
I	Individuals are assigned to clusters so that *within any cluster* all individuals are identical with respect to the variable of investigation. For example, if income is the variable, all adults on a block have identical incomes; or, if attitude on some topic is the variable, all individuals on a block have identical attitudes.
II	Individuals are randomly assigned to clusters without regard to any of their characteristics.
III	Each cluster has exactly the same mean on the variable of investigation as does the entire population.

As will be seen shortly, methods I and III represent two extremes, while II falls in between the two.

With this as background, it is possible to argue concerning the manner in which cluster sampling affects the precision of an estimate of a population mean. If the assignment to clusters is according to method II, then any random selection of clusters and any random selection of individuals from within the chosen clusters will be equivalent to drawing a simple random sample of the same total size from the entire population. That is, no gain or loss in precision, in comparison with random sampling, is effected in this situation.

Now consider the assignment to clusters according to method I. Since all elements within a cluster are identical, a single element provides complete information concerning the entire cluster. Therefore, if more than one element is taken from each sample cluster, the additional (over one) elements are "wasted" and the sampling variability of the resulting estimate will be larger than that which would be given by a simple random sample containing the same total number of elements. The extreme loss, relative to simple random sampling, is of course attained when each sample cluster is completely enumerated.

Finally, if method III is used for assignment, we need study only one cluster since this is equivalent to studying the entire population. An estimate of the population mean will have zero sampling variability if a 100 per cent sample is drawn from the selected cluster.

The natural clusters that must be used in the actual application of sampling procedures will never match exactly any one of the three preceding classifications. And yet it is reasonable to argue that a system of natural clusters, coupled with a variable whose population mean is to be estimated, will ordinarily fall between I and II. That is, elements within clusters defined on the basis of geographical proximity or organizational attachment will not be identical on most variables of interest, but they will tend to be more alike than if they had been randomly assigned to clusters. Thus people living on the same block will not differ widely on such characteristics as income or education, and these similarities may in turn induce like attitudes or opinions. As a consequence, estimates obtained from cluster samples will ordinarily be less precise than estimates produced from random samples of the same size.

As an illustration of these effects, consider the data of Tables 10.2 and 10.3. Those of Table 10.2 refer to clusters defined by taking a stated number of adjacent households from the same block in large cities of the United States. The comparisons of cluster sampling with simple random sampling are presented in terms of sample size. For example, suppose that one wishes to estimate the fraction of dwelling units in a large city that are owned by the occupant. This is the first item in Table 10.2. If clusters of 3 adjacent households are drawn randomly and each cluster is completely enumerated, a total sample of 130 households will be required to estimate the required fraction with the same precision as would be provided by a simple random sample of 100 households. Notice that the total size of sample needed to equal the precision of a random sample of 100 households increases as the size of cluster increases. Thus 530 households are required if the clusters contain 27 adjacent households The actual magnitude of these values cannot be predicted without performing computations, but preceding considerations would certainly lead

TABLE 10.2. SIZES OF CLUSTER SAMPLES OF ADJACENT HOUSEHOLDS REQUIRED TO PROVIDE ESTIMATES HAVING THE SAME PRECISION AS SIMPLE RANDOM SAMPLES OF 100 CASES
(Data computed for selected cities in the United States with 1940 populations of over 100,000)

Characteristic	Fraction of total population possessing characteristic	Size of sample for clusters containing:		
		3 adjacent households	9 adjacent households	27 adjacent households
Dwelling unit occupied by owner	.1967	130	240	530
Dwelling unit with rental value of $75 and over	.0174	180	330	520
Males aged 25–34 years	.0829	110	120	150
Males employed as craftsmen, foremen, and kindred workers	.0522	100	120	150

SOURCE: Hansen, Hurwitz, and Madow [2, vol. I, p. 597, table D-4].

to a prediction of this general type of behavior. Dwelling units located adjacent to one another on the same city block will tend to be alike with respect to owner occupancy or tenant occupancy. This clustering effect is not so pronounced for the item "males, aged 25–34 years." That is, men and women of various age groups are distributed more or less randomly among the various sections of a city.

Table 10.3 presents similar data for clusters defined by taking pairs of adjacent farms or by taking strings of eight consecutive farms along a road. Again it is to be expected that farms lying near one another will tend to be alike with respect to many characteristics and that cluster sampling will be less precise than random sampling for studying these characteristics. This is borne out by the data, with the exception of the variable "number of turkeys per farm" where random sampling and cluster sampling are much the same. Presumably this result arose because the farmers of the county used for the study did not specialize in growing turkeys, and the few farmers that did grow turkeys were more or less randomly scattered over the entire county.

Accepting the fact that the precision of estimates will often behave in the manner illustrated in Tables 10.2 and 10.3, we can now observe that the design of a cluster sample will often require achieving some sort of optimum balance between costs and precision. As one proceeds from simple random sampling, which may be regarded as the sampling of clusters of size 1, to the sampling and complete enumeration of larger and larger clusters, the total sample size needed to maintain equal precision will also increase. On the other hand, the cost per sample element

TABLE 10.3. SIZES OF CLUSTER SAMPLES OF ADJACENT FARMS REQUIRED TO PROVIDE ESTIMATES HAVING THE SAME PRECISION AS SIMPLE RANDOM SAMPLES OF 100 CASES
(Data computed for Morrow County, Ohio, in 1938)

Characteristic	Size of sample for clusters containing:	
	2 adjacent farms	8 consecutive farms along a road
Crop land harvested, acres	113	183
Wheat, acres	113	171
Soybeans, acres	103	123
Turkeys, number	100	97

SOURCE: Frederick F. Stephan, "The Morrow County Project," mimeographed paper prepared for the Committee on Sample Census of Agriculture, 1938.

will decrease because of decreasing travel costs and the like. It is sometimes possible to obtain approximate solutions for the point of optimum balance. For example, one can assume that there is a fixed amount of money available to carry out a particular survey and can then obtain the cluster size that will produce estimates having the greatest possible precision.

10.2.3. *Further Aspects of Cluster Sampling.* Two difficulties not explicitly mentioned in the preceding discussion will be encountered in the design of most cluster samples. These are (1) the natural clusters that are readily available for use (counties, cities, blocks, employers, professional and social organizations, and the like) will each contain a large number of population elements, and, as is indicated by the contents of Tables 10.2 and 10.3, complete enumeration of the sample clusters will produce estimates that have much greater sampling variability than can be tolerated for a given expenditure of time and money; and (2) these natural clusters will not all contain exactly the same number of population elements, and the probability model of Table 10.1 must be changed to take this into account.

These two difficulties can sometimes be overcome simultaneously if it is possible to justify the allocation of large resources to the definition of small areal clusters. Such an undertaking is feasible only if the resulting system of clusters can be used in the design of a large number of different sample surveys. An example of this approach is provided by the Master Sample of Agriculture, developed jointly by Iowa State College, U.S. Bureau of Agricultural Economics, and U.S. Bureau of the Census. The materials available through the Master Sample of Agriculture are described by King and Jessen [6]. They consist primarily of county maps, for each county in the United States, on which small areal sampling units have been defined in terms of easily identifiable natural boundaries.

If one is not able to define small equal-sized clusters, then it is possible to select a sample of clusters and to choose random samples of elements from within each of the chosen clusters. The technique of *subsampling* was mentioned in the preceding discussion, but the data of Tables 10.2 and 10.3 refer only to complete enumeration. The general effect of subsampling will be to decrease the influence of similarity among elements within the same cluster and hence to increase the precision of $\bar{x}_{cl}$ for fixed total size of sample. Here again a search may be made for an optimum solution, namely, that combination of number of clusters and number of elements from each cluster which will provide the "best" estimate for a given expenditure.

Finally, we observe that many new possibilities for theory and practice are opened up if the clusters are not of equal size. For example, the probability model may specify "large" selection probabilities for "large" clusters and "small" probabilities for "small" clusters. A wide variety of probability models that are useful for dealing with unequal-sized clusters will be found in references [1] and [2]. Also, a very usable "how to do it" account of these procedures for obtaining a two-stage sample in a city has been given by Kish [7].

10.3. Stratified Simple Random Sampling

10.3.1. *Definition.* Sample surveys are always conducted for the purpose of adding information to the store of knowledge that *already exists* relative to some defined population. Rarely, if ever, is one faced with the problem of sampling a population about which absolutely nothing is known. And yet neither of the two probability-model sampling procedures discussed thus far makes any use of this prior information in producing a "better" or more "representative" sample. Simple random sampling of course requires a list of population elements, but this list merely defines the population and tells nothing about the characteristics of the population. Similarly, cluster sampling requires only a listing of groups of population elements and, in general, produces estimates that are less precise than would be provided by simple random samples of the same size. The question immediately arises as to whether known characteristics of the population can be used to increase the precision of a sample estimate over that provided by simple random sampling, assuming no increase in sample size. In this section we shall discuss briefly one probability-model sampling procedure that can sometimes be used in achieving this result.

Information from routine records or previous statistical investigations, including censuses, often supplies the means for identifying subgroups within a population, including the approximate number or fraction of elements contained in each of the subgroups. For example, the decennial census gives the number of adult men and women in the population, the number of adults living in cities of various sizes and the number living in rural areas, the number living in the various geographic regions of the United States, the number employed in the different occupations, and so on; similarly, the personnel records of a plant will permit the identification of groups of workers which differ as to length of service, general type of duty (hourly paid, supervisory, or professional), or specific kind of work performed.

Obviously it is not possible to classify elements into subgroups on the basis of the variable that is to be investigated in a specific survey. The required information will not be available until after the study has been completed. However, one can sometimes argue on the basis of general knowledge that the subgroups of a defined system will in some sense differ with respect to the subject of investigation. Thus the planners of a preelection survey would know that the fraction of Southerners who will vote Democratic will almost certainly be different from the fraction of Northerners who will vote Democratic. Or, suppose a government agency is going to conduct a survey of farms in the United States to estimate the amount of land planted to wheat for a given year. This agency would know that the average number of acres per farm in a Northeastern state such as Vermont will be quite different from the average number of acres for a Middle Western state like Kansas.

If it is surmised that the subgroups of a defined system will differ with respect to the subject of an investigation, and if the fraction of the population falling in each of these subgroups is known from a previous study, then it seems reasonable that some controlled representation of these groups in the sample would be profitable in terms of increasing the precision of sample estimates. Simple random sampling affords no such control since all possible samples have an equal chance of being drawn and some of these may not even contain a single element from an important subgroup (see Table 10.4). One form of probability-model procedure which does permit control over the representation of subgroups is *stratified simple random sampling.* From a formal point of view stratified simple random sampling is defined as follows:

1. The population is divided into a set of mutually exclusive and exhaustive subgroups or *strata.*
2. A simple random sample of elements is selected from *each* stratum.
3. An estimate is prepared separately for *each* stratum.
4. These estimates are *combined* in some prescribed manner to provide an estimate for the entire population.

Notice that stratified sampling is quite different in form from cluster sampling, even though both techniques make use of population subgroups. Cluster sampling requires one to select a sample of groups, while stratified sampling requires that *each* subgroup be represented in the final sample.

The fact that stratified simple random sampling does conform to a probability model can be easily illustrated by means of a hypothetical example. Consider a plant in which there is one professional worker, two individuals in supervisory capacities, and three hourly-paid workers. The professional worker will be designated as P, the two supervisors as S_1 and S_2, and the three hourly-paid workers as H_1, H_2, and H_3. A stratified simple random sample of size 3 might then be obtained by specifying that a single worker is to be drawn at random from each of the three groups or strata. (Since there is only one professional worker, he must be included in every sample.) Table 10.4 gives the probability model for simple random sampling and for the stated form of stratified simple random sampling. Any sample that does not have an element from each of the three strata is assigned a zero probability, and those samples that do satisfy this requirement are assigned equal probabilities. Since the definition of stratified simple random sampling implies that the sample from one stratum is independent of the samples drawn from any other stratum, these probabilities can also be obtained by direct multiplication. Thus

$$\begin{aligned}\Pr(PS_1H_3) &= \Pr(P) \cdot \Pr(S_1) \cdot \Pr(H_3) \\ &= 1(\tfrac{1}{2})(\tfrac{1}{3}) \\ &= \tfrac{1}{6}\end{aligned}$$

Although the contents of Table 10.4 illustrate the probability-model character of stratified simple random sampling, they have not yet taken into account the final two points contained in the definition of this form of sampling, namely, that an estimate is prepared separately for each stratum and that these estimates are then combined in some prescribed manner to provide an estimate for the entire population. For example, suppose we wished to estimate the average monthly income of the six workers in the hypothetical example. It would clearly not be appropriate to take the simple average income of the three workers in the sample since the single hourly-paid worker represents himself and two others while the single professional worker represents only himself.

TABLE 10.4. COMPARISON OF PROBABILITY MODELS FOR SIMPLE RANDOM SAMPLING AND FOR STRATIFIED SIMPLE RANDOM SAMPLING
($N = 6$, $n = 3$; strata are (P), (S_1S_2), and $(H_1H_2H_3)$; a single element is drawn at random from each stratum.)

All possible samples of size 3	Probabilities under simple random sampling	Probabilities under stratified simple random sampling
PS_1S_2	1/20	0
PS_1H_1	1/20	1/6
PS_1H_2	1/20	1/6
PS_1H_3	1/20	1/6
PS_2H_1	1/20	1/6
PS_2H_2	1/20	1/6
PS_2H_3	1/20	1/6
PH_1H_2	1/20	0
PH_1H_3	1/20	0
PH_2H_3	1/20	0
$S_1S_2H_1$	1/20	0
$S_1S_2H_2$	1/20	0
$S_1S_2H_3$	1/20	0
$S_1H_1H_2$	1/20	0
$S_1H_1H_3$	1/20	0
$S_1H_2H_3$	1/20	0
$S_2H_1H_2$	1/20	0
$S_2H_1H_3$	1/20	0
$S_2H_2H_3$	1/20	0
$H_1H_2H_3$	1/20	0
Total...............	1	1

The general probability model for stratified simple random sampling provides the solution to this problem of form of estimate. Assume that we wish to estimate the population mean μ of a variable x. Let $\bar{x}_1$ be the mean of the simple random sample drawn from the first stratum and let W_1 be the fraction of the entire population contained in the first stratum; let $\bar{x}_2$ be the sample mean for the second stratum and W_2 the fraction of the entire population in this stratum; . . . ; and let $\bar{x}_L$ be the mean of the last stratum and W_L the fraction of the entire population in the last stratum. Then it can be shown that one should use the weighted sample mean $\bar{x}_{st}$ as an estimate of the unknown population mean where

$$\bar{x}_{st} = W_1\bar{x}_1 + W_2\bar{x}_2 + \cdots + W_L\bar{x}_L$$
$$= \sum_{i=1}^{L} W_i\bar{x}_i \qquad (10.1)$$

If (10.1) is applied in the example of Table 10.4, we see that the estimate for any one of the six possible stratified simple random samples should be of the form

$$(1/6)P + (1/3)S + (1/2)H$$

where P, S, and H stand, respectively, for the value of the variable associated with the sample professional worker, the sample supervisory worker, and the sample hourly-paid worker.

The final evaluation of a probability-model sampling procedure must be made on the basis of some of the characteristics of sampling distributions generated by repeated application of the procedure. The probability model for stratified simple random sampling leads to the following results:

1. $\bar{x}_{st}$, as defined in (10.1), is an unbiased estimate of the unknown population mean. That is, $E(\bar{x}_{st}) = \mu$.

2. It is possible to obtain from the model an expression for the variance of $\bar{x}_{st}$, that is, $\sigma^2_{\bar{x}_{st}}$ or $V(\bar{x}_{st})$, and to estimate this variance from a sample. See Cochran [1, chap. 5].

3. If, in addition, the sample drawn from each stratum is large, the sampling distribution of $\bar{x}_{st}$ can be approximated by a normal distribution and one can therefore compute confidence limits for μ.

The preceding discussion has left unanswered two major questions. These are: (1) How should the total sample be divided or allocated among the strata? (2) What gains, over simple random sampling, can be expected through the use of stratification in any particular situation? Answers to these two questions are related to one another, and they will be considered next.

10.3.2. *Comparison of the Precision of Stratified Simple Random Sampling with That of Simple Random Sampling.* The relative precision of estimates provided by these two forms of sampling will be illustrated in terms of a small finite population where the value of the variable attached to each element in the population is known. The data to be used are given in Table 10.5, and they represent the contract monthly rent paid by each of eight families living in a large city. The mean and variance of this population are given at the foot of the table. This is clearly an artificial example because the population is very small and because the values of the variable are assumed to be known. Nevertheless, the observations we make from these data concerning stratification can be carried over into a general theory which is applicable in "real" situations of the type discussed earlier. With respect to the illustrative variable "rent," there is at least one important instance where stratified sampling is used to determine the average monthly contract rent in large cities. Contract monthly rent is one of the components of the Consumer Price Index of the Bureau of Labor Statistics, and current data are obtained on a sampling basis [8].

Suppose that the average rent in the population of Table 10.5 is to be estimated from samples of size 4. We know from the contents of Chap. 9

TABLE 10.5. CONTRACT MONTHLY RENT PAID BY EACH OF EIGHT FAMILIES COMPRISING A SMALL FINITE POPULATION
(Data are hypothetical.)

Family designation	Rent, dollars
A	16
B	28
C	34
D	36
E	42
F	48
G	58
H	90

$\mu = \$44$
$\sigma^2 = 505.14 \text{ (dollars)}^2$

that the precision of an estimate may be evaluated in terms of the variance of the sampling distribution of that estimate. Furthermore, if simple random samples of size 4 are drawn without replacement from this population, the variance of the sample mean $\bar{x}$ can be obtained by using (9.2) and its value is approximately 63. This variance of the mean of simple random samples is the standard relative to which we shall compare the precision of estimates produced by various forms of stratified simple random sampling and will henceforth be taken as equal to 100.

Our earlier and more general discussion implied that stratification is of benefit if strata are defined so that they are in some sense different from one another with respect to the variable under investigation. This can be easily accomplished in the present example because all values of the variable are known. Let us arbitrarily assume that two strata of equal size are to be used and that they will be defined so that their respective means are as different as possible. These strata are identified in Table 10.6. The four elements having the smallest values of the variable are in stratum I, and the four having the largest values are in stratum II.

With the sample size fixed at 4 and with the strata given, the only additional factor at our disposal is the allocation of the sample among the strata. The most obvious and intuitively appealing form of allocation is *proportional allocation.* That is, if 10 per cent of the population is in a particular stratum, then 10 per cent of the sample will be allocated to that stratum; or, if 70 per cent of the population is in a stratum, 70 per cent of the sample will be drawn from that stratum. In the present example, 50 per cent of the population is in each of the two strata and proportional allocation therefore requires taking a simple random sample of size 2 from stratum I and a simple random sample of size 2 from stratum II. This specification of stratified simple random sampling gives

TABLE 10.6. IDENTIFICATION OF STRATA FOR HYPOTHETICAL POPULATION OF TABLE 10.5

Stratum	Elements	Values of the variable	Strata means	Strata variances
I	*A, B, C, D*	16, 28, 34, 36	28.5	81
II	*E, F, G, H*	42, 48, 58, 90	59.5	457

36 possible samples of size 4 that can be drawn, as opposed to the 70 samples that are possible with simple random sampling. These 36 samples are listed in Table 10.7, and the value of $\bar{x}_{st}$, as computed from (10.1), is given for each sample.

Before investigating the distribution of these values it is of interest to examine the exact character of the estimate $\bar{x}_{st}$. Taking a single sample, say *ADFH*, expression (10.1) says to take the mean for elements *A* and *D* and the mean for elements *F* and *H* and to weight these means in accordance with the fraction of elements in the respective strata. Thus we have

$$\begin{aligned}\bar{x}_{st} &= W_{\text{I}}\bar{x}_{\text{I}} + W_{\text{II}}\bar{x}_{\text{II}} \\ &= \left(\frac{1}{2}\right)\left(\frac{16 + 36}{2}\right) + \left(\frac{1}{2}\right)\left(\frac{48 + 90}{2}\right) \\ &= (1/4)(16 + 36 + 48 + 90) \\ &= 47.5\end{aligned}$$

Notice that the general expression given by (10.1) reduced in this case to taking the mean of the four sample elements without regard to the strata from which they came. This procedure can always be used, no matter how many strata there may be or how the population is divided among the strata, as long as *proportional allocation* is used. That is, the sample mean, computed without weighting, is an unbiased estimate of the population mean. (Some error is introduced if the sample is small and if it is not possible to achieve an exact match between the fraction of the population in a stratum and the fraction of the sample in that stratum.) Sampling and estimation procedures that have this property are called *self-weighting*, and the use of such procedures often aids one in avoiding computational difficulties that can arise in analyzing sample results.

Each of the samples in Table 10.7 has by definition an equal probability of being the one selected in a particular drawing. Therefore the mean

TABLE 10.7. THE POSSIBLE STRATIFIED SIMPLE RANDOM SAMPLES OF SIZE 4 WITH PROPORTIONAL ALLOCATION
(Strata are given in Table 10.6.)

Sample	$\bar{x}_{st}$	Sample	$\bar{x}_{st}$
ABEF	33.5	*BCEF*	38.0
ABEG	36.0	*BCEG*	40.5
ABEH	44.0	*BCEH*	48.5
ABFG	37.5	*BCFG*	42.0
ABFH	45.5	*BCFH*	50.0
ABGH	48.0	*BCGH*	52.5
ACEF	35.0	*BDEF*	38.5
ACEG	37.5	*BDEG*	41.0
ACEH	45.5	*BDEH*	49.0
ACFG	39.0	*BDFG*	42.5
ACFH	47.0	*BDFH*	50.5
ACGH	49.5	*BDGH*	53.0
ADEF	35.5	*CDEF*	40.0
ADEG	38.0	*CDEG*	42.5
ADEH	46.0	*CDEH*	50.5
ADFG	39.5	*CDFG*	44.0
ADFH	47.5	*CDFH*	52.0
ADGH	50.0	*CDGH*	54.5

$$\mu_{\bar{x}_{st}} = E(\bar{x}_{st}) = 44.0$$
$$\sigma^2_{\bar{x}_{st}} = V(\bar{x}_{st}) = 33.62$$

and variance of the sampling distribution of $\bar{x}_{st}$ can be obtained by assigning a probability of $1/36$ to each of the 36 samples and applying formulas (4.6) and (5.5).* Their respective values are 44.0 and 33.62, as given

* The variance of $\bar{x}_{st}$ can also be obtained in terms of the separate strata variances. Let W_1 be the fraction of the population in the first stratum, N_1 the number of elements in the first stratum, n_1 the size of the simple random sample drawn from the first stratum, and σ_1^2 the variance of the variable x in the first stratum. Appropriate subscripts identify the corresponding quantities for the other strata. Then it can be shown that

$$V(\bar{x}_{st}) = W_1^2\left(1 - \frac{n_1}{N_1}\right)\frac{\sigma_1^2}{n_1} + W_2^2\left(1 - \frac{n_2}{N_2}\right)\frac{\sigma_2^2}{n_2} + \cdots + W_L^2\left(1 - \frac{n_L}{N_L}\right)\frac{\sigma_L^2}{n_L}$$
$$= \sum_{i=1}^{L} W_i^2\left(1 - \frac{n_i}{N_i}\right)\frac{\sigma_i^2}{n_i} \qquad \textbf{(10.2)}$$

at the foot of Table 10.7. Since $E(\bar{x}_{st}) = 44.0$ and $\mu = 44.0$ (see Table 10.5), this example provides a numerical demonstration of the statement made in Sec. 10.3.1 that $\bar{x}_{st}$ is an unbiased estimate of the population mean. Now let us investigate the relative precision of $\bar{x}_{st}$. Taking the variance of the mean of a simple random sample of size 4 (63.14) as 100 per cent, we see that $V(\bar{x}_{st})$ is equal to 53 per cent. That is, the variance of the estimate has almost been cut in half by using stratified simple random sampling with proportional allocation. And we still have exactly the same sample size!

An intuitive argument for this gain in precision can be obtained by examining the samples listed in Table 10.7. By controlling the representation of the two strata in the sample, zero probabilities are being assigned to some of the samples that can be drawn under simple random sampling. It so happens in the present example that the eliminated samples are those having the most extreme values for their arithmetic means. For example, the smallest value of $\bar{x}_{st}$ in Table 10.7 is 33.5, and the largest value is 54.5. With simple random sampling, we might draw samples with still smaller or larger means such as the following: *ABCD* ($\bar{x} = 28.5$), *ABCE* ($\bar{x} = 33.0$), *DFGH* ($\bar{x} = 58.0$), or *EFGH* ($\bar{x} = 59.5$). The gain in precision of stratified simple random sampling over simple random sampling arises from this elimination of samples with very small or very large arithmetic means. If the samples eliminated by a particular set of strata do not tend to have especially large or small arithmetic means, then it might be inferred from the foregoing that stratified simple random sampling with proportional allocation will not offer much gain in precision over simple random sampling.

The results that have just been obtained from a particular numerical example can be formalized by working with the appropriate probability models. It is then possible to prove the following:

1. The precision of $\bar{x}_{st}$, computed from stratified simple random samples with *proportional allocation*, cannot ordinarily be less than the precision of the mean of a simple random sample of the same size. (Exceptions can occur because of the finite-population corrections, but these exceptions are rarely met in practice.) The two precisions will be equal when the population means of all strata are identical.

2. The greater the differences in the population means of the strata, the greater will be the gain in the precision of stratified simple random sampling (with proportional allocation) over that of simple random sampling.

Notice that the second point gives a guide for the definition of strata. They should be determined so that their means are as different as possible. Naturally one does not have the actual values of the variable available for performing this division into strata, but some related variable for which data are known may serve almost as well. Thus areas of a city can almost certainly be distinguished, perhaps on the basis of past census data, so that the rental levels in the areas will be different.

Proportional allocation of a sample among strata has a certain amount of intuitive appeal. Furthermore, it can be justified on the objective grounds that it will ordinarily increase the precision of sample estimates. Proportional allocation, however, is clearly not the only type that should be considered. As a matter of fact, a close examination of the samples listed in Table 10.7 will show that the appearance or nonappearance of element H (which has an extreme value of the variable) in a sample

TABLE 10.8. THE POSSIBLE STRATIFIED SIMPLE RANDOM SAMPLES OF SIZE 4 WITH DISPROPORTIONATE ALLOCATION

(Strata are given in Table 10.6; one element is drawn at random from stratum I and three elements at random from stratum II.)

Sample	$\bar{x}_{st}$	Sample	$\bar{x}_{st}$
AEFG	32⅔	*CEFG*	41⅔
AEFH	38	*CEFH*	47
AEGH	39⅔	*CEGH*	48⅔
AFGH	40⅔	*CFGH*	49⅔
BEFG	38⅔	*DEFG*	42⅔
BEFH	44	*DEFH*	48
BEGH	45⅔	*DEGH*	49⅔
BFGH	46⅔	*DFGH*	50⅔

$$\mu_{\bar{x}_{st}} = E(\bar{x}_{st}) = 44.0$$
$$\sigma^2_{\bar{x}_{st}} = V(\bar{x}_{st}) = 24.71$$

has a very pronounced effect on the value of the sample mean. This suggests that we use some form of *disproportionate* allocation which will guarantee that element H, or any other "extreme" element, appear in a greater fraction of the possible samples. There is only one way of doing this with the strata as defined and that is to draw a single element at random from stratum I and three elements at random from stratum II. There are 16 samples of this form, and they are listed in Table 10.8. The value of $\bar{x}_{st}$ must now be computed by the strict use of (10.1). That is, a stratified simple random sample with disproportionate allocation is not self-weighting. As an illustration, consider sample *CFGH*. We have

$$\begin{aligned}\bar{x}_{st} &= W_{\text{I}}\bar{x}_{\text{I}} + W_{\text{II}}\bar{x}_{\text{II}} \\ &= \left(\frac{1}{2}\right)(34) + \left(\frac{1}{2}\right)\left(\frac{48 + 58 + 90}{3}\right) \\ &= 49\tfrac{2}{3}\end{aligned}$$

The values of $\bar{x}_{st}$ for the other samples are given in Table 10.8.

The values of $E(\bar{x}_{st})$ and $V(\bar{x}_{st})$ can be computed by using (4.6) and (5.5) and the fact that each of the possible samples has a probability of $\frac{1}{16}$. As shown at the foot of the table, $E(\bar{x}_{st}) = 44.0$ and $V(\bar{x}_{st}) = 24.71$. Again taking the variance of the mean of a simple random sample of size 4 as 100 per cent, we find that $V(\bar{x}_{st})$ is now equal to 39 per cent. Thus the precision of the estimate has been increased over that provided by proportional allocation. A heuristic justification of this result is as follows. It is more difficult to estimate the mean of stratum II than it is to estimate the mean of stratum I because the elements of stratum II are more variable among themselves. A measure of this difference in internal variability is provided by the strata variances as presented in Table 10.6. Therefore more of the sample should be taken from stratum II than is required by proportional allocation. The payoff in this argument is in the observed reduction of the variance of $\bar{x}_{st}$ from 53 per cent for proportional allocation to 39 per cent for this form of disproportionate allocation.

Once it has been recognized that some form of disproportionate allocation may give a more precise estimate than does proportional allocation, it seems natural to inquire whether there is a best form of allocation. It can be shown that there does exist an allocation which provides estimates with the greatest possible precision, and this is known as *optimum allocation*. Optimum allocation is defined in the following manner. As before, let $W_1, W_2, \ldots, W_L$ be the fractions of the population contained in the various strata, and in addition, let $\sigma_1, \sigma_2, \ldots, \sigma_L$ be the standard deviations of the variable x in these strata. Then if n is the total sample size and n_1 is the number of elements to be selected at random from the first stratum, the optimum value of n_1 is

$$n_1 = n\left(\frac{W_1\sigma_1}{W_1\sigma_1 + W_2\sigma_2 + \cdots + W_L\sigma_L}\right)$$

Similarly, the optimum size of sample to be drawn from the hth stratum is

$$n_h = n\left(\frac{W_h\sigma_h}{\sum_{i=1}^{L} W_i\sigma_i}\right) \tag{10.3}$$

Going back to the two strata of Table 10.6, these expressions become

$$n_1 = 4\left[\frac{(\frac{1}{2})9}{(\frac{1}{2})9 + (\frac{1}{2})(21.4)}\right] = 1.2$$

and

$$n_2 = 4\left[\frac{(\frac{1}{2})(21.4)}{(\frac{1}{2})9 + (\frac{1}{2})(21.4)}\right] = 2.8$$

Since the sample sizes have to be whole numbers, we take $n_1 = 1$ and $n_2 = 3$, and this is the allocation used in Table 10.8.

In concluding this demonstration of the effects of stratified simple random sampling on the precision of an estimate of an unknown population mean, there are two important points that should be made. These are:

1. The gains are possible only because information is available concerning the population. For proportional allocation, we need to be able to define strata that will have different means, we need to know the exact fraction of the population that falls in each stratum—if errors exist in the fractions used, the estimate $\bar{x}_{st}$ will be biased—and we must be able to make an advance classification of elements into strata so that a random sample of known size can be selected from each. For optimum allocation, we need in addition to know the standard deviations of the separate strata. The saying that "you can't obtain something for nothing" most certainly holds here.

2. It is possible, through faulty allocation, to increase the variance of $\bar{x}_{st}$ over the variance of the mean of a random sample of the same size. For example, suppose that the sampling procedure with the strata of Table 10.6 was to take three elements at random from stratum I and a single element at random from stratum II. Although the expected value of $\bar{x}_{st}$ is still equal to 44.0, the variance of $\bar{x}_{st}$ now becomes 138 per cent of the variance of the mean of a random sample of the same size. That is, the mere fact that one is using stratified simple random sampling does not guarantee an increase in precision.

10.3.3. *Further Remarks on Stratification.* Thus far stratified random sampling has been viewed only as a device for increasing the precision of an estimate of a population mean. Surveys are frequently not so much concerned with obtaining an estimate of some population characteristic as they are in providing data for making comparisons among the separate strata. Stratification then serves as a technique for assuring that an adequate sample size will be obtained from each stratum. Thus simple-random-sampling theory might be used to determine the sample size within each stratum without reference to the precision of a population estimate. If a later combination of the strata results is required, formula (10.1) will still apply since this provides an unbiased estimate of the population mean no matter how the sample is allocated among strata.

It is also important to recognize that cost and administrative considerations may influence the definition of strata or the allocation of the sample among strata. For example, the cost of obtaining data may vary greatly from stratum to stratum (as from urban strata to rural strata). Reducing the size of the sample in a high-cost stratum may permit a large enough increase in the size of sample in low-cost strata to reduce the variance of the estimate for the population. Strata can also be used as administrative units. Large-scale surveys extending over a wide geographic area may require a number of field offices which can handle the hiring, training, and supervision of interviewers. Under such circumstances it is usually possible to define geographic strata so that a field office can be located in each stratum and to it can be assigned the responsibility for the field work conducted in that stratum. This type of design is particularly essential when a permanent field organization is required for repeated surveys.

Finally, we emphasize that it is not enough to state that a sample is "representative" and therefore "good" simply because of the controlled representation of the various strata. The probability model exhibited in Table 10.4 was obtained by requiring that a simple random sample be drawn from each stratum. No subjective factors were permitted to influence the selection of elements from within a stratum. Judgment of course enters into the definition of strata, but poor judgment in this respect merely means that the estimate is not as precise as it might otherwise have been. The use of judgment in this manner does not invalidate the unbiased nature of the estimates, nor does it prohibit us from estimating the precision of an estimate from an actual sample.

There do exist sampling procedures which have controlled representation of the various strata, but with the interviewer exercising his own judgment in the final selection of respondents. These are usually called *quota* sampling procedures since the interviewer is only told to obtain a quota of men, a quota of women, a quota of people over 40, and so on. If each interviewer were actually able to act as a random selecting device, then quota sampling would be very close to stratified simple random sampling with proportional allocation. Unfortunately, it is impossible to demonstrate that this is or is not the case. The situation can be summarized by the following statement, taken from the account of the investigation of the preelection polls of 1948 [9]:

> The available evidence is not adequate to measure the extent to which quota sampling contributed to the systematic errors of the 1948 election predictions. Examination of the sample educational distributions suggests that there was a considerable systematic error in most of the quota samples. The major portion of such contributions would probably be found in the selection of respondents by interviewers, and not in the determination of regional quotas or in the selection of sampling points (and the assignment of their quotas). The mere fact that these measurements cannot be made is in itself a serious weakness of the method. Users of quota samples should recognize their shortcomings, and the implicit difficulty of defending their use when the accuracy of forecasts is important.

10.4. Systematic Sampling

There are many instances in which a sample is to be drawn from a population in which the elements have a certain natural ordering. Thus names are arranged alphabetically on a list and houses are ordered with respect to position on a block. When such an ordering exists there is an immediately obvious method of selection which has not yet been treated. This consists of counting off along the ordered elements and taking every kth element (where k represents any positive integer) into the sample. This procedure is called *systematic sampling.* Ordinarily, *the number of the element at which the counting is to start is obtained by choosing a random number between* 1 *and* k.

There are several rather intuitive advantages associated with the use of systematic selection. In particular, the mechanical details of selection are simpler than for random sampling, where tables of random numbers must be used, and the sample elements appear to be evenly spread over the entire population. The appeal of an "even spread" is especially strong when the ordering is on a geographic basis. Unfortunately, intuition is not necessarily a good guide in deciding upon a sampling procedure, and so we need to examine a bit more closely the characteristics of systematic sampling.

Provided that the sample size is an exact divisor of the population size, it is very easy to exhibit the probability model for systematic sampling. For example, suppose there are 12 dwelling units on a block which appear in the order

$$ABCDEFGHIJKL$$

If a systematic sample of size 3 is desired, then the sampling interval (the k of our general definition) should be 4 (population size/sample size). Since one of the first four elements is chosen at random and every fourth element thereafter is drawn into the sample, the following four samples are assigned equal probabilities:

$$AEI,\ BFJ,\ CGK,\ DHL$$

All other possible samples of size 3—and there are 217 of them—are assigned a probability of zero.

It can be shown from this probability model that the arithmetic mean of a systematic sample is an unbiased estimate of the population mean. However, it is extremely difficult to draw any general conclusions concerning the relative precision of the mean of a systematic sample as compared to the mean of a simple random sample of the same size. This difficulty arises from the fact that systematic sampling may be like simple random sampling, or like stratified simple random sampling with proportional allocation, or like cluster sampling, depending upon the manner in which the list is prepared. A brief discussion of each of these possibilities follows:

1. If the list from which the sample is drawn represents a random ordering of the population elements, systematic and simple random sampling are identical. It is frequently assumed that an alphabetic listing of names will give such a random ordering, although even here one must take care since members of the same family will be listed together.

2. If the list is prepared by groups—for example, the addresses on each block listed together or the workers in a plant listed by department—and the elements within a group are randomly ordered, then systematic selection will be equivalent to stratified simple random sampling with proportional allocation. This, of course, assumes that the sampling interval is small enough so that every group is represented in the sample.

3. If the list is set up so that there is some type of relationship between elements k units apart, then it is most helpful to look upon systematic sampling as a form of cluster sampling. In this case, the mean of a systematic sample is apt to be a much less precise estimate of the population mean than is the mean of a simple random sample of the same size. For example, if one went down a street taking every tenth house and there happened to be 10 houses on a block, then the sample would contain either all corner houses or no corner houses. This might lead to large variability between sample means for certain types of investigations.

The principal shortcoming of systematic sampling is that a single sample provides no evidence as to which of these three views is appropriate, and one may be dealing with the last situation without even recognizing it. Hence systematic sampling should be used with care.

10.5. Complex Sample Designs

Four fundamental probability-model sampling procedures have now been described. These are simple random sampling, cluster sampling, stratified simple random sampling, and systematic sampling. These procedures, together with a wide variety of modifications, provide the basic tools of the sample designer. However, the goals set forth in Sec. 10.1 can rarely be achieved through the use of any single one of them. It is usually necessary to fit them together to provide a well-integrated sample design, and this can be done in such a way that the final product is still a probability-model procedure.

A simple illustration of the foregoing might arise if we wished to draw a sample of adults in a large city. A possible procedure might be:

1. Clusters of adults, probably those living on the same city block, would be defined.

2. The clusters would be grouped into strata on the basis of characteristics like geographic location, average rent paid, and size.

3. A simple random sample of clusters would be selected within each stratum.

4. A systematic sample of households would be chosen from each sample cluster.

5. A single adult would be chosen at random from within each sample houschold.

Notice that this sample design proceeds in stages: first, a sample of blocks; then, a sample of households from the chosen blocks; and finally, a sample of one adult from each selected household.

Complex sample designs are usually *multistage*, and each stage may offer an opportunity to use stratification in conjunction with some form of probability-model procedure for sampling within strata. Hence the number of conceivable probability-model procedures that could be used in most situations is enormous. No simple classification is possible, and it is therefore the responsibility of the individual or organization reporting the results of a survey to describe the sampling procedure in precise detail. No person using the data should be satisfied with less. An excellent account of a highly complex sample design, that employed by the Current Population Survey of the U.S. Bureau of the Census, is given by Hansen, Hurwitz, and Madow [2, vol. I, chap. 12B]. The Current Population Survey is a sample survey, carried out on a monthly basis, that provides a large volume of current data concerning the characteristics of the entire population of the United States.

10.6. The Execution of a Sampling Plan

No matter how well a sample is designed, the final results are conditioned by the quality of its execution. If deviations from the original design occur in the field work, this will mean that deductions made from the probability model will be approximations to what actually holds in practice. The accuracy of these approximations can be judged, if at all, only by careful examination of all the major phases of the survey operation. We shall now examine briefly a few of the deviations from design that may occur in the sampling of human populations.

The ordinary use of cluster sampling calls for the definition on a map of small areal units and the selection of a sample of these units. Field workers are then sent to the chosen clusters, either to enumerate the dwelling units in each cluster or to enumerate and select a sample of respondents. Errors can creep into this operation in a number of ways. The boundaries of the sample cluster can be misinterpreted; dwelling units can be omitted from the enumeration; or errors can be made in carrying out the sampling instructions. It is difficult to predict what the effects of such errors will be, but it is easy to see that they may have some effect. For example, omitted dwelling units will tend to be those lying in unnoticed alleys, those located in the rear of commercial or industrial establishments, and the like, and the occupants of such dwelling units will have different characteristics from those people who live in more accessible locations.

While the household or dwelling is the ultimate unit in some surveys such as those of housing, labor force, or consumer behavior, it is not suitable for studies of attitudes and similar individual variables, since an interview with one member of the household may influence the responses of other members that might be interviewed in turn. Hence the final step in sampling may be the selection of one individual at random in each household. Following the determination of the exact individuals who are to be interviewed, it is next necessary to interview them. This, as past experience well demonstrates, is not a trivial problem. Urban surveys ordinarily find only 30 to 40 per cent of a sample of *individuals* at home on the first call. Even if the interviewer is permitted to obtain the information from any responsible person who is at home in a dwelling unit in his sample, only 60 to 70 per cent of the assigned interviews are completed on the first call. Since those individuals who are relatively difficult to contact differ in many respects from those who are relatively easy to contact (more frequently male, younger, better educated, have fewer children), one must balance the risk of having serious biases enter the sample results if one substitutes an interview with some other person who is at home against the greater cost of obtaining an interview by making callbacks. Some illustrative data on this problem are furnished by Hilgard and Payne [10].

Even after individuals have been contacted for interview, sample losses may still be expected, primarily because of refusals. The loss of people who refuse to be interviewed can also have a marked effect on sample results since refusals are more frequent in some groups, such as older females and the foreign-born, than others.

10.7. Survey Design

The ultimate results obtained by a sample survey are affected not only by the sample design and its execution but also by such other factors as interviewing techniques, response errors, techniques of analysis, and the like. These elements were emphasized and discussed in Chap. 2. Combining all these elements, we may then say that the *survey design* is a complete plan of action for obtaining information about a population. It specifies the means by which the information is to be collected and processed, including the sampling procedure and the measurement devices that are to be used. The details of the survey design are to be arrived at primarily through a joint consideration of the use which is to be made of the data, the resources which are available, and the adequacy of the sampling and measurement procedures. Thus we see that the sampling procedure is only one portion of the total survey design and must therefore be looked upon in its relationship to the whole rather than in isolation.

The relationships between the various portions of a survey design, and the decisions which this forces one to make, are well illustrated by an excellent account given by Marks, Mauldin, and Nisselson [11] of the design of the Post-enumeration Survey of the U.S. Bureau of the Census. The purpose of the Post-enumeration Survey was to measure error in the 1950 censuses. The following quotation sets the problem of survey design:

> Although the Post-enumeration Survey possessed certain distinctive features peculiar to its purposes and content, its history illustrates the complex decision process involved in survey planning. This decision process requires the balancing of cost and accuracy considerations both within a particular aspect of a survey and among different aspects. The over-all cost of a survey and the accuracy of the estimates the survey provides are resultants of the costs and accuracies associated with a variety of processes—sample selection, interviewing, coding, tabulating, etc. Since most surveys have a fixed over-all expenditure limit, increased expense to improve the accuracy of one phase will usually require curtailed expenditure and lowered accuracy in some other phase—for example, may require a decrease in sample size with consequent increase in variance. Between survey processes a balance is necessary; to achieve a reduction of 5 per cent in coding error at the expense of decreased checking of the interviewing with an attendant increase of 10 per cent in interviewing error is hardly desirable, particularly if the interviewing error is already larger than the coding error. Within a given survey process—for example, interviewing—an increased expenditure for training must be balanced against an increase in expenditures for supervision, interviewer salaries, more detailed questions, etc., in terms of the relative effect of the various expenditures on the quality of the interview output.

EXERCISES

10.1. Using the strata and values of the variable given in Table 10.6, determine the mean and variance of $\bar{x}_{st}$ if three elements are drawn at random from stratum I and a single element is drawn at random from stratum II.

10.2. Cluster sampling and stratified sampling are often combined in the same sample design. As an illustration, consider a finite population of 12 elements which are grouped in clusters in the following manner: (ABC), (DEF), (GHI), and (JKL). Suppose these clusters are placed in two strata:

Stratum I: (ABC), (DEF)
Stratum II: (GHI), (JKL)

A possible sampling procedure might then be to draw a single cluster at random from each stratum and to then select two elements at random from the chosen clusters. Demonstrate that this is a probability-model sampling procedure by using the enumeration techniques of Secs. 10.2.1 and 10.3.1. What would be the sample quantity that one would use with this sampling procedure to estimate the population mean?

10.3. A recent study of advertising media (see Table 3.5 for reference) wished to interview a "representative" sample of the United States noninstitutional population 10 years of age and over. Respondents were to be asked whether they had seen a recent issue of certain magazines, had listened to a recent broadcast of selected radio programs, or had seen a recent showing of a number of television programs. The first stage in the sample design specified that a sample of metropolitan areas and counties was to be chosen. The nonmetropolitan counties, which can be viewed as clusters of individuals, were stratified by geography, population density, per cent of persons engaged in agriculture, per cent of homes with refrigeration, and per cent of women in the labor force. (The metropolitan areas were also stratified, but with some different criteria.) Why do you think these quantities were employed as stratification criteria?

10.4. Under a quota-sampling procedure the interviewers make the final selection of respondents. Even though these selections are subject to controls on the sex distribution, age distribution, and the like, there is no guarantee that the interviewers will act as random selecting devices within these strata. List some of the ways in which you think that interviewers might depart from random sampling.

REFERENCES CITED

1. Cochran, William G.: *Sampling Techniques*, John Wiley & Sons, Inc., New York, 1953.
2. Hansen, Morris H., William N. Hurwitz, and William G. Madow: *Sample Survey Methods and Theory*, vols. I and II, John Wiley & Sons, Inc., New York, 1953.
3. U.S. Department of Labor: *Mobility of Molders and Coremakers, 1940–1952*, Bulletin 1162, 1954.
4. Gray, P. G., and J. Corlett: "Sampling for the Social Survey," *Jour. Roy. Stat. Soc.*, A, **113**:150–206 (1950).
5. Shaul, J. R. H.: "Sampling Surveys in Central Africa," *Jour. Amer. Stat. Assoc.*, **47**:239–254 (1952).
6. King, A. J., and R. J. Jessen: "The Master Sample of Agriculture," *Jour. Amer. Stat. Assoc.*, **40**:38–56 (1945).
7. Kish, Leslie: "A Two-stage Sample of a City," *Amer. Sociol. Rev.*, **17**:761–769 (1952).
8. U.S. Department of Labor: *Techniques of Preparing Major BLS Statistical Series*, Bulletin 1168, 1954.
9. Mosteller, Frederick, et al.: *The Pre-election Polls of 1948*, Bulletin 60, Social Science Research Council, New York, 1949.
10. Hilgard, Ernest H., and Stanley F. Payne: "Those Not at Home: Riddle for Pollsters," *Pub. Op. Quart.*, **8**:254–261 (1944).
11. Marks, Eli S., W. Parker Mauldin, and Harold Nisselson: "The Post-enumeration Survey of the 1950 Census: A Case History in Survey Design," *Jour. Amer. Stat. Assoc.*, **48**:220–243 (1953).

CHAPTER 11

χ^2 PROCEDURES FOR QUALITATIVE DATA

11.1. The Basic Goodness-of-fit Problem for Qualitative Variables

Techniques were developed in Chap. 8 for treating some problems of statistical inference that arise in studying dichotomous variables. Although most qualitative variables have more than two natural categories, they can always be reduced to a dichotomy if one is willing to focus attention on some single category of the variable and to group all the remaining categories in a second, or residual, category (see Sec. 8.1.1 for illustrations). Frequently this focus on a single category is not desired. The investigator wishes to study *simultaneously* more than two categories, and the approach of Chap. 8 is no longer directly applicable. A form of analysis will be presented in this chapter that is often useful for treating problems of this type.

The general model with which we shall be working is specified as follows. There exists a large or infinite population in which each element can be classified into one and only one category of a defined qualitative variable. Let k stand for the total number of categories of the variable that are to be distinguished in the analysis, and let these separate categories be designated by 1, 2, . . . , k. The population is completely described with respect to this variable by the fractions of the population falling in the various categories. Let p_1 be the fraction of the population falling in the first category, p_2 the fraction in the second category, . . . , and p_k the fraction in the kth and last category. Since the categories are mutually exclusive and exhaustive, $\sum_{i=1}^{k} p_i = 1$. Under most circumstances, the values of the p_i's are unknown, and information concerning them can be obtained only from a sample.

If a single element is drawn at random from the foregoing population, the probability is p_1 that the element falls in the first category of the variable, p_2 that it falls in the second category, . . . , and p_k that it falls in the last category. We shall consider samples of n elements, drawn so that:

1. The probability of obtaining an element in a particular category, say the ith, is equal to p_i for each of the n drawings.
2. The drawings are independent in the probability sense. That is, the appearance of a particular category of the variable on any one or more drawings has no influence on the categories that appear in any of the remaining drawings.

These conditions define the *multinomial probability model*, which is merely an extension of the binomial model as described in Sec. 8.1.2. The results to be presented here are derived from this multinomial model.

The procedures to be given in this chapter will often be applied to simple random samples drawn without replacement from finite populations. As in the case of the binomial model, some care must be exercised in making such applications. It was noted in Sec. 8.1.3 that simple random sampling—where the elements are drawn one by one without replacement from a finite population—does not satisfy the conditions required by the binomial probability model. The probability of obtaining an S on a particular draw is not a constant but depends upon the number of S's and F's that were obtained in previous draws. Similarly, in cases where the multinomial model would be used, the probability that a particular element of the sample falls in a stated category of the variable is not a constant but depends upon what has happened in the previous draws. However, it was observed in the discussion accompanying Table 8.8 that the binomial model will give a "very close" approximation to what is actually true if the sample size is small relative to the population size. This same restriction on the relative size of the sample is required when the multinomial model is used to describe results obtained from simple random samples drawn without replacement from a finite population. If the samples are drawn with replacement, the multinomial model holds exactly.

The basic goodness-of-fit problem for qualitative variables can now be phrased in the following terms. Consider a population and a qualitative variable with k categories. There exists, on the basis of an a priori analysis of the situation, some hypothesis which specifies the exact fraction of the entire population in each of the categories; that is, the null hypothesis specifies the exact values of $p_1, p_2, \ldots, p_k$. A sample of size n is drawn from the population in such a manner that the multinomial model applies, at least to a close approximation. The number of sample elements falling in each category of the variable is observed. Let u_1 be the number falling in the first category, u_2 the number in the second, . . . , and u_k the number in the last. The sum of the observed frequencies $\sum_{i=1}^{k} u_i$ is equal to n since each element in the sample falls in one and only one category of the variable. Recognizing that the values of the u_i's will vary from sample to sample, we wish an objective procedure for deciding whether the observed distribution of the sample elements among the categories is such as might be expected if the hypothesized values of the p_i's are actually the correct ones.

One kind of situation in which the foregoing type of analysis might be applied is illustrated by the data of Table 11.1. These data were obtained from an examination of 4 years of work experience of each of 289 individuals who were employed in 1947 by the Park Gate Iron and

TABLE 11.1. NUMBER OF INDUSTRIAL ACCIDENTS OCCURRING IN THE DIFFERENT HOURS OF AN 8-HR SHIFT

(Data obtained from an examination of 4 years of work experience of each of 289 individuals who were employed in 1947 by the Park Gate Iron and Steel Company, England, and who "stayed" with the company for the following 4 years)

Hour of the shift	Number of accidents u_i	Fraction of accidents u_i/n	Hypothesized fraction of accidents p_i
1	20	.124	.125
2	24	.149	.125
3	14	.087	.125
4	18	.112	.125
5	22	.137	.125
6	21	.130	.125
7	22	.137	.125
8	20	.124	.125
Total.....	161	1.000	1.000

SOURCE: J. M. M. Hill and E. L. Trist, "Changes in Accidents and other Absences with Length of Service," *Human Relations*, **8**:121–152 (1955).

Steel Company, England, and who "stayed" with the company for the following 4 years. For present purposes, data have been abstracted from the cited source which show the hour (of an 8-hour shift) at which each of a number of industrial accidents occurred. That is, the occurrence of an industrial accident is regarded as a single trial of a conceptually repeatable experiment. Each accident is classified according to the hour of the shift at which it occurred. Then p_1 is the probability that an accident occurs in the first hour, p_2 the probability for the second hour, . . . , and p_8 the probability for the eighth and last hour. The true values of these p_i's are unknown. One might possibly expect that the p_i's for the later hours would be higher than for the earlier hours—because of increasing fatigue—but there is no way to use this hunch in arriving at exact values of the p_i's. We therefore state as an exploratory hypothesis: the probability of an accident occurring in a particular hour of the shift is the same for all hours of the shift. That is, we hypothesize that $p_1 = p_2 = \cdots = p_8 = \frac{1}{8}$. Table 11.1 shows that there are differences between these hypothesized probabilities and the fractions of accidents that were actually observed. This, in and of itself, is no reason for rejecting the hypothesis. The hypothesis can be correct, and deviations will still arise because of sampling fluctuations. The problem is to devise a test that will indicate whether the deviations are large enough to lead one to suspect the correctness of the hypothesis.

There is another way of viewing this basic goodness-of-fit problem. Suppose someone draws what is purportedly a simple random sample from a population in which the fraction of elements falling in the various categories of some qualitative variable are known. Thus the United States census gives the fraction of the entire adult population of the United States which has achieved stated levels of educational attainment. Provided that the sample size is small relative to the population size, a comparison of the observed frequencies in the categories with the results predicted by the multinomial model then provides a test of whether this model is appropriate or not. This in turn provides a test of whether the sampling procedure actually conformed to the model for simple random sampling.

11.2. A Criterion for Testing Goodness of Fit

A test of whether the hypothesized values of $p_1, p_2, \ldots, p_k$ are correct must depend in some way upon a comparison of these values with their corresponding observed frequencies or fractions. It seems reasonable to ask for a criterion or measure of agreement which has the following properties:

1. Since no single category of the variable is to receive special attention, the criterion should be in some sense a measure of total agreement over all categories.

2. It should be possible to separate the possible values of the criterion into two classes, such that values in one class indicate "good" agreement while values in the second class indicate "poor" agreement, and to define objectively what is meant by "good" and "poor."

Experience has shown that a criterion with these properties can be obtained by comparing the observed frequency in each category of the variable with the *expected*, or *average*, number of sample elements that would fall into the category if the hypothesis were true. It can be proved that if n drawings are made from the population, the probability being p_1 that any drawing gives a result that is classified in the first category of the variable, then the *expected number* of observations that will fall in the first category is np_1. This is the result that would be obtained if an indefinitely large number of sets of n drawings were made and the average number of elements falling in the first category were computed. Similar statements hold for each of the other categories of the variable. Thus the criterion will depend upon a comparison of the observed frequencies $u_1, u_2, u_3, \ldots, u_k$ and the corresponding expected frequencies $np_1, np_2, \ldots, np_k$.

A criterion which is frequently employed in making this over-all comparison of observed and expected frequencies is:

$$X^2 = \frac{(u_1 - np_1)^2}{np_1} + \frac{(u_2 - np_2)^2}{np_2} + \cdots + \frac{(u_k - np_k)^2}{np_k}$$

$$= \sum_{i=1}^{k} \frac{(u_i - np_i)^2}{np_i} \tag{11.1}$$

The calculation of this criterion can be iilustrated by considering its application to the data of Table 11.1. The null hypothesis was that these data constitute a sample from a population in which

$$p_1 = p_2 = \cdots = p_8 = 1/8$$

Since $n = 161$, the expected frequency for each hour of the shift is $(161)(1/8) = 20.125$. Therefore the value of X^2 is found to be

$$X^2 = \frac{(20 - 20.125)^2}{20.125} + \frac{(24 - 20.125)^2}{20.125} + \cdots + \frac{(22 - 20.125)^2}{20.125} + \frac{(20 - 20.125)^2}{20.125}$$

$$= .001 + .845 + 1.715 + .175 + .224 + .063 + .224 + .001$$

$$= 3.248$$

We shall learn shortly how to interpret this and other specific values of X^2.

It is clear that very small values of X^2 correspond to very close agreement between the observed and expected frequencies and that larger and larger values indicate poorer and poorer agreement. But this is not enough. An objective procedure is needed for distinguishing between "large" and "small" values. Such a procedure can be obtained by considering the behavior of X^2 when repeated samples are drawn from a population for which the hypothesized values of the p_i's are actually the correct ones. In short, we desire the sampling distribution of X^2 under the assumption that the null hypothesis is true.

The exact sampling distribution of X^2 can be obtained without too much difficulty if the sample size and the number of categories of the variable are both small. For example, consider the situation where $k = 3$, $n = 6$, and $p_1 = p_2 = p_3 = 1/3$. The expected frequency for each category is 2. It is now possible to enumerate all the ways in which a sample of six elements can be distributed among the three categories of the variable. There are 28 such possibilities, and these are exhibited in Table 11.2. The value of X^2 is given for each triplet of observed frequencies (u_1,u_2,u_3). Thus if $(u_1,u_2,u_3) = (0,1,5)$,

$$X^2 = \frac{(0-2)^2}{2} + \frac{(1-2)^2}{2} + \frac{(5-2)^2}{2} = 7$$

Also presented are the probabilities of obtaining any specified triplet of observed frequencies. That is, $\Pr(u_1,u_2,u_3)$* is the fraction of times that

TABLE 11.2. THE POSSIBLE WAYS THAT A SAMPLE OF SIZE 6 CAN BE DISTRIBUTED AMONG THREE CATEGORIES OF A VARIABLE
(The values of X^2 and of $\Pr(u_1,u_2,u_3)$ were computed under the assumption that $p_1 = p_2 = p_3 = 1/3$.)

u_1	u_2	u_3	X^2	$\Pr(u_1,u_2,u_3)$	u_1	u_2	u_3	X^2	$\Pr(u_1,u_2,u_3)$
2	2	2	0	.1235	0	2	4	4	.0206
1	2	3	1	.0823	2	0	4	4	.0206
2	1	3	1	.0823	2	4	0	4	.0206
2	3	1	1	.0823	0	4	2	4	.0206
1	3	2	1	.0823	4	0	2	4	.0206
3	1	2	1	.0823	4	2	0	4	.0206
3	2	1	1	.0823	0	1	5	7	.0082
1	1	4	3	.0412	1	0	5	7	.0082
1	4	1	3	.0412	1	5	0	7	.0082
1	1	4	3	.0412	0	5	1	7	.0082
0	3	3	3	.0274	5	0	1	7	.0082
3	0	3	3	.0274	5	1	0	7	.0082
3	3	0	3	.0274	0	0	6	12	.0014
					0	6	0	12	.0014
					6	0	0	12	.0014
									1.0001

the triplet (u_1,u_2,u_3) would appear if an indefinitely large number of samples of size 6 were drawn from the defined population.

Now the exact values of u_1, u_2, and u_3 are of no particular interest in the present situation, except as they determine a probability and a value of X^2. We observe that there are six different values of X^2 given in

* The multinomial probability model leads to formal expressions for these probabilities. Thus it can be shown that for samples of size n,

$$\Pr(u_1, u_2, \ldots, u_k) = \frac{n!}{u_1!u_2! \cdots u_k!} p_1^{u_1}p_2^{u_2} \cdots p_k^{u_k} \qquad (11.2)$$

Notice that this reduces to the binomial probability given by (8.1) if $k = 2$.

Table 11.2 and that the probability of any one of these can be computed by using the law of addition for mutually exclusive events, (7.4). Thus

$$\begin{aligned} \Pr(X^2 = 0) &= \Pr(2,2,2) = .1235 \\ \Pr(X^2 = 12) &= \Pr(0,0,6) + \Pr(0,6,0) + \Pr(6,0,0) \\ &= .0014 + .0014 + .0014 \\ &= .0042 \end{aligned}$$

The entire sampling distribution of X^2 can be obtained in this manner and is presented in Table 11.3.

TABLE 11.3. THE SAMPLING DISTRIBUTION OF X^2
($n = 6$, $k = 3$, $p_1 = p_2 = p_3 = \frac{1}{3}$)

X^2		$\Pr(X^2)$
0		.123
1		.494
3		.206
4		.123
7		.049
12		.004
	Total	.999

This sampling distribution can be used to define test procedures for judging whether a sample of size 6, drawn in accordance with the multinomial model, comes from a population in which $p_1 = p_2 = p_3 = \frac{1}{3}$. For example, we might do the following:

1. Draw a sample and observe the triplet (u_1,u_2,u_3).
2. Compute X^2 in accordance with (11.1).
3. If X^2 is 7 or greater, *reject* the hypothesis that $p_1 = p_2 = p_3 = \frac{1}{3}$.
4. If X^2 is less than 7, *accept* the hypothesis that $p_1 = p_2 = p_3 = \frac{1}{3}$.

The probability that the hypothesis will be rejected when it is actually true is equal to

$$\Pr(X^2 = 7 \mid p_1 = p_2 = p_3 = \tfrac{1}{3}) + \Pr(X^2 = 12 \mid p_1 = p_2 = p_3 = \tfrac{1}{3}) = .053$$

Thus a test procedure has been defined that has a level of significance of 5.3 per cent. If the hypothesis were rejected when X^2 is 4 or greater, then the level of significance would be 17.6 per cent. Similarly, if it were rejected only when $X^2 = 12$, the level of significance would be 0.4 per cent. The difficult task of distinguishing between "large" and "small" values of X^2 has been translated into the problem of setting the level of significance of the test procedure. This conforms to the general line of approach developed in Chaps. 8 and 9.

Although the foregoing simple example outlines the reasoning and the details underlying a goodness-of-fit test procedure, it does not provide a practical means for solving most problems. The computations required to obtain the exact sampling distribution of X^2 become overwhelming if n is at all large. Fortunately, it seems reasonable to expect that a useful approximation to this sampling distribution can be found if it is recognized that approximate critical values of X^2 for the case $k = 2$ can be obtained from results given in Chap. 9.

Consider a situation where each element in the population is either an S or an F, and let p be the fraction of S's for the entire population. Suppose we wish to test, on the basis of a random sample drawn from the population, the null hypothesis that $p = p_0$ against the two-sided alternative that $p \neq p_0$. It was shown in Sec. 9.4.2 that if n is large enough to permit one to approximate the sampling distribution of u/n with a normal distribution, then a test procedure having a 5 per cent level of significance is specified as follows:

1. If the observed u/n is between $p_0 - 1.96\sqrt{p_0q_0/n}$ and $p_0 + 1.96\sqrt{p_0q_0/n}$, accept the hypothesis.
2. If u/n is less than $p_0 - 1.96\sqrt{p_0q_0/n}$ or greater than $p_0 + 1.96\sqrt{p_0q_0/n}$, reject the hypothesis.

The acceptance and rejection regions of this test can be written in other forms. Thus saying that u/n is less than $p_0 - 1.96\sqrt{p_0q_0/n}$ or greater than $p_0 + 1.96\sqrt{p_0q_0/n}$ is exactly the same as saying that $[(u/n) - p_0]$ is less than $-1.96\sqrt{p_0q_0/n}$ or greater than $+1.96\sqrt{p_0q_0/n}$. This in turn is the same as saying that

$$\frac{[(u/n) - p_0]}{\sqrt{p_0q_0/n}}$$

is less than -1.96 or greater than $+1.96$. Finally, we observe that by squaring all quantities, we eliminate minus signs and obtain the following alternative statement of the test procedure:

1. Accept the hypothesis if

$$\frac{[(u/n) - p_0]^2}{p_0q_0/n} < (1.96)^2 = 3.84$$

2. Reject the hypothesis if

$$\frac{[(u/n) - p_0]^2}{p_0q_0/n} > (1.96)^2 = 3.84$$

But it can now be demonstrated by simple algebraic manipulation that

$$\frac{[(u/n) - p_0]^2}{p_0q_0/n} = \frac{[u - np_0]^2}{np_0} + \frac{[(n - u) - nq_0]^2}{nq_0}$$

and the right hand side of this equality is the criterion X^2 for the case of a variable with two categories. The observed frequency in the S category is u, and np_0 is the expected frequency if the hypothesis is true; the observed frequency in the second category is $(n - u)$, and nq_0 is the corresponding expected frequency.

Other critical values of X^2 for the case $k = 2$ can be obtained by squaring the appropriate multipliers in Table 9.3. Thus rejecting the hypothesis when $X^2 > (1.282)^2 = 1.64$ gives a two-sided test procedure with a level of significance of 20 per cent; rejecting when $X^2 > (1.645)^2 = 2.71$ gives a level of significance of 10 per cent; and rejecting when

$$X^2 > (2.576)^2 = 6.64$$

gives a level of significance of 1 per cent. Thus the problem of obtaining critical values of X^2 for the case $k = 2$ is completely solved if n is large enough to permit one to approximate the sampling distribution of u/n with a normal distribution. We shall now present approximate solutions for cases where k is greater than 2.

11.3. The χ^2 Approximation to the Distribution of the Test Criterion

Table 11.3 gives the exact sampling distribution of X^2 for particular values of n, k, and the hypothesized p_i's. This distribution has been used in determining tests of goodness of fit with known levels of significance. A way will now be described of obtaining approximate critical values of X^2 for situations where it is not possible to compute the exact sampling distribution.

The contents of Fig. 11.1 illustrate the form of solution that will be considered. The exact sampling distribution of Table 11.3 is represented by the histogram in this figure, each cell of the histogram corresponding to one of the six possible values of X^2. Although the area of the cell corresponding to a particular value of X^2 is proportional to the probability associated with that value, notice that it has not been possible to center all these cells on the values of X^2. The dividing points have arbitrarily been taken halfway between successive values of X^2. Thus the cell associated with $X^2 = 4$ runs from 3.5 (halfway to the next lower value of X^2, 3) to 5.5 (halfway to the next higher value of X^2, 7). The first cell is centered at $X^2 = 0$ and the last cell at $X^2 = 12$. The level of significance for a test procedure with the critical value of X^2 set at 7, for example, can be obtained from this figure by determining the relative area of the histogram which falls to the right of 5.5 (halfway to the next lower value of X^2, 4).

Thus far nothing has been added which makes it any easier to obtain critical values of X^2. But now consider the possibility of approximating the histogram by a smooth curve such as is drawn in Fig. 11.1, where the total area under the curve is set equal to 1 and where the fraction of the area lying to the right of any given point is known. Actually, the smooth curve of Fig. 11.1 is not just a curve drawn in "by eye" but is a special theoretical probability distribution known as the χ^2 (read chi square) *distribution with 2 degrees of freedom.* The total area under this smooth curve is equal to 1, and tables are available which show the fraction of the total area falling to the right of any given point [1, table 7]. A comparison of the exact and approximate probabilities of obtaining a value

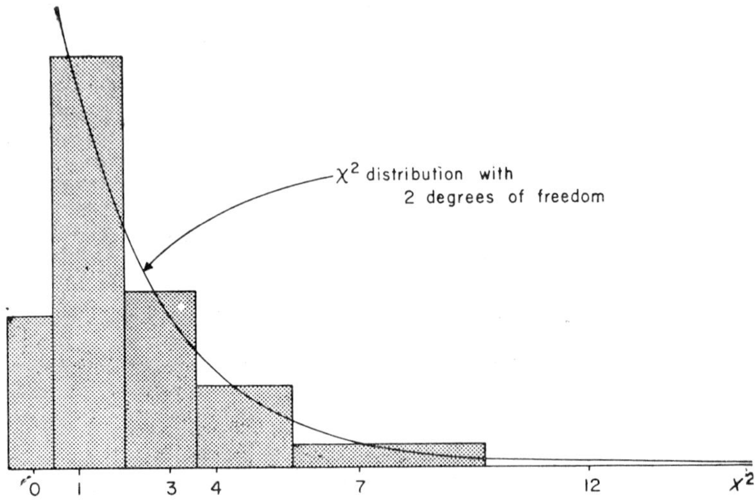

FIG. 11.1. Graphical representation of the exact sampling distribution of X^2 and a smooth-curve approximation. ($n = 6$, $k = 3$, $p_1 = p_2 = p_3 = \frac{1}{3}$.)

of X^2 equal to or greater than some stated value is given in Table 11.4. The agreement is quite remarkable considering that the total sample size is only 6 for the present example.

It might be expected that the approximation provided by the χ^2 distribution would become better and better as the size of sample increases. The principal effect of increasing the sample size will be to increase the number of possible values of X^2, thus making the cells of the histogram narrower and narrower and decreasing the amount of "jump" that occurs between adjacent cells. It can be proved that this is actually the case, and the following is a formal statement of the result:

> Suppose that repeated samples of size n are drawn in accordance with the multinomial probability model from a population in which p_1, p_2, and p_3 are the true fractions of elements falling in the three mutually exclusive and exhaustive categories of a qualitative variable. If n is sufficiently large, the sampling distribution of X^2, as defined by (11.1), can be closely approxi-

mated by a χ^2 distribution with 2 degrees of freedom. The quality of the approximation improves as n increases.

Notice that the result imposes no restrictions on the values of p_1, p_2, and p_3—although they must of course add to 1—and that with n sufficiently large, it is not necessary to make the "halfway correction" that was used in Fig. 11.1 and Table 11.4. Critical values of χ^2 for use in determining goodness-of-fit test procedures when n is large will be found in Table 11.5.

TABLE 11.4. A COMPARISON OF THE EXACT SAMPLING DISTRIBUTION OF X^2 WITH THE APPROXIMATION PROVIDED BY THE χ^2 DISTRIBUTION WITH 2 DEGREES OF FREEDOM
($n = 6$, $k = 3$, $p_1 = p_2 = p_3 = 1/3$)

Value of X^2	Exact probability that X^2 will be equal to or greater than given value	Probability from χ^2 approximation (distribution read halfway to next lower value of X^2)
0	1.000	1.000
1	.877	.778
3	.382	.368
4	.176	.174
7	.053	.064
12	.004	.009

The discussion up to this point has been focused entirely on qualitative variables with three categories. Since X^2 is the sum of k positive terms when the variable has k categories, it is to be expected that a large number of different approximating distributions will be required, one for each particular value of k. A solution to this problem is provided by the fact that a χ^2 distribution with 2 degrees of freedom, hereafter abbreviated as χ^2(2 df), is only one out of a family of χ^2 distributions. Thus there are χ^2 distributions with 1 df, 2 df, 3 df, ... Three of these distributions,* χ^2(2 df), χ^2(4 df), and χ^2(6 df), are presented for illustrative purposes in Fig. 11.2. It can be proved that:

Let repeated samples of size n be drawn in accordance with the multinomial probability model from a population in which

* The equation of a χ^2 distribution with ν (read nu) degrees of freedom is given by

$$f(\chi^2) = \frac{1}{2^{\nu/2}\left(\frac{\nu}{2} - 1\right)!} (\chi^2)^{(\nu/2)-1} e^{-(1/2)\chi^2}$$

where χ^2 = particular value of variable
ν = number of degrees of freedom
e = constant approximately equal to 2.7183

$p_1, p_2, \ldots, p_k$ are the true fractions of elements falling in the k categories of a qualitative variable. If n is sufficiently large, the sampling distribution of X^2, as defined by (11.1), can be closely *approximated* by a χ^2 *distribution* with $(k - 1)$ *degrees of freedom*. The quality of the approximation improves as n increases.

χ^2 distributions are also useful in solving other problems* that arise in statistical inference.

In order to make use of the foregoing results, a table of critical values of χ^2 is needed for various numbers of degrees of freedom. Such a table

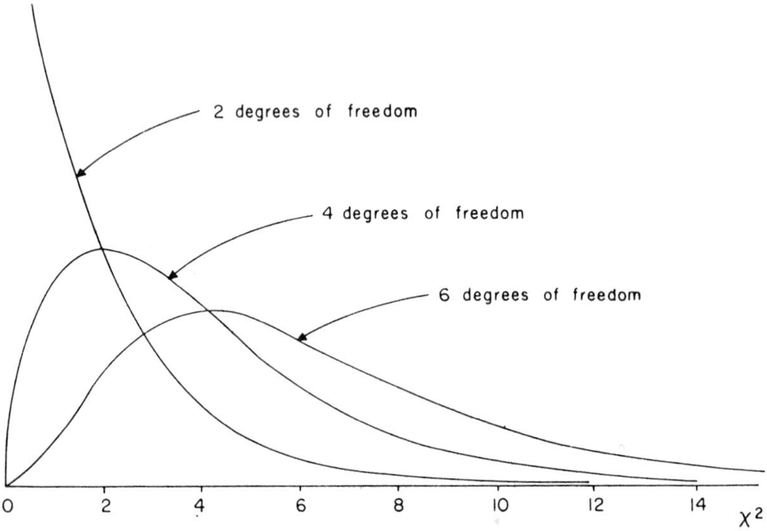

FIG. 11.2. χ^2 distributions for 2, 4, and 6 degrees of freedom.

is given in [1, table 8], and a portion of this table is reproduced in Table 11.5. These values can now be used to judge the significance of an observed value of X^2, provided that n is sufficiently large to permit the use of the χ^2 approximation. Consider, for example, the data of Table 11.1. Since the variable has eight categories in this instance, the sampling distribution of X^2 should be approximated by a χ^2 distribution

* One of the most fundamental applications of the χ^2 distribution is the following. Consider a variable x which has a normal distribution with variance σ^2. If random samples of size n are drawn from this population and if s^2 is the sample variance, as defined by (5.3), then the quantity

$$\frac{(n-1)s^2}{\sigma^2}$$

has a χ^2 distribution with $(n - 1)$ degrees of freedom. This result leads, among other things, to confidence limits and tests of hypotheses for the population variance σ^2.

with $(8 - 1) = 7$ df. Suppose the hypothesis that

$$p_1 = p_2 = \cdot \cdot \cdot = p_8 = 1/8$$

is to be tested at the 10 per cent level of significance. From Table 11.5, $\chi^2_{.10}(7 \text{ df})$ is equal to 12.017. The hypothesis will be rejected if X^2 is

TABLE 11.5. SOME CRITICAL VALUES OF χ^2
(The probability is α of obtaining a value of χ^2 greater than the tabular value.)

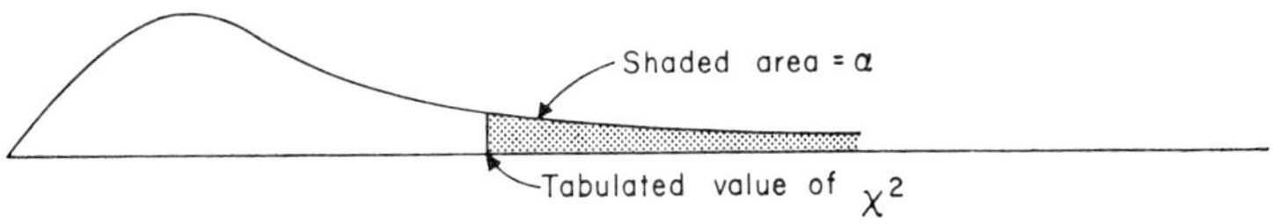

df	α				df	α			
	0.25	0.10	0.05	0.01		0.25	0.10	0.05	0.01
1	1.323	2.706	3.841	6.635	21	24.935	29.615	32.670	38.932
2	2.773	4.605	5.991	9.210	22	26.039	30.813	33.924	40.289
3	4.108	6.251	7.815	11.345	23	27.141	32.007	35.172	41.638
4	5.385	7.779	9.488	13.277	24	28.241	33.196	36.415	42.980
5	6.626	9.236	11.070	15.086	25	29.339	34.382	37.652	44.314
6	7.841	10.645	12.592	16.812	26	30.434	35.563	38.885	45.642
7	9.037	12.017	14.067	18.475	27	31.528	36.741	40.113	46.963
8	10.219	13.362	15.507	20.090	28	32.620	37.916	41.337	48.278
9	11.389	14.684	16.919	21.666	29	33.711	39.088	42.557	49.588
10	12.549	15.987	18.307	23.209	30	34.800	40.256	43.773	50.892
11	13.701	17.275	19.675	24.725	40	45.616	51.805	55.758	63.691
12	14.845	18.549	21.026	26.217	50	56.334	63.167	67.505	76.154
13	15.984	19.812	22.362	27.688	60	66.981	74.397	79.082	88.379
14	17.117	21.064	23.685	29.141	70	77.577	85.527	90.531	100.425
15	18.245	22.307	24.996	30.578	80	88.130	96.578	101.879	112.329
16	19.369	23.542	26.296	32.000	90	98.650	107.565	113.145	124.116
17	20.489	24.769	27.587	33.409	100	109.141	118.498	124.342	135.807
18	21.605	25.989	28.869	34.805					
19	22.718	27.204	30.144	36.191					
20	23.828	28.412	31.410	37.566					

SOURCE: Ref. 1, table 8. This abridged table is published here with the permission of the authors and their publishers.

greater than 12.017 and accepted if X^2 is less than 12.017. The computed value of X^2 is 3.248 (see Sec. 11.2), and the hypothesis is therefore accepted.

As a slightly different application of the χ^2 test of goodness of fit, consider the data of Table 11.6. These data were obtained in a national survey conducted in May, June, and July, 1954, under the auspices of the Fund for the Republic. The purpose of the survey was to examine "in some depth the reactions of Americans to two dangers. One, from the Communist conspiracy outside and inside the country. Two, from those who in thwarting the conspiracy would sacrifice some of the very liberties which the enemy would destroy." The observed frequencies in this table give the distribution of answers obtained from 109 presidents of local bar associations (in a sample of cities of 10,000 to 150,000 population) in response to the question quoted in the title of Table 11.6.

TABLE 11.6. DISTRIBUTION OF ANSWERS GIVEN BY A SAMPLE OF LOCAL BAR ASSOCIATION PRESIDENTS IN RESPONSE TO THE QUESTION "Sometimes a man says he refuses to answer questions about Communism because he does not want to be forced to testify against his former friends. Should he be punished very severely, severely, not too severely or not at all?" (Sample of cities of 10,000 to 150,000 population)

Response category	u_i	p_i^*	np_i	$(u_i - np_i)^2$	$(u_i - np_i)^2/np_i$
Very severely..........	5	.06	6.5	2.25	0.35
Severely..............	41	.24	26.2	219.04	8.36
Not too severely.......	31	.33	36.0	25.00	0.69
Not at all.............	21	.22	24.0	9.00	0.38
Don't know...........	11	.15	16.3	28.09	1.72
Total...............	109	1.00	109	......	$X^2 = 11.50$

* These are the fractions obtained from a national cross-section sample of 4,933 adults. Since the sample is large, these are here assumed to be the true fractions for the entire population.

SOURCE: Samuel A. Stouffer, *Communism, Conformity, and Civil Liberties*, p. 234, Copyright © 1955 by Samuel A. Stouffer, reprinted by permission of Doubleday & Company, Inc., New York.

Viewing these 109 presidents as a random sample from the population of all presidents of local bar associations for cities of 10,000 to 150,000 population, suppose one would like to test the hypothesis that the distribution of responses in this population is exactly the same as in the population of all adults in the United States. The cited source gives the distribution obtained from a national cross-section sample of 4,933 adults. Since this sample is large, the fractions obtained are here assumed to be the true fractions for the entire adult population. The variable has five categories, and the distribution of the test criterion is therefore approximated by the distribution of $\chi^2(4 \text{ df})$. Assuming a test at the 5 per cent level of significance is desired, $\chi^2_{.05}(4 \text{ df})$ is read from Table 11.5 as 9.488. The computation of X^2 is summarized in Table 11.6, and its value is found to be 11.50. Therefore the hypothesis is rejected at the 5 per cent level of significance.

A particular χ^2 distribution is identified by the number of degrees of freedom. The preceding discussion has arbitrarily stated that when the qualitative variable has k categories the sampling distribution of X^2 for a basic goodness-of-fit test procedure can be approximated by a χ^2 distribution with $(k - 1)$ degrees of freedom. A rough interpretation of the use of the phrase "degrees of freedom" is as follows: We have assumed that repeated samples of size n are being drawn from the population. It is therefore necessary that

$$u_1 + u_2 + \cdot \cdot \cdot + u_k = n$$

Because of this restriction on the sum of the observed frequencies, only $(k - 1)$ of them are free to vary independently. Once any $(k - 1)$ have been fixed, the value of the kth frequency is automatically determined. This means that X^2 is the sum of only $(k - 1)$ independent chance quantities. This fact was formally demonstrated at the end of Sec. 11.2 where it was shown that X^2, originally computed as the sum of two terms for a variable with $k = 2$, could be reduced to a single term.

11.4. Some Additional Comments on the Test Procedure

The χ^2 distribution provides a close approximation to the sampling distribution of X^2 if n is sufficiently large. The practical problem is, of course, to assign some quantitative meaning to the phrases "close approximation" and "sufficiently large." It is extremely difficult to make precise yet general statements concerning these matters, and we shall merely summarize briefly the general recommendations given by Cochran [2] in a recent review paper on the χ^2 test of goodness of fit. The criterion used by Cochran in examining available evidence with respect to the "close approximation" phrase was to compare the exact probability with the corresponding probability as read from a χ^2 table that a value of X^2 equal to or greater than some stated value would be obtained, this comparison being made at $\alpha = .05$ and $\alpha = .01$ in the χ^2 table. He states, "A disturbance is regarded as unimportant if when the P is 0.05 in the χ^2 table, the exact P lies between 0.04 and 0.06, and if when the tabular P is 0.01, the exact P lies between 0.007 and 0.015."

The recommendations given in [2] are based not upon n but upon the smallest expected frequency, that is, the smallest value of np_i. For the problem to which we have thus far applied the χ^2 test of goodness of fit, they read as follows:

1. *1 degree of freedom.* This applies only to the binomial case discussed at the end of Sec. 11.2. (Another type of problem where X^2 has 1 degree of freedom will be discussed in the next section.) If the smallest expected value is 5 or greater, then the χ^2 distribution can be used. If it is less than 5, then the techniques developed in Chap. 8 can be applied.

2. *2 or more degrees of freedom and all expected values less than 5.* A correction is usually required. One computes either the exact distribution of X^2, as was done in Table 11.2, or a corrected value of X^2, X_c^2. X_c^2 is the point halfway between the observed value of X^2 and the next *smallest* possible value of X^2. If X_c^2 is larger than the χ^2 tabular value, the hypothesis is rejected. Otherwise, it is accepted. This latter technique is known as the *correction for continuity*, and the underlying rationale is illustrated in Fig. 11.1 and Table 11.4.

3. *2 or more degrees of freedom and some expected values greater than 5.* Use the regular χ^2 tabular values without correcting X^2 for continuity.

For more detailed discussion of these points, the reader should consult [2].

The level of significance of the χ^2 test of goodness of fit has exactly the same interpretation as for all the previously discussed tests of statistical hypotheses. That is, it is the probability of rejecting, when it is actually true, the hypothesis that $p_1, p_2, \ldots, p_k$ have some stated set of values. Even though the point has not been stressed, it is also a fact that this hypothesis may be accepted when it is false. That is, one may commit a Type II error. However, the χ^2 test is usually applied where there is no clear-cut alternative hypothesis in mind on the basis of which to compute the probability of a Type II error. The test can reject for any pattern of deviations of the true fractions from the hypothesized values of $p_1, p_2, \ldots, p_k$ and not just for some specified pattern of deviations. It is for this reason that the test is often called a "general purpose" test. If one has some particular alternative or set of alternatives in mind, then it is frequently possible to devise a test that is in some sense "better" in detecting instances where the alternative is really true.

Also related to this consideration of alternative hypotheses is the matter of total sample size n. Provided that n is large enough so that the sampling distribution of X^2 can be closely approximated by a χ^2 distribution, the critical value of χ^2 required to define a test procedure with stated level of significance depends only on k, the number of categories, and not at all upon the value of n. This, however, does not mean that the test procedure behaves in the same manner for a large sample as for a small sample. It can be shown that for any fixed set of deviations between the true and hypothesized values of the p_i's and for any fixed level of significance, the probability of rejecting the hypothesis approaches 1 as the sample size increases indefinitely. Thus with small samples the test may not have a very good chance of detecting large errors in the hypothesized values of the p_i's, while for large samples it may almost certainly reject the hypothesis for very small and possibly uninteresting deviations between the true and hypothesized values of the p_i's. The size of sample should therefore play an important role in evaluating the result of a χ^2 test of goodness of fit.

The χ^2 test of goodness of fit has thus far been applied only to situations where the hypothesized values of the p_i's were completely specified in advance of examining the data that were to be used in testing the hypothesis. Sometimes the hypothesis only gives the functional form of the p_i's, and their actual values are determined by estimating certain parameters from the observed data. The χ^2 test of goodness of fit can still be used, but the appropriate number of degrees of freedom is no longer equal to $(k - 1)$ and must be adjusted to the particular problem. We shall discuss an example by way of illustration.

TABLE 11.7. TESTING THE HYPOTHESIS THAT THE DISTRIBUTION OF NUMBER OF LIMESTONE PEBBLES IN SAMPLES OF 10 PEBBLES IS A BINOMIAL DISTRIBUTION
(Lake Michigan beach gravel; 100 samples, each consisting of 10 pebbles)

Number of limestone pebbles per sample	Number of samples u_i	p_i^*	np_i	$(u_i - np_i)^2$	$(u_i - np_i)^2/np_i$
0	0	.001	0.1	0.01	0.10
1	1	.010	1.0	0.00	0.00
2	6	.044	4.4	2.56	0.58
3	7	.117	11.7	22.09	1.89
4	23	.205	20.5	6.25	0.30
5	26	.246	24.6	1.96	0.08
6	21	.205	20.5	0.25	0.01
7	12	.117	11.7	0.09	0.01
8	3	.044	4.4	1.96	0.45
9	1	.010	1.0	0.00	0.00
10	0	.001	0.1	0.01	0.10
Total............	100	1.000	100.0	.....	$X^2 = 3.52$

* These probabilities were obtained by using expression (8.1). The value of p is unknown but can be estimated from the data as (total number of limestone pebbles in all samples combined)/(total number of pebbles in all samples combined) = 499/1,000 = .499.

SOURCE: W. C. Krumbein, "Applications of Statistical Methods to Sedimentary Rocks," *Jour. Amer. Stat. Assoc.*, **49**:51–66, table 3 (1954).

The data of Table 11.7 refer to the variable "number of limestone pebbles in a sample of 10 pebbles." They were obtained by taking 100 samples, each sample consisting of 10 pebbles, from a small area of Lake Michigan beach. The question at issue is, Are the limestone pebbles randomly distributed among the other types of pebbles? The alternative might be that limestone pebbles tend to be clustered together more than would be expected on a random basis. If the limestone pebbles are randomly distributed among the others, then the probability of obtaining some exact number, say 6, of limestone pebbles in a sample of 10 pebbles is given by

$$\frac{10!}{6!4!}\,p^6q^4$$

where p is the fraction of limestone pebbles in a very large population of pebbles. That is, the distribution of number of limestone pebbles follows a binomial distribution [see expression (8.1)]. The value of p is unknown, but it can be estimated from the data as

$$\frac{\text{Total number of limestone pebbles in all samples combined}}{\text{Total number of pebbles in all samples combined}}$$

$$= \frac{499}{1{,}000} = .499$$

The p_i of Table 11.7 were computed by using expression (8.1) with the value of p set at .499.

X^2 can now be computed in the ordinary manner, and its value is found to be 3.52. However, we now need to know how many degrees of freedom to associate with the approximating χ^2 distribution. The variable has 11 categories in this instance, and one's first reaction might be to use the $(k - 1)$ rule, thus approximating with χ^2(10 df). But the hypothesized p_i's depend upon the observed data to the extent that the value of a single parameter p has been estimated from the data. This will tend to make the "fit" better than if the value of p had been specified by the hypothesis. It can be shown that one should use $\chi^2(k - 2)$ degrees of freedom as the approximating distribution. One degree of freedom is subtracted because of the fixed sample size (100), and a second degree of freedom is subtracted because a single parameter has been estimated from the data. Assuming a test at the 5 per cent level of significance, $\chi^2_{.05}$ (9 df) is read from Table 11.5 as 16.919. Since X^2 is considerably less than 16.919, the hypothesis of a binomial distribution, or of a random distribution of limestone pebbles, is accepted.

Although it will not be illustrated here, the foregoing principle for determining degrees of freedom can be applied generally. If one wishes to test the goodness of fit of some theoretical distribution, the number of degrees of freedom of X^2 will be equal to the number of categories of the variable less 1 degree of freedom for the fixed sample size and less 1 degree of freedom for each parameter of the distribution that must be estimated from the data before numerical values of the p_i's can be specified.

11.5. Testing the Hypothesis of Independence in a 2 × 2 Contingency Table

Numerous illustrations of the joint classification of an element of a population or sample with respect to two qualitative variables were presented in Chap. 7. For example, Exercise 7.2 shows how each of a national sample of 390 individuals who had voted in the 1948 Presidential election was jointly classified with respect to the variable "time at which a decision was made on a candidate" and "candidate for whom the individual voted." For this particular illustration, and for others that could be considered, the manner in which the classification on one variable "goes with," or is "related to," or is "associated with" the classification on the second variable is of great aid in understanding the phenomenon under study. Thus it would mean one thing to a political scientist if the same fraction of Dewey and Truman voters had made their final decision on a candidate in the last 2 weeks before the election and quite a different thing if one of these fractions were larger than the other. In analyzing data of this type, one always asks first whether there exists any association between the two variables. If the presence of association is confirmed, one can attempt to measure the degree of association and can try to explain the reasons for the association. This section (which deals with problems where each of the two qualitative variables has two categories) and the following section (where each of the two variables will be allowed to have an arbitrary number of categories) will present χ^2 tests for determining whether or not two qualitative variables are associated with one another.

As an example of the cross classification of two dichotomous variables, consider the data of Table 11.8. These data were obtained as part of a study of retirement. The authors describe the study as follows:

> The survey of the rural aged reported here is part of the Cornell Study of Occupational Retirement. One of the research projects in this study was a nationwide urban survey in 1952 and 1953 of a representative cross-section of approximately 1000 men and 200 women 60 years of age and over; in this project the focus of interest was upon problems of adjustment to old age and retirement. To check on the widely held assumption that aging in rural areas presents few, if any, of the problems of adjustment found in urban areas, a survey of 249 males 60 years of age and over, resident in the villages and open country of 6 predominantly rural counties in New York State, was conducted in the summer of 1952.

The data of Table 11.8 show the joint classification of 94 retired individuals according to "mode of retirement" and "occupation prior to retirement," and they bear on the "widely held assumption that aging in rural areas presents few, if any, of the problems of adjustment found in urban areas." We shall be concerned here with the problem of determining whether these two specific qualitative variables are associated with one another, and this is pertinent since retirement presumably requires less adjustment if it occurs gradually.

Although it is extremely difficult to devise satisfactory measures of the degree of association between two qualitative variables, and some comments on this problem will be found in Sec. 11.7, it is relatively easy to specify what is meant by the absence of association. We shall say that there is no association between two qualitative variables, A and B, if the probability that a randomly selected element falls in a particular category of the A variable, say A_i, *and* in a particular category of the B variable, say B_j, is equal to the product of the probability that the ele-

TABLE 11.8. JOINT CLASSIFICATION OF RETIRED RURAL RESIDENTS ACCORDING TO OCCUPATION AND MODE OF RETIREMENT
(Data obtained from a random sample of males 60 years of age and over, resident in the villages and open country of six predominantly rural counties in New York State)

Mode of retirement	Occupation		Total
	Retired farm operators	Retired nonfarmers	
Retired abruptly	9	53	62
Retired gradually	14	18	32
Total	23	71	94

SOURCE: Philip Taietz, Gordon F. Streib, and Milton L. Barron, *Adjustment to Retirement in Rural New York State*, Bulletin 919, Table 16, Cornell University Agricultural Experiment Station, Ithaca, N.Y., 1956.

ment falls in A_i and the probability that it falls in B_j. That is

$$\Pr(A_i \text{ and } B_j) = \Pr(A_iB_j) = \Pr(A_i)\Pr(B_j) \tag{11.3}$$

for all possible values of i and all possible values of j. In the terminology of Sec. 7.5, we are saying that there is no association between the two variables if they are independent. From a practical viewpoint, this means that a knowledge of the A category of an element does not aid in predicting the B category, and vice versa. A χ^2 procedure for testing the hypothesis that two dichotomous qualitative variables are independent will now be presented.

The hypothesis of independence in relation to the contents of Table 11.8 states that the four cell frequencies describe a sample drawn from a population of retired persons for which

$$\begin{aligned}\Pr(A_1B_1) &= \Pr(A_1)\Pr(B_1)\\ \Pr(A_1B_2) &= \Pr(A_1)\Pr(B_2)\\ \Pr(A_2B_1) &= \Pr(A_2)\Pr(B_1)\\ \Pr(A_2B_2) &= \Pr(A_2)\Pr(B_2)\end{aligned}$$

The symbol A_1 refers to "retirement occurred abruptly," A_2 to "retirement occurred gradually," B_1 to "retired farm operator," and B_2 to "retired nonfarmer." The first problem is to use these multiplicative probability expressions in specifying an expected frequency for each of the four cells of the table. The criterion X^2 can then be computed as in the preceding sections, and a "large" value of X^2 will lead to rejecting the hypothesis of independence.

The hypothesis that $\Pr(A_iB_j) = \Pr(A_i)\Pr(B_j)$ does not in itself provide the numerical values which are required for computing expected frequencies. However, it is possible to circumvent this difficulty by estimating $\Pr(A_i)$ and $\Pr(B_j)$ from the observed data. Since the 94 individuals of Table 11.8 constitute a random sample of all retired males living in the defined six-county area, we may estimate $\Pr(A_1)$ by $(62/94)$, $\Pr(A_2)$ by $(32/94)$, $\Pr(B_1)$ by $(23/94)$, and $\Pr(B_2)$ by $(71/94)$. These estimates provide estimates of $\Pr(A_iB_j)$ if the hypothesis is in fact true. Finally, expected frequencies are obtained by multiplying these estimated probabilities by the total size of sample, 94. Thus the estimated expected frequency for the A_1B_1 cell is given by

$$\left(\frac{62}{94}\right)\left(\frac{23}{94}\right)94 = \frac{(62)(23)}{94} = 15.2$$

Notice that this gives a very simple computational procedure for obtaining the expected frequencies. The expected frequency for any cell is

TABLE 11.9. COMPUTATION OF EXPECTED FREQUENCIES FOR DATA OF TABLE 11.8

Mode of retirement	Occupation		Total
	Retired farm operators	Retired nonfarmers	
Retired abruptly	$\frac{(23)(62)}{94} = 15.2$	$\frac{(71)(62)}{94} = 46.8$	62
Retired gradually	$\frac{(23)(32)}{94} = 7.8$	$\frac{(71)(32)}{94} = 24.2$	32
Total	23	71	94

equal to the product of the totals for the row and column in which the cell falls, divided by the total sample size n. The computations of expected frequencies are summarized in Table 11.9. The value of X^2 can now be computed by using (11.1) and is equal to

$$\frac{(9 - 15.2)^2}{15.2} + \frac{(53 - 46.8)^2}{46.8} + \frac{(14 - 7.8)^2}{7.8} + \frac{(18 - 24.2)^2}{24.2}$$
$$= \frac{(-6.2)^2}{15.2} + \frac{(6.2)^2}{46.8} + \frac{(6.2)^2}{7.8} + \frac{(-6.2)^2}{24.2}$$
$$= 2.53 + 0.82 + 4.93 + 1.59 = 9.87$$

The sampling distribution of X^2, under the hypothesis of independence, will be approximated by a χ^2 distribution with an appropriate number of degrees of freedom. One's first impulse might be to carry over the argument of Sec. 11.3 and use 3 df since there are four cells. This, however, is not correct since the probabilities for obtaining the expected frequencies have been *estimated* from the observed data. We have actually estimated two probabilities, namely $\Pr(A_1)$ and $\Pr(B_1)$. An estimate of $\Pr(A_2)$ is automatically determined by the estimate of $\Pr(A_1)$ since $\Pr(A_1)$ and $\Pr(A_2)$ must sum to 1. Similarly, an estimate of $\Pr(B_2)$ is automatically determined by the estimate of $\Pr(B_1)$ since $\Pr(B_1)$ and $\Pr(B_2)$ must sum to 1. The general rule stated in the last paragraph of Sec. 11.4 can now be applied. The number of degrees of freedom associated with X^2 is equal to the number of categories (4), less one (1) degree of freedom for the fixed sample size, less one degree of freedom for each parameter (2) that is estimated from the data. Thus X^2, used for testing the hypothesis of independence in a 2×2 table has 1 degree of freedom. Reference to Table 11.5 shows that $\chi^2_{.01}$ (1 df) is 6.635. Since the observed value of X^2, 9.87, is larger than 6.635, the hypothesis of independence is rejected at the 1 per cent level of significance. Observe that we have only rejected the hypothesis of "no association." Nothing has been said about the type or strength of association that may exist between the two variables.

The general 2×2 table, of which Table 11.8 is only a particular example, is of the form illustrated in Table 11.10.

TABLE 11.10. GENERAL 2×2 TABLE

Categories of variable A	Categories of variable B		Total
	B_1	B_2	
A_1	u_{11}	u_{12}	$u_{1\cdot}$
A_2	u_{21}	u_{22}	$u_{2\cdot}$
Total.........	$u_{\cdot 1}$	$u_{\cdot 2}$	n

The observed frequencies u are identified with double subscripts where the first subscript gives the category of variable A and the second subscript gives the category of variable B. The total number of observations falling in category A_1 (without regard to category in variable B) is $u_{1\cdot}$ and the total number of observations falling in category B_2 (without regard to category in variable A) is $u_{\cdot 2}$.

The test procedure just discussed views this general table in the following manner. A random sample of size n is drawn from a specified population, and $\Pr(A_iB_j)$ is the probability of an observation falling in the

A_iB_j cell. The hypothesis to be tested on the basis of the observed values of the u_{ij} is that of independence, namely, that $\Pr(A_iB_j)$ can be written as $\Pr(A_i)\Pr(B_j)$. This is not the only way of viewing a 2×2 table, and, as a matter of fact, one of these other ways was discussed in some detail in connection with the analysis of Table 9.9. We may regard category B_1 as identifying one population and B_2 as identifying a second population. A random sample of size $u_{\cdot 1}$ is drawn from population B_1, and an *independent* random sample of size $u_{\cdot 2}$ is drawn from population B_2. The hypothesis to be tested states that the fraction of A_1's in population B_1 is the same as the fraction of A_1's in population B_2. A large sample test of this hypothesis was illustrated on the data of Table 9.9 where A_1 corresponds to a "yes" response and A_2 to the combination of "no" and "no opinion" responses. Actually, these two views of a 2×2 table are not as different as might at first appear. It can be shown that *if the sample size is sufficiently large:*

1. The criterion X^2 is computed in exactly the same manner for both cases.
2. The sampling distribution of X^2 can be approximated by $\chi^2(1 \text{ df})$ in both cases.
3. The X^2 test is identical with the large sample, *two-sided* test, given in Sec. 9.5, of the hypothesis that there is no difference between two population fractions.

If the sample size is small, the $\chi^2(1 \text{ df})$ distribution may not give a very close approximation to the sampling distribution of X^2. Under such circumstances, one usually attempts to find the exact sampling distribution of X^2 and to use this in setting up the hypothesis-testing procedure. This was illustrated in Tables 11.2 and 11.3 for one particular example. In the case of the general 2×2 table of Table 11.10, this approach leads to difficulty. It is not possible to determine numerically the exact sampling distribution of X^2 for either of the two interpretations of this table that have thus far been set forth. However, there exists still a third interpretation that does lead, under the hypothesis of no association between A and B, to an exact sampling distribution for X^2. This interpretation requires that we think of drawing repeated samples from the population in such a manner that each sample has exactly the same *marginal frequencies* as does the observed table. That is, each sample has the same values of n, $u_{\cdot 1}$, $u_{\cdot 2}$, $u_{1\cdot}$, and $u_{2\cdot}$. Tables have been prepared on the basis of this exact sampling distribution to cover situations where $u_{1\cdot}$ and $u_{2\cdot}$ are both less than or equal to 15 [1, table 38]. These tables should be used to test independence in a 2×2 table whenever they are applicable in terms of sample size and distribution of the sample among the four cells, even though one may feel that data arose from one of the two previous views of the 2×2 table.

It can be shown that if n is sufficiently large, the criterion X^2 is still the appropriate quantity for testing independence with this third view of the 2×2 table. Furthermore, the sampling distribution of X^2 can still be approximated by χ^2(1 df). If any of the expected frequencies in such a table are less than 5, and if the exact tests given in [1] do not apply because of the sample size and the distribution of the sample among the four cells, then it is probably advisable to *correct X^2 for continuity* (see [2]). In the present case of the 2×2 table, this correction is very easily applied. We shall illustrate this correction with the data of Tables 11.8 and 11.9, although such a correction is probably not required in this case. The computation of X^2 was

$$X^2 = \frac{(-6.2)^2}{15.2} + \frac{(6.2)^2}{46.8} + \frac{(6.2)^2}{7.8} + \frac{(-6.2)^2}{24.2}$$

Notice that each difference between an observed and expected frequency is the same. (This will always be true for a 2×2 table.) A value of X^2, corrected for continuity, is obtained by *decreasing the absolute value of each difference by 0.5*. Thus

$$\begin{aligned} X_c^2 &= \frac{(5.7)^2}{15.2} + \frac{(5.7)^2}{46.8} + \frac{(5.7)^2}{7.8} + \frac{(5.7)^2}{24.2} \\ &= 2.14 + .69 + 4.17 + 1.34 \\ &= 8.34 \end{aligned}$$

This is still significant at the .01 level.

11.6. Testing the Hypothesis of Independence in an $r \times c$ Contingency Table

The methods of Sec. 11.5 may be immediately extended to include situations in which the qualitative variable A has any arbitrary number of categories, say r, and the variable B has any arbitrary number of categories, say c. The joint classification (with respect to the two variables) of the elements of a random sample of size n drawn from a defined population may then be described by means of a contingency table having r rows and c columns. Table 11.11 gives the general case. The observed number of elements falling in the ith category of variable A *and* in the jth category of variable B is denoted by u_{ij}, the total number of elements falling in the ith category of variable A (without regard to variable B) by $u_{i\cdot}$, and the total number of elements falling in the jth category of variable B (without regard to variable A) by $u_{\cdot j}$. Again we wish to test the hypothesis of independence, namely, that expression (11.3) holds for all possible values of i and all possible values of j.

Since the values of $\Pr(A_i)$ and $\Pr(B_j)$ are not specified by the hypothesis, they must be estimated from the observed data before expected frequencies can be computed for the rc cells of the table. These probabilities may be estimated from the marginal totals of the table. Thus we can estimate $\Pr(A_i)$ by $u_{i\cdot}/n$ and $\Pr(B_j)$ by $u_{\cdot j}/n$. As a consequence, the expected frequency for the A_iB_j cell, under the hypothesis of independence, is given by

$$\left(\frac{u_{i\cdot}}{n}\right)\left(\frac{u_{\cdot j}}{n}\right)n = \frac{u_{i\cdot}u_{\cdot j}}{n} \tag{11.4}$$

The value of X^2 can now be computed by applying formula (11.1), where the summation extends over the rc cells in the table.

TABLE 11.11. GENERAL $r \times c$ CONTINGENCY TABLE

Categories of variable A	Categories of variable B						Total
	B_1	B_2	$\cdot$	B_j		B_c	
A_1	u_{11}	u_{12}	$\cdot$	u_{1j}	$\cdot$	u_{1c}	$u_{1\cdot}$
A_2	u_{21}	u_{22}	$\cdot$	u_{2j}	$\cdot$	u_{2c}	$u_{2\cdot}$
$\cdot$	$\cdot$	$\cdot$	$\cdot$	$\cdot$	$\cdot$	$\cdot$	$\cdot$
A_i	u_{i1}	u_{i2}	$\cdot$	u_{ij}	$\cdot$	u_{ic}	$\cdot$
$\cdot$	$\cdot$	$\cdot$	$\cdot$	$\cdot$	$\cdot$	$\cdot$	$\cdot$
A_r	u_{r1}	u_{r2}	$\cdot$	u_{rj}	$\cdot$	u_{rc}	$u_{r\cdot}$
Total.........	$u_{\cdot 1}$	$u_{\cdot 2}$	$\cdot$	$\cdot$	$\cdot$	$u_{\cdot c}$	n

The number of degrees of freedom to be associated with X^2 is determined by applying the general rule of Sec. 11.4. We have estimated $(r - 1)$ probabilities for the A variable and $(c - 1)$ probabilities for the B variable. For each variable, the number of estimated probabilities is one less than the number of categories because of the fact that

$$\Pr(A_1) + \cdots + \Pr(A_r) = 1$$

and $\Pr(B_1) + \Pr(B_2) + \cdots + \Pr(B_c) = 1$. Therefore the number of degrees of freedom for X^2 is equal to the number of categories (rc), less one (1) degree of freedom for the fixed sample size, less one degree of freedom for each estimated A probability $(r - 1)$, less one degree of freedom for each estimated B probability $(c - 1)$. Hence the number of degrees of freedom for X^2 is

$$rc - (r - 1) - (c - 1) - 1 = (r - 1)(c - 1) \tag{11.5}$$

The application of these techniques can be illustrated in terms of the data of Table 11.12. These data were obtained from a random sample of 321 nonstewards, drawn from among the members of Local 688 of the Teamsters Union in St. Louis, Mo., in 1949. Among other things, each

TABLE 11.12. ANSWERS TO THE QUESTION
"Are you willing to picket to help some other shop get organized or get a raise in pay?"
CLASSIFIED ACCORDING TO NUMBER OF YEARS IN THE UNION
(Random sample of 321 nonstewards, drawn from among the members of Local 688 of the Teamsters Union in St. Louis, Mo., 1949)

Response to question	Number of years in the union				Total
	Less than 1	1–3	4–10	Over 10	
Yes	27	54	137	28	246
No	14	18	34	3	69
Don't know and no answer	3	2	1	0	6
Total	44	74	172	31	321

SOURCE: Arnold M. Rose, *Union Solidarity*, p. 77, University of Minnesota Press, Minneapolis, 1952.

TABLE 11.13. COMPUTATION OF EXPECTED FREQUENCIES FOR THE DATA OF TABLE 11.12

Response	Number of years in the union				Total
	B_1	B_2	B_3	B_4	
A_1	$\frac{(44)(246)}{321} = 33.7$	$\frac{(74)(246)}{321} = 56.7$	$\frac{(172)(246)}{321} = 131.8$	$\frac{(31)(246)}{321} = 23.8$	246
A_2	$\frac{(44)(69)}{321} = 9.5$	$\frac{(74)(69)}{321} = 15.9$	$\frac{(172)(69)}{321} = 37.0$	$\frac{(31)(69)}{321} = 6.6$	69
A_3	$\frac{(44)(6)}{321} = 0.8$	$\frac{(74)(6)}{321} = 1.4$	$\frac{(172)(6)}{321} = 3.2$	$\frac{(31)(6)}{321} = 0.6$	6
Total	44	74	172	31	321

respondent was asked the question, "Are you willing to picket to help some other shop get organized or get a raise in pay?" We shall here be concerned with whether the responses to this question can be considered as independent of the number of years that an individual had belonged to the union. The computation of expected frequencies, under the hypothesis of independence, is summarized in Table 11.13. The computation of X^2 is:

$$
\begin{aligned}
X^2 = {} & \frac{(27 - 23.7)^2}{33.7} + \frac{(54 - 56.7)^2}{56.7} + \frac{(137 - 131.8)^2}{131.8} + \frac{(28 - 23.8)^2}{23.8} \\
& + \frac{(14 - 9.5)^2}{9.5} + \frac{(18 - 15.9)^2}{15.9} + \frac{(34 - 37.0)^2}{37.0} + \frac{(3 - 6.6)^2}{6.6} \\
& + \frac{(3 - 0.8)^2}{0.8} + \frac{(2 - 1.4)^2}{1.4} + \frac{(1 - 3.2)^2}{3.2} + \frac{(0 - 0.6)^2}{0.6} \\
= {} & 15.44
\end{aligned}
$$

The number of degrees of freedom associated with X^2 is, by (11.5), $(r - 1)(c - 1) = (3 - 1)(4 - 1) = 6$. Suppose a test at the 1 per cent level of significance is desired. From Table 11.5, we obtain

$$\chi^2_{.01}(6 \text{ df}) = 16.812$$

Since the observed value of X^2, 15.44, is less than this critical value, we do not reject, at the 1 per cent level of significance, the hypothesis of independence. This hypothesis would, however, be rejected at the 5 per cent level of significance since $\chi^2_{.05}(6 \text{ df}) = 12.592$.

An $r \times c$ contingency table may also arise in quite a different way from the one which we have just described. Suppose the symbols $B_1, B_2, \ldots, B_c$ are used to denote c different populations and that a random sample of size $u_{\cdot 1}$ is drawn from population B_1, an independent random sample of size $u_{\cdot 2}$ is drawn from $B_2, \ldots$, and an independent random sample of size $u_{\cdot c}$ is drawn from population B_c. That is, we *fix* the total sample size at n and we *fix* the number of elements that is drawn from each of the populations B_1 up to B_c. Under these circumstances, one is usually interested in testing the hypothesis of *homogeneity of distributions.* That is, the hypothesis to be tested is that the fraction of A_1's is the same in each of the populations, that the fraction of A_2's is the same, . . . , and that the fraction of A_r's is the same. The criterion X^2 can be used to test this hypothesis, and if the sample size is sufficiently large, the sampling distribution of X^2 can still be approximated by a χ^2 distribution with $(r - 1)(c - 1)$ degrees of freedom.

The foregoing view of an $r \times c$ contingency table is the appropriate one if a stratified simple random sample is drawn from a population, the separate strata being identified as $B_1, B_2, \ldots, B_c$. Thus the data of Table 11.12 would be viewed in this manner if the members of the group sampled had been first divided into the four stated "length of membership" categories and a random sample of 44 drawn from the "less than 1 year" group, an independent random sample of 74 drawn from the "1–3 year" group, . . . , and an independent random sample of 31 drawn from the "over 10 year" group. Even though the number of elements to be drawn from each of the four stated "length of membership" categories is not fixed in advance and a simple random sample is drawn from the entire population, X^2 can still be used to test for homogeneity of distributions. We simply regard the observed table as one out of an indefinitely large number of tables, each of which has exactly the same number of elements in the $B_1, \ldots, B_c$ categories as does the observed table.

As a final point, we may ask under what circumstances the sampling distribution of X^2 can be "closely" approximated by a χ^2 distribution. Cochran [2, 7] suggests that if relatively few expectations are less than 5 (say 1 cell out of 5 or more, or 2 cells out of 10 or more), a minimum expectation of 1 is allowable in computing X^2.

11.7. Measuring the Degree of Association in Contingency Tables

The preceding two sections have discussed the use of the criterion X^2 for testing the hypothesis of independence or of homogeneity of distributions on the basis of observed contingency tables. If the hypothesis is rejected, we are in effect saying that there exists some type of association between the two qualitative variables A and B. Many attempts have been made to devise measures of the degree or strength of association between two qualitative variables, the usual goal being to obtain a dimensionless quantity that has the value 0 if A and B are independent and the value 1 if there is "perfect" association. Most of these measures depend directly upon the value of X^2 since a "small" value of X^2 indicates independence and larger and larger values indicate greater and greater departures from independence. Some of the measures that have been proposed are:

Mean-square contingency:

$$\phi^2 = \frac{X^2}{n}$$

Pearson's coefficient of mean-square contingency:

$$C = \sqrt{\frac{X^2}{n + X^2}} = \sqrt{\frac{\phi^2}{1 + \phi^2}}$$

Tschuprow's coefficient of contingency:

$$T^2 = \frac{\phi^2}{\sqrt{(r-1)(c-1)}}$$

A comprehensive account of these traditional measures of association will be found in Yule and Kendall [3, chaps. 2, 3].

These traditional measures suffer from a variety of defects, including the fact that adequate knowledge of their sampling distributions is not available. However, their chief defect is probably that they do not indicate in any way the nature of the departure from independence. Since they depend upon X^2, this behavior can perhaps best be illustrated by returning to our treatment of the test for independence. By analogy with the discussion contained in Sec. 11.4, this test will not be at all selective in terms of the patterns of deviations from independence for which it rejects. This can be seen by examining a bit more closely the data of Table 11.12 with reference to the fraction of "yes" answers given by union members in each of the four "length of membership" categories. These fractions are .61 for the "less than 1 year" category, .73 for the "1–3 year" category, .80 for the "4–10 year" category, and .90 for the "over 10 year" category. There appears to be a steady increase in the fraction of "yes" answers as the length of membership increases. And yet the value of X^2, and therefore the value of any measure of association based on X^2, would be exactly the same if one made any arbitrary interchange of the headings of the columns of Table 11.12.

This difficulty arises from the fact that the categories of variable B have a definite pattern of organization, or ordering, and that X^2 does not take this organization into account. As Fisher [4, p. 90] has stated, "To measure the degree of association it is necessary to have some hypothesis as to the nature of the departure from independence to be measured." A number of specialized measures of association, which do permit one to take into account the patterning (or lack of patterning) of the categories of one or both variables, have been suggested by Goodman and Kruskal [5], and the interested reader will find it instructive to refer to this paper. As experience with the use of these measures accumulates and as knowledge is obtained concerning their sampling distributions, it is to be expected that they (or some suitable variants) will provide increasingly useful tools for the social scientist.

EXERCISES

11.1. One of the basic theoretical probability distributions (in addition to the binomial and normal distributions) that has important applications in statistics is the Poisson distribution. This distribution can often be used to describe the number of occurrences of some phenomenon, such as calls on a telephone exchange or the appearance of cars at a toll gate, in a stated interval of time. A description of the Poisson distribution, together with many applications, will be found in Feller [6, pp. 115–123 and chap. 17].

As an application of the Poisson distribution, consider the following set of data. They were obtained by observing the number of taxicab arrivals at Euston Station, England, in 1947, in 1-minute intervals (intervals observed between 9:45 and 10:00 A.M.). The predicted probabilities were obtained from the Poisson distribution.

Data for Exercise 11.1

Number of taxicab arrivals in a 1-minute interval	Observed number of 1-minute intervals (u_i)	Predicted fraction* of intervals on basis of Poisson distribution (p_i)
0	18	.259
1	18	.349
2	14	.236
3	7	.106
4	3	.036
5	0	.010
6	0	.002
7	0	.000
Total..........	60	.998

* In order to obtain these predicted fractions, *one* parameter of the Poisson distribution must be estimated from the data.

Source: David G. Kendall, "Some Problems in the Theory of Queues," *Jour. Roy. Stat. Soc.*, B, **13**:151–185 (1951).

Use the X^2 criterion to test the hypothesis that the observed data were drawn from a population that can be described by a Poisson distribution. (In determining the number of degrees of freedom, refer to the footnote on the table.)

11.2. The National Bureau of Economic Research conducted in 1939 a study designed to serve two purposes: "first, to analyze the significance of some credit factors generally considered important by credit men, to discern which of these factors have been associated in the past with bad loans, and to determine whether or not this information can be used to predict the course of future transactions; second, to experiment with the use of statistical methods that may be applied to the problem of credit risk selection." The basic data were obtained by asking certain lending institutions to select samples of "good" loans and "bad" loans from among those loans made in the period 1935 to 1938. The following table shows the classification of good loans and bad loans, made by a sample of industrial-banking companies, according to whether the borrower did or did not possess life insurance:

Borrower possessed:	Loan classified as: Good	Bad	Total
Life insurance............	284	249	533
No life insurance..........	66	93	159
Total..................	350	342	692

Source: David Durand, *Risk Elements in Consumer Instalment Financing*, p. 64, National Bureau of Economic Research, New York, 1941.

a. Use the criterion X^2 to test the hypothesis that the fraction of borrowers possessing life insurance was the same for both good and bad loans. State the number of degrees of freedom, the significance level used, and whether the hypothesis is accepted or rejected.

b. If the hypothesis of similar fractions is rejected, examine the data in the table and state whether you feel the criterion of "life insurance possessed" would be of much value in assessing a potential borrower.

11.3. During and toward the close of World War II the Research Branch of the Information and Education Division in the War Department was concerned with the problem of predicting what kind of employment men would seek upon discharge from the Army, as well as how great the demand would be for further schooling. The following table shows the relation between "plans to return to previous employer as expressed at separation from the Army" and "extent of return within 3 to 4 months after discharge." These data were obtained from interviews and follow-ups with a sample of individuals separated from the Army in December, 1945. All these men had worked full time for an employer just before entering the Army.

DATA FOR EXERCISE 11.3

Status 3 to 4 months after discharge*	Plans expressed at separation*			Total
	Definitely return	Probably or possibly return	Plan not to return	
Had returned to previous employer or were planning to return......	496	344	86	926
Tried to return and could not get job back; or asked to return but decided not to return..........	36	96	63	195
Made no effort to return.........	66	185	331	582
Total.........................	598	625	480	1,703

* The categories given here represent a grouping of some of the categories given in the source.

SOURCE: Samuel A. Stouffer et al., *The American Soldier*, vol. IV, *Measurement and Prediction*, p. 633, Princeton University Press, Princeton, N.J., 1950.

a. Test the hypothesis that the variable "plans expressed at separation" is independent of the variable "status 3 to 4 months after discharge."

b. If one wished to measure the degree of association between these two variables, explain why a measure based on the value of X^2 would not be completely satisfactory.

11.4. Exercise 7.2 shows how each individual in a national sample of 390 individuals who had voted in the 1948 Presidential election was jointly classified with respect to the variable "time at which a decision was made on a candidate" and "candidate for whom the individual voted." Test the hypothesis that these two qualitative variables are independent in the population from which the sample was selected.

11.5. As part of an Office of Naval Research project on "A Program of Research on the Fundamental Problems of Organizing Human Behavior," the Survey Research Center of the University of Michigan carried out a special study in the summer of 1947 in the home office of the Prudential Insurance Company. A number of matched pairs of sections or work groups carrying out clerical operations were selected for the study, one member of each pair being a "high productivity" section and the other a "low productivity" section. The employees in these sections were asked the question "How well does your supervisor explain the new jobs or methods that come along?" The following distributions of responses were obtained:

Type of section	Response to question				Total
	Very well	Fairly well	Not so well	Not ascertained	
Employees in high sections	77	46	59	27	209
Employees in low sections	103	61	29	17	210
Total	180	107	88	44	419

SOURCE: Daniel Katz, Nathan Maccoby, and Nancy C. Morse, *Productivity, Supervision and Morale in an Office Situation*, pt. I, p. 19, University of Michigan, Survey Research Center, Ann Arbor, Mich., 1950.

Use the criterion X^2 to test the hypothesis that the distribution of responses to the question is the same for employees in high sections as for employees in low sections. State the number of degrees of freedom, the significance level used, and whether the hypothesis is accepted or rejected.

11.6. It was observed in Sec. 10.6 that sample surveys often encounter difficulty in contacting all respondents who have been designated for interview. One of the ways of investigating the effect of not interviewing the "difficult to contact" is to compare the characteristics of individuals interviewed in their homes on the first call with the characteristics of individuals interviewed on the second call, on the third call, and so on. The following table presents data of this type obtained from a sample survey conducted among adults living in Elmira, N.Y., in June, 1948.

1944 Presidential vote	Individuals interviewed on:			Total
	First call	Second call	Third or later call	
Roosevelt	138	122	95	355
Dewey	124	115	85	324
Didn't vote	90	67	75	232
Too young	25	25	30	80
Other	14	9	14	37
Total	391	338	299	1,028

SOURCE: Philip J. McCarthy, *Sampling Procedure for* 1948 *Voting Study*, mimeographed, Cornell University, Ithaca, N.Y., 1949.

Use the X^2 criterion to test the hypothesis that the distribution of responses to the question on "1944 Presidential vote" is the same for individuals interviewed on the first call as for individuals interviewed on the second call and for individuals interviewed on the third or later call.

REFERENCES CITED

1. Pearson, E. S., and H. O. Hartley: *Biometrika Tables for Statisticians*, vol. I, Cambridge University Press, New York, 1954.
2. Cochran, William G.: "The χ^2 Test of Goodness of Fit," *Annals Math. Stat.*, **23**:315–345 (1952).
3. Yule, G. Udny, and M. G. Kendall: *An Introduction to the Theory of Statistics*, 14th ed., Hafner Publishing Company, New York, 1950.
4. Fisher, R. A.: *Statistical Methods for Research Workers*, 11th ed., Hafner Publishing Company, New York, 1950.
5. Goodman, Leo A., and William H. Kruskal: "Measures of Association for Cross Classifications," *Jour. Amer. Stat. Assoc.*, **49**:732–764 (1954).
6. Feller, William: *An Introduction to Probability Theory and Its Applications*, vol. 1, John Wiley & Sons, Inc., New York, 1950.
7. Cochran, W. G., "Some Methods for Strengthening the Common χ^2 Tests," *Biometrics*, **10**:417–451 (1954).

CHAPTER 12

THE LINEAR ASSOCIATION OF TWO QUANTITATIVE VARIABLES

12.1. The Association of Two Quantitative Variables

The association between pairs of qualitative variables was treated in Chap. 11, but the preceding work on quantitative variables has been restricted to the consideration of only a single variable at a time. It is clear that the elements of most populations can be classified simultaneously with respect to more than one quantitative variable. Thus numerical values can be attached to an adult male for such variables as age, height, weight, intelligence quotient, school attainment, or school attainment of his father; a worker in a plant may be characterized by weekly wage, hours worked in a week, rate of production, or length of experience; a family has a size, a total yearly income, or an average total weekly expenditure for food; and a manufacturing firm may be assigned numerical values for number of employees, total production, total sales in a year, or dividends paid on stock. Problems that arise when quantitative variables are considered in *pairs* will be the principal concern in this chapter. That is, each element of a population or sample is measured simultaneously on two quantitative variables, say x and y, and the basic data therefore consist of *pairs of measurements*. Our major purpose will be to present some methods that may be applied in studying the way in which the values of the variable x are associated with, or related to, the values of the variable y. Several brief illustrations follow of the use that can be made of the association between two quantitative variables.

As a first illustration, let us consider a pair of quantitative variables between which association obviously exists. If one studies families in the United States with respect to total income and expenditure for food, it is to be expected that families with large incomes will spend more for food, on the average, than will families with small incomes. However, a home economist, nutrition expert, or agricultural marketing specialist will wish to go further than this and secure a concise description of the manner in which expenditure for food changes with family income. For example, they will want to know if a 10 per cent increase in income will be accompanied by less than, by more than, or by exactly a 10 per cent increase in expenditure for foods. They will also want to know how an increase in amount spent for food will be distributed among various competing types of foods. Such items of information are of value in planning research on human nutrition, in devising educational campaigns on family nutrition habits, and in setting up agricultural marketing programs. A very detailed investigation of these and many other facets of the food consumption of urban families in the United States is reported in a publication of the U.S. Department of Agriculture [1].

Another illustration of the use of association between two quantitative variables can be obtained in the field of personnel selection. If the personnel department of a large plant hires and trains a more or less random selection of all individuals who apply for particular positions, it will find that some are successful and that others perform in an unsatisfactory manner. Since it is costly to train workers who will ultimately be classed as unsatisfactory, it is natural to attempt to eliminate at least a portion of the poor workers before any effort has been invested in their training. The answer may partially lie in the administration of certain psychological tests. That is, a worker would be classified according to two variables, namely, *test performance* and *job performance.* In this illustration, unlike the previous one, it would probably not be possible to ignore the question, Does there actually exist some type of association between the two variables? It would be necessary to study a sample of workers who had taken the tests and had then worked in the plant, determine whether any association existed between test performance and job performance, describe this association, and attempt to use it in selecting future employees. Thus one might hire prospective workers with "high" test performance and pass over those with "low" test performance. This approach could not be expected to provide a perfect means of selection, but it might improve considerably over what would be accomplished with unplanned or random selection. The actual determination and evaluation of a selection program has to be based upon a very careful plan of test construction and of collecting and analyzing data. An account of the development and evaluation of one selection program is described by Arbous [2]. Similar problems of selection arise, for example, in determining admission policies of colleges and universities, in choosing persons to receive specialized training in the Armed Forces of the United States (aircraft pilots, electronic technicians, and the like), and in identifying potential executives for large corporations.

As implied in the discussion of these two examples, any statistical investigation which involves a study of association between two quantitative variables must provide answers to one or more of the following questions:

1. Does there actually exist some type of association between the two variables?
2. What is the form of the association?
3. How can the degree of the association be measured?
4. How can the association be used in solving some stated practical or theoretical problem?

Each of these questions will receive some attention during the course of this chapter.

12.2. The Linear-regression Model

12.2.1. *The Specification of the Model.* There are many different ways in which two quantitative variables can be associated with one another, but exact techniques of analysis and evaluation depend upon choosing a particular *model of association*. The *simple linear-regression model* which will be considered in this section is only a single example of a model of association, but experience has demonstrated that it—and other closely related models—can be frequently applied to describe the association between two quantitative variables.

A very simple example will introduce the principal features of the linear regression model. Suppose that one takes many automobile trips and that each trip is characterized by the time spent on the road and by the total distance covered. If it is possible to maintain a constant speed, say 40 miles per hour, then the exact amount of time required to make a trip of a stated distance will be known. If y stands for the time required (in hours) and x stands for the distance to be covered (in miles), the values of y and x for a trip are associated with one another by the equation $y = (1/40)x = .025x$. Since the graph of this equation is a straight line, as illustrated by the solid line in Form A of Fig. 12.1, we say that there is *linear association* between the two quantitative variables y and x.

There are several points that should be noted concerning Form A of Fig. 12.1. The possible values of x, distance to be covered, are entered on the horizontal scale of the diagram. The reason for putting the variable "distance covered" on the horizontal scale is simply that most trips are planned to cover some stated distance and that one wishes to determine from this the amount of time that will be required. The variable whose value is *fixed* in advance will be called the *independent* variable, and it will always be indicated by the symbol x. The variable whose value is to be determined from a fixed value of the independent variable will be called the *dependent* variable, and it will always be indicated by the symbol y. Values of the dependent variable will appear on the vertical scale of a graph. Observe that the line $y = .025x$ goes through the origin of the graph, that is, the point whose x coordinate is 0 and whose y coordinate is 0, and this mirrors the fact that no time is required to cover zero distance. The number .025 which multiplies x in the

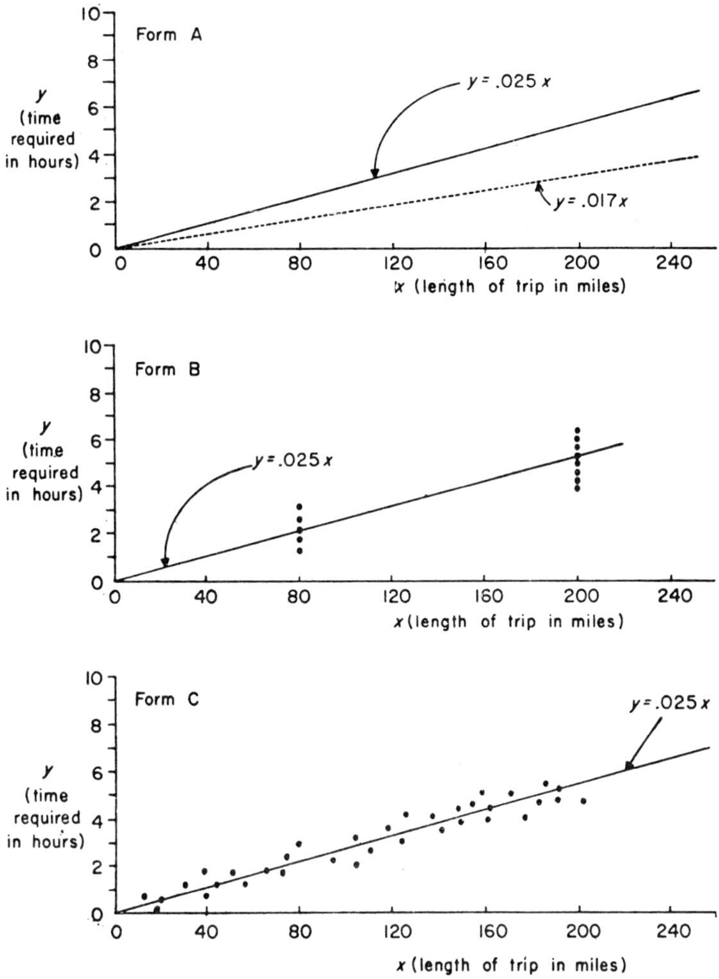

FIG. 12.1. Association between time required for an automobile trip and distance covered. (Hypothetical data; average time required to make a large number of trips of length x assumed to be $x/40$.)

equation of the line is called the *slope* of the line, and it is the amount by which time increases when the distance is increased by one unit, namely 1 mile. Thus the dashed line in Form A of Fig. 12.1 indicates the association for a constant speed of 60 miles per hour. The equation of this line is $y = (1/60)x = .017x$.

The discussion thus far has assumed that a motorist is able to maintain a constant speed of 40 miles per hour. Of course exactly the same result will be obtained if he is able to compensate for periods in which he travels less than 40 miles per hour with periods in which he travels more than 40 miles per hour, the compensation being such as to always make his average speed *for any single trip*—distance covered divided by time required—exactly 40 miles per hour. But such perfect compensation is rarely possible. There are many factors which operate on a more or less chance basis that will influence the average rate of speed for a trip, even though a conscious effort is made to hold this at 40 miles per hour. These factors will cause the time required for going some stated distance to vary from one trip to the next. Thus rough roads, poor weather conditions, heavy traffic, and mechanical difficulties would tend to increase the time required to travel a stated distance, while good roads and weather, coupled with light traffic, would decrease this time. As a result of these chance factors, the times required to make five trips, each of 80 miles, might vary as indicated by the five dots shown at $x = 80$ on Form B of Fig. 12.1; or the times required to make eight trips, each of 200 miles, might differ among themselves as indicated by the eight dots at $x = 200$. In showing such variation, we are implicitly assuming that the deviations of the observed times from $x/40$ are *independent* of one another. If each trip of a given distance covered exactly the same stretch of road, under similar weather and traffic conditions, all observations might be over or under the time that would be required at an average speed of 40 miles per hour. In statistical terminology, these observations would not be independent of one another.

Form C of Fig. 12.1 shows what might result if a record were kept during a given period of time of all trips made in that period, each trip being identified as to distance traveled and time required. The important things to notice about both Form B and Form C are:

1. The observed points tend to be clustered or concentrated about the line $y = .025x$, and this arises as a result of the conscious effort to average 40 miles per hour on each trip.
2. The observed points spread out both above and below the line $y = .025x$ as a result of factors which operate on more or less a chance basis and which influence the time required to make a trip of some stated distance.

This variability about the line is a natural phenomenon and must play an important role in any analysis of this type. For example, a bus company would have to set its scheduled time between two cities with full cognizance of the variability in actual times. Setting the scheduled time too low would mean many "late" arrivals and a large number of dissatisfied customers; setting it too high would place it in a bad light in comparison with its competitors.

The foregoing observations concerning variability about the line $y = .025x$ can be formalized in the following manner. Let us focus attention on trips of some *fixed* distance, say x_0, where a conscious effort is made to drive in such a manner that each trip requires $x_0/40$ hours. We can think of an indefinitely large or infinite population of such trips, each of which is characterized by a value of the variable y, that is, by the time required to make the trip. The variable y will have some distribution in this population of trips. Designate the mean and variance of this distribution by $\mu_{y \cdot x_0}$ and $\sigma^2_{y \cdot x_0}$, respectively. As before, the symbol μ is used for a population arithmetic mean or average, and the symbol σ^2 for a population variance. Therefore, $\mu_{y \cdot x_0}$ is to be read as "the population mean of y when the value of x is fixed at x_0" and $\sigma^2_{y \cdot x_0}$ as "the population variance of y when the value of x is fixed at x_0." By the preceding analysis, it seems reasonable to equate $\mu_{y \cdot x_0}$ and the value on the line of association, $.025x_0$. Thus the line of association $y = .025x$ is really a *line of means*, and its equation can be more precisely written as $\mu_{y \cdot x} = .025x$. Such a line of means is often called the *line of regression* of y (the variable whose values are influenced by chance factors) on x (the variable whose values are known and can be fixed in advance). This particular linear-regression model is presented graphically in Fig. 12.2. Figure 12.2 is to be viewed as a three-dimensional drawing, where the curves at $x = x_1$ and $x = x_2$ lie in planes forming right angles with the plane of the paper. These curves represent the distribution of y for x fixed at x_1 and for x fixed at x_2, respectively. A similar distribution could be drawn for any other fixed value of x.

The preceding hypothetical example has indicated some of the main ideas underlying the regression approach to the problem of association between two quantitative variables x and y. That is, we think of a family of populations, each population being identified by a fixed and known value of the independent variable x. The elements within a given population may vary with respect to their associated values of the dependent variable y, even though they all have exactly the same value of the variable x. The mean and variance of the distribution of y values in this population are denoted by $\mu_{y \cdot x}$ and $\sigma^2_{y \cdot x}$. The purpose of a regression analysis is to describe the manner in which $\mu_{y \cdot x}$ depends upon the value of x, to determine the values of $\sigma^2_{y \cdot x}$, and to use this information in solving stated problems.

In the hypothetical example concerning the distance and time of an automobile trip, the values of $\mu_{y \cdot x}$ were *forced* to lie on a known straight line passing through the origin, and this line was called the *regression line of y on x*. In actual investigations of association, there is of course no guarantee that such linear association will exist. As a matter of fact, any of the following situations may arise:

1. The values of $\mu_{y \cdot x}$ may fall on, or very close to, a straight line which may or may not pass through the origin.

2. The values of $\mu_{y \cdot x}$ may tend to lie on some curve, in which case we speak of *curvilinear* regression.

3. There may be no well-defined relation between the values of $\mu_{y \cdot x}$ and the corresponding value of x.

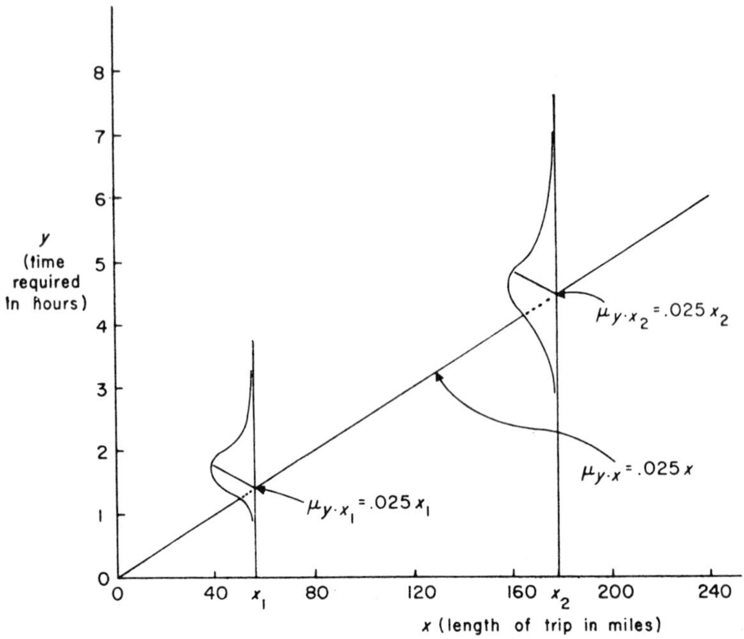

FIG. 12.2. Linear regression model showing association between time required for an automobile trip and distance covered. (Average time required to make a large number of trips of length x assumed to be $x/40$; the curves represent the distribution of y values for fixed values of x.)

We shall now discuss an actual set of data which illustrates some of these points.

The study [1], cited in Sec. 12.1, presents data on the association between total income (after taxes) and money value of food consumed by urban families of two or more persons in the United States. These data were obtained in a nationwide survey of 1,558 urban housekeeping families conducted in the spring of 1948. A portion of these data is given in Table 12.1. In order to show the association between family income and value of food consumed, we have plotted in Fig. 12.3 the seven points determined by the average income within a class interval and the average money value of food consumed per week by families falling in that class interval. The independent variable x is total yearly family income, and its values are plotted on the horizontal scale of the graph. The dependent variable y is average money value of food consumed per week per family, and its values are plotted on the vertical scale of the graph. If one now considers a population of urban housekeeping families, all of whom have approximately the same total yearly

TABLE 12.1. FAMILY INCOME (AFTER TAXES) AND MONEY VALUE OF FOOD CONSUMED
(Urban housekeeping families of two or more persons in the United States, 1947)

Yearly income, dollars	Number of families	Average income, within class interval	Average total money value of food consumed*	Average money value of food consumed per week*
Under 1,000	53	610	675	13.0
1,000–1,999	204	1,555	817	15.7
2,000–2,999	410	2,505	1,085	20.9
3,000–3,999	351	3,485	1,261	24.2
4,000–4,999	167	4,421	1,452	27.9
5,000–7,499	154	5,861	1,506	29.0
7,500 and over	72	11,766	2,090	40.2

* Includes food purchased for consumption at and away from home, value of food produced and consumed at home, and value of food received as a gift or as part of compensation for services rendered.

SOURCE: Ref. 1, table 25.

income, he will find that these families will vary with respect to money value of food consumed per week because of such factors as size of family, section of the country, food preferences of members of the family, and the like. However, one can consider the average money value of food consumed per week by these families, $\mu_{y \cdot x}$, and it is *approximations* to these averages that are plotted in Fig. 12.3. These values are only approximations because they were derived from a sample. Thus the plotted points are approximations to the regression of money value of food consumed per week on total yearly family income.

If attention is now restricted to those groups of families having a total yearly income of less than \$7,500—that is, to the first six plotted points counting from the left—it would appear that the regression of y on x is very nearly linear. A straight line has been drawn "by eye" in Fig. 12.3, and the deviations of the observed points from this line are relatively small. Notice that this line does not go through the origin, as was the case in Figs. 12.1 and 12.2. The general equation of a straight line can be written in the form $y = a + bx$, where the quantity a is called the *y intercept* of the line, and, as before, the quantity b is called the *slope* of the line. The value of a is equal to the height, either positive or negative, at which the line cuts the y axis. This is indicated in Fig. 12.3, where the value of a is shown to be 11.5. The value of b is the amount by which y increases when x is increased by one unit. This is indicated

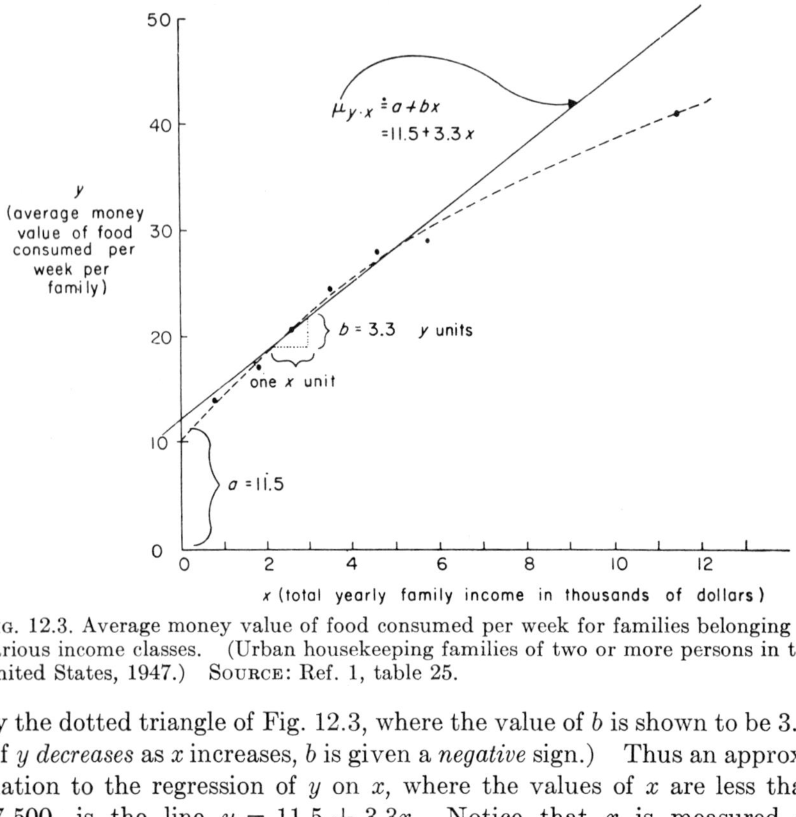

FIG. 12.3. Average money value of food consumed per week for families belonging to various income classes. (Urban housekeeping families of two or more persons in the United States, 1947.) SOURCE: Ref. 1, table 25.

by the dotted triangle of Fig. 12.3, where the value of b is shown to be 3.3. (If y *decreases* as x increases, b is given a *negative* sign.) Thus an approximation to the regression of y on x, where the values of x are less than \$7,500, is the line $y = 11.5 + 3.3x$. Notice that x is measured in thousands of dollars.

It is clear from an examination of Fig. 12.3 that, although a straight line provides a reasonable "fit" to the points where family income is less than \$7,500, the one remaining point ($x = 11.8$ thousands of dollars) indicates a marked departure from the straight line. This is in accordance with one's natural feeling that there is a limit to the amount of food that a family can consume, irrespective of the total amount of income that the family receives. It would therefore appear that the association of y and x over the entire range of possible values of x is best described by a regression curve such as is drawn with dashes in Fig. 12.3. (The cited source for the data of Table 12.1 and Fig. 12.3 used a regression curve in studying the association between family income and expenditure for food.) Even though association may actually be curvilinear, it is often possible to approximate portions of a curve with a straight line as is done in Fig. 12.3 for values of x falling below $7,500.

The preceding discussion and examples have introduced the concepts that are required in defining the simple linear-regression model that will form the basis for the remaining work of this chapter. The simple linear-regression model is defined as follows:

1. There exists a family of populations, each population being identified by a fixed and known value of the independent variable x. The elements within a given population may vary with respect to their associated values on the dependent variable y, even though they all have exactly the same value of the variable x.

2. Let $\mu_{y \cdot x}$ be the average value of y in the population which is identified by a fixed and known value of the variable x. We shall deal with situations in which the relation between the various values of $\mu_{y \cdot x}$ may be closely approximated by a straight line, that is, situations in which

$$\mu_{y \cdot x} = \alpha + \beta x \tag{12.1}$$

The y intercept of this population regression line is α (read alpha), and the slope of the regression line is β (read beta). (Note the use of the Greek letters α and β to indicate that these are population values of the y intercept and slope of the regression line. The corresponding Roman letters a and b were used in Fig. 12.3 to refer to sample estimates of these population quantities.) If this approximate linear association does not hold for all values of x, attention will be restricted to the range of possible values of x for which it can be expected to hold.

3. The variances of the distributions of y values may also depend on the fixed and known value of x. Thus the general notation for these variances is $\sigma^2_{y \cdot x}$. The simple linear-regression model requires that these variances all be equal. That is, the techniques to be presented require that

$$\sigma^2_{y \cdot x} = \sigma^2 \tag{12.2}$$

for all values of x.* (Note that this would probably not be true for the hypothetical situations pictured in Figs. 12.1 and 12.2. The times required for extremely short trips would probably not vary much among themselves simply because of the limitation on distance to be covered. Also, the variances for long trips might be small because the longer dis-

* This assumption of equal variances about the regression line is sometimes known as the assumption of homoscedasticity.

tances would allow the effects of the various chance factors to offset one another. However, at some intermediate distances the variances might be larger than for either short or long trips.)

In addition to the foregoing requirements, it is often necessary to place certain restrictions upon the manner in which a sample of pairs of observations is obtained and upon the shape or form of the distributions of y for fixed values of x. These restrictions will be described as they are needed.

Working within the framework of this simple linear-regression model, we shall now provide answers to such questions as (1) How do we obtain data with which to study the linear association between two quantitative variables y and x? (2) How do we estimate the equation of the regression line from an observed sample of data? (3) How do we measure and estimate variability about the regression line? (4) How do we use this information in solving stated practical and theoretical problems?

12.2.2. *Estimating the Equation of the Regression Line.* The first problem to be faced in investigating the linear association between two quantitative variables is that of obtaining an approximation to the regression line of y on x. The basic data to be used in estimating this regression line will consist of n pairs of values of x and y, each pair being the values of x and y for a single element drawn from the population under study. These n pairs of observations will be designated as (x_1,y_1), (x_2,y_2), . . . , (x_n,y_n), where (x_1,y_1) are the values of x and y for the first sample element and, in general, (x_i,y_i) are the values of x and y for the ith sample element.

These pairs of observations (x_i,y_i) may be obtained in either of the two ways illustrated in Forms B and C of Fig. 12.1. Thus if the method illustrated in Form B is used, *two* or more values of the variable are first chosen and a simple random sample is then selected from each of the subpopulations identified by these fixed values of x. The value of y is determined for each sample element. (At least two different values of x are required to obtain an estimate of the equation of the regression line.) If the method illustrated in Form C is used, one merely selects a random sample of elements from the combined x populations and determines the values of x and y for each sample element.

The preceding requirements on sample selection are often summarized in the following terms. The value of y for the ith sample element y_i can be expressed as

$$y_i = \alpha + \beta x_i + d_i$$

where d_i is the amount by which y_i deviates from the mean $\mu_{y \cdot x_i}$ of all elements having a value of x equal to x_i. A sample of pairs (x_i,y_i) should be drawn in such a manner that the d_i's are *independent* of one another. A hypothetical situation in which this would not hold true has been described in connection with the discussion of Form B of Fig. 12.1. Another example of such dependence can be stated in terms of the data on food consumption and family income presented in Fig. 12.3. Suppose the regression line of money value of food consumed on family income is desired for the population of *all* urban housekeeping families of two or more persons in the United States. Suppose, however, that in selecting a sample of families in a particular income class, say $3,000 to $3,999, only families having five or more persons are drawn into the sample. The deviations of the food expenditures of these families from the average expenditure of *all* families in the $3,000 to $3,999 income range would not be independent. They would tend to show too high a fraction of large and positive (above the average) deviations as compared to the population as a whole.

As an illustration of the actual data that are obtained in investigating the linear association between two quantitative variables, consider the contents of Table 12.2. These data were obtained during the course

TABLE 12.2. OBSERVATIONS ON SPEED AND DENSITY OF TRAFFIC FLOW THROUGH A VEHICULAR TUNNEL

(Each pair of observations refers to a 5-minute interval of traffic flow in the fast lane of the Lincoln Tunnel, a back-up of vehicles existing at the tunnel entrance during each observation period.)

5-min interval (P.M.)	Average density, vehicles/mi x_i	Average speed, mi/hr y_i
4:30–4:35	69	16.8
4:35–4:40	88	14.8
4:40–4:45	65	19.1
4:45–4:50	84	14.9
5:10–5:15	63	21.6
5:30–5:35	53	25.7
6:05–6:10	82	16.8
6:10–6:15	53	25.1
6:15–6:20	70	19.7
6:20–6:25	34	31.8

SOURCE: Ref. 3, table 1, p. 154.

of an investigation of the flow of traffic through the Lincoln Tunnel, which connects New York City and New Jersey under the Hudson River. The author of the study states [3]:

In the flow of traffic through vehicular tunnels, it is shown that an approximate linear relationship between speed and density of traffic exists, whereby, as the average speeds of vehicles increase, the density of the traffic stream decreases. The relationship is used to establish a series of theoretical equations that, in turn, permit the determination of optimum volumes under given traffic composition.

Although these data by themselves provide no obvious reason for taking one or the other of the two variables as the independent variable, the further analyses in which the regression line was to be used led the

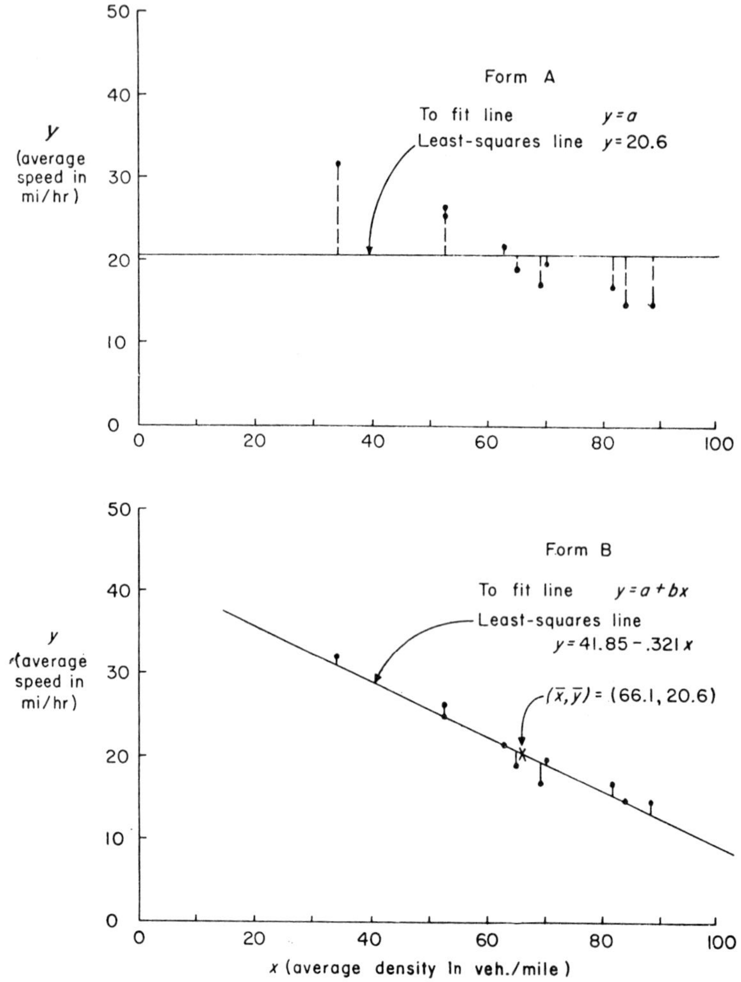

FIG. 12.4. Scatter diagram of observations on speed and density of traffic flow through a vehicular tunnel. (Each point refers to a 5-minute interval of traffic flow in the fast lane of the Lincoln Tunnel.) SOURCE: Ref. 3, table 1, p. 154.

author to choose density of traffic as the independent variable. We shall conform to this choice in the following discussion.

The 10 pairs of observations given in Table 12.2 are shown graphically in Fig. 12.4. This plot of individual observations is called a *scatter diagram*, and the preparation of such a diagram is always the first step in analyzing the association between two quantitative variables. If there is no a priori reason for assuming linear association between two variables, an examination of the scatter diagram will permit one to make a fairly reliable judgment concerning the presence or absence of linear association in particular instances. The points in Fig. 12.4 seem to indicate clearly the presence of linear association.

In estimating the equation of the regression line $y = \alpha + \beta x$ from a set of observations, one's first impulse might be to draw a line through the points "by eye." The difficulty is that this procedure could give quite different lines if it were independently repeated a number of times on the same set of points. Moreover, this procedure provides no basis for evaluating the characteristics of the estimates that are obtained for α and β. It is necessary to define what is meant by a "best-fitting" line, where the definition is such that (1) it leads, for each set of observed points, to a unique line; and (2) it is possible to evaluate the characteristics of the estimates that are obtained for α and β. We shall here define a best-fitting line in terms of the *criterion of least squares.*

In order to introduce the criterion of least squares, let us first take the somewhat artificial view presented in Form A of Fig. 12.4 of the present problem. Suppose we wish to determine the best-fitting *horizontal* line for these observed points. The equation of a horizontal line is of the form $y = a$, the slope being zero, and we wish to specify the "best" value for the constant a. One horizontal line is drawn in Form A of Fig. 12.4, and the deviations of the observed values of y from this line, for fixed values of x, are indicated by the vertical dashed lines. The value of a is to be chosen so that these deviations are as small as possible. Clearly a horizontal line cannot be a best-fitting line for this observed set of points, as may be seen by comparing the vertical deviations in Form A with those in Form B, but the criterion of least squares has exactly the same meaning for a horizontal line as for the general straight line.

If one merely considers the deviations as they appear in Form A, obtains the sum of deviations for all points, and attempts to make this sum of deviations as small as possible, difficulty is encountered in dealing with the plus and minus signs of the deviations. (This is the same difficulty that was discussed in Chap. 5 in connection with defining measures of variability for the distribution of a single quantitative variable.) We avoid this difficulty by treating the *squared* deviations about the horizontal line. Formalizing these considerations, we have:

> The best-fitting horizontal line, according to the *criterion of least squares,* is that line $y = a$ for which the sum of squared vertical deviations about the line takes on its smallest possible (least) value.

The sum of squared vertical deviations about the line $y = a$, for the 10 observed points given in Table 12.2, can be written as

$$\begin{aligned}\Sigma(y_i - a)^2 &= (y_1 - a)^2 + (y_2 - a)^2 + \cdots + (y_{10} - a)^2 \\ &= (16.8 - a)^2 + (14.8 - a)^2 + \cdots + (31.8 - a)^2\end{aligned}$$

Thus we wish to find that number a for which this sum of squares, hereafter abbreviated SS, assumes its smallest possible value. The fact that such a number exists can be easily illustrated by computing the SS for a few different values of a and presenting the results graphically. For example:

Line $y = 17$:
$$\begin{aligned}SS &= (16.8 - 17)^2 + (14.8 - 17)^2 + \cdots + (31.8 - 17)^2 \\ &= 402.53\end{aligned}$$
Line $y = 19$:
$$\begin{aligned}SS &= (16.8 - 19)^2 + (14.8 - 19)^2 + \cdots + (31.8 - 19)^2 \\ &= 297.33\end{aligned}$$
Line $y = 20.63$:
$$\begin{aligned}SS &= (16.8 - 20.63)^2 + (14.8 - 20.63)^2 + \cdots + (31.8 - 20.63)^2 \\ &= 270.76\end{aligned}$$
Line $y = 22$:
$$\begin{aligned}SS &= (16.8 - 22)^2 + (14.8 - 22)^2 + \cdots + (31.8 - 22)^2 \\ &= 289.53\end{aligned}$$
Line $y = 25$:
$$\begin{aligned}SS &= (16.8 - 25)^2 + (14.8 - 25)^2 + \cdots + (31.8 - 25)^2 \\ &= 461.73\end{aligned}$$

These five values of $\Sigma(y_i - a)^2$ are plotted in Fig. 12.5, and a smooth curve has been drawn through the resulting points. (The jagged lines on the ordinate and abscissa indicate that the scales have not been started at zero.)

An examination of Fig. 12.5 shows that the quantity $\Sigma(y_i - a)^2$ depends very markedly on the magnitude of a and that the least value of the SS is attained when a is approximately equal to 21. If a is either larger or smaller than 21, then the SS becomes larger than it is in the neighborhood of $a = 21$. The exact value of a at which the SS attains its minimum is provided by the following theorem:*

> Consider n values of a variable, $y_1, y_2, \ldots, y_n$, and form the sum of squared deviations of these values from an arbitrary constant a. This sum of squared deviations $\Sigma(y_i - a)^2$ takes on its smallest possible value when a is set equal to $\bar{y}$. As before, $\bar{y}$ designates the arithmetic mean of the y_i's.

* The proof of this result is easily obtained by means of differential calculus. The derivative of $\Sigma(y_i - a)^2$ with respect to a is set equal to zero, and the solution of the resulting equation is $a = \bar{y}$.

Thus the minimum of the curve in Fig. 12.5 occurs at $a = \bar{y} = 20.63$, and the horizontal line in Form A of Fig. 12.4 has been drawn at this value.

The criterion of least squares can also be used to determine the equation of a best-fitting line, where the slope of the line is not arbitrarily set equal to zero as in the preceding discussion. This situation is illustrated in Form B of Fig. 12.4. The general equation of a line is of the form $y = a + bx$. The criterion of least squares states that those values of a and b should be selected which minimize the sum of squared vertical

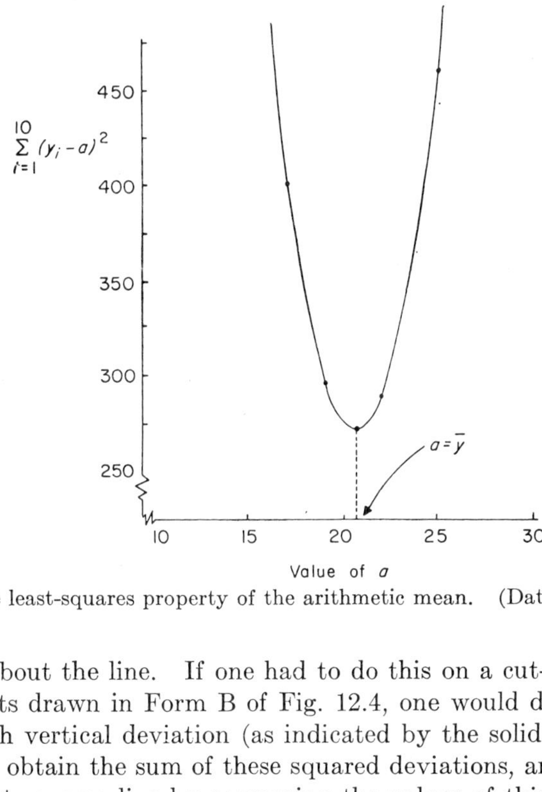

FIG. 12.5. The least-squares property of the arithmetic mean. (Data used are from Table 12.2.)

deviations about the line. If one had to do this on a cut-and-try basis for the points drawn in Form B of Fig. 12.4, one would draw in a line, measure each vertical deviation (as indicated by the solid vertical lines in Form B), obtain the sum of these squared deviations, and attempt to find the least-squares line by comparing the values of this SS among a large number of combinations of possible values of a and b. This would obviously be a very difficult task. However, a formal approach leads quickly to the desired solution.

Consider an observed point, say (x_i, y_i). Remembering that the value of x_i is regarded as fixed and known, the y coordinate of the point on the general line $y = a + bx$ that corresponds to this fixed value of x is obtained by substituting the fixed value of x in the equation of the line, that is, $(a + bx_i)$. The deviation of the observed y value from this *line value* is $[y_i - (a + bx_i)]$. Thus the leftmost point in Form B of

Fig. 12.4 has coordinates (34, 31.8). The line value, corresponding to an x of 34, is $(a + b34)$, and the vertical deviation of the observed y value from this line value is $[31.8 - (a + b34)]$. We wish to choose those values of a and b which minimize the sum of squared vertical deviations about the line, that is, which minimize $\Sigma[y_i - (a + bx_i)]^2$.

With this statement of the problem, the following result can be obtained:*

> Consider n pairs of observations, $(x_1,y_1), (x_2,y_2), \ldots, (x_n,y_n)$. Form the sum of squared vertical deviations of the y values from an arbitrary line $y = a + bx$. This sum of squared deviations $\Sigma[y_i - (a + bx_i)]^2$ takes on its smallest possible value when
>
> $$b = \frac{n\Sigma x_i y_i - (\Sigma x_i)(\Sigma y_i)}{n\Sigma x_i^2 - (\Sigma x_i)^2} \tag{12.3}$$
>
> and $$a = \bar{y} - b\bar{x} \tag{12.4}$$
>
> The value of b is computed first from (12.3), and this value is used in (12.4) to obtain the value of a.

The application of formulas (12.3) and (12.4) to the data of Table 12.2 is illustrated in Table 12.3. Observe particularly the computation of the quantity $\Sigma x_i y_i$. Written out, this quantity becomes

$$\Sigma x_i y_i = x_1 y_1 + x_2 y_2 + \cdots + x_n y_n$$

That is, one first obtains the product of *each* pair of associated x and y values and then obtains the sum of these *cross products*. Also note that when the appropriate quantities are substituted into the formula for b, the result has a negative sign. This means that the value of y *decreases* by .321 units when x is increased by one unit, and this negative slope must be taken into account in computing the value of a. This computation is

$$\begin{aligned} a &= \bar{y} - b\bar{x} \\ &= 20.63 - (-.321)(66.1) \\ &= 20.63 + 21.22 \\ &= 41.85 \end{aligned}$$

* Formulas (12.3) and (12.4) are obtained by the use of differential calculus. The partial derivatives of $\Sigma[y_i - (a + bx_i)]^2$ with respect to a and b are each set equal to zero, thus leading to two equations in the two unknowns a and b. These two normal equations are

$$\begin{aligned} na + (\Sigma x_i)b &= \Sigma y_i \\ (\Sigma x_i)a + (\Sigma x_i^2)b &= \Sigma x_i y_i \end{aligned}$$

Formulas (12.3) and (12.4) are obtained by solving these normal equations for a and b.

The occurrence of two minus signs together makes the resulting product have a plus sign. Thus the least-squares line for the data of Table 12.2 is $y = 41.85 - .321x$. This is the line drawn in Form B of Fig. 12.4. Finally, we draw attention to the fact that with an automatic calculating machine it is not necessary to record each individual value of x_i^2 and x_iy_i as is done in Table 12.3. Such a machine will *cumulate* the results of the individual operations, and one need only record Σx_i^2 and Σx_iy_i. (See discussion accompanying Table 5.3.)

TABLE 12.3. COMPUTATION OF THE SLOPE AND y INTERCEPT OF A LEAST-SQUARES LINE
(Data taken from Table 12.2)

Average density, vehicles/mi x_i	Average speed, mi/hr y_i	Squares of x values x_i^2	Cross products of x and y values x_iy_i
69	16.8	4,761	1,159.2
88	14.8	7,744	1,302.4
65	19.1	4,225	1,241.5
84	14.9	7,056	1,251.6
63	21.6	3,969	1,360.8
53	25.7	2,809	1,362.1
82	16.8	6,724	1,377.6
53	25.1	2,809	1,330.3
70	19.7	4,900	1,379.0
34	31.8	1,156	1,081.2
$\Sigma x_i = 661$ $\bar{x} = 66.1$	$\Sigma y_i = 206.3$ $\bar{y} = 20.63$	$\Sigma x_i^2 = 46,153$	$\Sigma x_iy_i = 12,845.7$

$$b = \frac{n\Sigma x_iy_i - (\Sigma x_i)(\Sigma y_i)}{n\Sigma x_i^2 - (\Sigma x_i)^2}$$
$$= \frac{(10)(12,845.7) - (661)(206.3)}{(10)(46,153) - (661)^2}$$
$$= \frac{-7,907.3}{24,609} = -.321$$
$$a = \bar{y} - b\bar{x} = 20.63 - (-.321)(66.1)$$
$$= 41.85$$

In plotting a straight line on a graph, it is usually not convenient to work directly with the slope and y intercept of the line. Since a straight line is determined by any *two* points on the line, it is only necessary to select any two arbitrary and convenient values of x, substitute these values into the equation of the line, and thus obtain the corresponding values of y *on the line*, plot these two points, and draw the line through them. Furthermore, *the least-squares line* computed in accordance with formulas (12.3) and (12.4) *always passes through the point determined by the arithmetic mean of the x values and the arithmetic mean of the y values,* that is, the point whose coordinates are $(\bar{x},\bar{y})$. This is illustrated in Form B of Fig. 12.4 where the least-squares line passes through the point $(\bar{x},\bar{y}) = (66.1, 20.6)$.

12.2.3. *Estimating the Variability about the Regression Line.* The linear-regression model was defined in Sec. 12.2.1 in terms of the line of means of y's for fixed values of x, $\mu_{y \cdot x} = \alpha + \beta x$ and in terms of the variance σ^2 of the y's about their respective means. Estimates of the y intercept and slope of this regression line have been obtained by applying the least-squares criterion. The problem of estimating σ^2 will now be discussed.

Consider a particular observed pair of values of x and y, say (x_i,y_i). The observed value of y_i is viewed as a random sample of size 1 from a population having a mean of $\alpha + \beta x_i$ and a variance of σ^2. The mean of this population is unknown, but the linear regression of y on x permits the use of the information provided by all n observations in estimating this mean. That is, we estimate $\mu_{y \cdot x_i}$ by $\bar{y}_{x_i} = a + bx_i$, where a and b are the least-squares estimates of α and β as determined in the preceding section. The deviation of the observed value of y from this estimated mean is therefore $(y_i - \bar{y}_{x_i}) = [y_i - (a + bx_i)]$. There are n of these deviations, one for each observed point (x_i,y_i), and they are shown for one set of data as solid vertical lines in Form B of Fig. 12.4. An appropriate average of the squares of these deviations will be used to estimate σ^2.

The quantity $s^2_{y \cdot x}$ will be taken as an estimate of σ^2 where

$$
\begin{aligned}
s^2_{y \cdot x} &= \frac{1}{n-2} \sum (y_i - \bar{y}_{x_i})^2 \\
&= \frac{1}{n-2} \sum [y_i - (a + bx_i)]^2 \\
&= \frac{1}{n-2} \{[y_1 - (a + bx_1)]^2 + [y_2 - (a + bx_2)]^2 \\
&\qquad + \cdots + [y_n - (a + bx_n)]^2\} \qquad (12.5)
\end{aligned}
$$

This is nothing more than the average of squared deviations about the least-squares regression line, a divisor* of $(n - 2)$ being used in obtaining the average. It can be shown that if an indefinitely large number of samples, each consisting of n pairs of observations (x_i,y_i), are drawn from a population in which the linear-regression model applies and if the

* The rationale for the $(n - 2)$ divisor is related to that set forth in Sec. 11.4 for use in determining the number of degrees of freedom to be associated with a value of X^2. Although there are n independent observations, two quantities are estimated from the data, namely, the values of α and β, before the sum of squared deviations about the regression line can be computed. It can be shown from this that $\Sigma[y_i - (a + bx_i)]^2$ is the sum of $(n - 2)$ independent squared terms.

quantity $s_{y \cdot x}^2$ is computed for each sample, then the long-run average of all these values of $s_{y \cdot x}^2$ will be equal to σ^2. In other words, $s_{y \cdot x}^2$ is an *unbiased* estimate of σ^2. The square root of $s_{y \cdot x}^2$ is often called the *standard error of estimate*. The computation of $s_{y \cdot x}^2$ and $s_{y \cdot x}$ is illustrated in Table 12.4 with the data of Table 12.2. The value of $s_{y \cdot x}^2$ is found to be 2.08, and that of $s_{y \cdot x}$, 1.44. (The use of these quantities will be discussed in the next section.) Note that the sum of algebraic deviations

TABLE 12.4. COMPUTATION OF THE VARIANCE ABOUT THE ESTIMATED REGRESSION LINE $s_{y \cdot x}^2$

(Data taken from Table 12.2; equation of estimated regression line, from Table 12.3, is $y = 41.85 - .321x$.)

x_i	y_i	$\bar{y}_{x_i} = 41.85 - .321x_i$	$(y_i - \bar{y}_{x_i})$	$(y_i - \bar{y}_{x_i})^2$
69	16.8	19.70	−2.90	8.4100
88	14.8	13.60	+1.20	1.4400
65	19.1	20.98	−1.88	3.5344
84	14.9	14.89	−0.01	0.0001
63	21.6	21.63	−0.03	0.0009
53	25.7	24.84	+0.86	0.7396
82	16.8	15.53	+1.27	1.6129
53	25.1	24.84	+0.26	0.0676
70	19.7	19.38	+0.32	0.1024
34	31.8	30.94	+0.86	0.7396
Total..	....	.....	−0.05	16.6475

$$s_{y \cdot x}^2 = \frac{\Sigma(y_i - \bar{y}_{x_i})^2}{n - 2} = \frac{16.6475}{8} = 2.0809$$

$$s_{y \cdot x} = 1.44$$

from the regression line, $\Sigma(y_i - \bar{y}_{x_i})$, is very close to zero. This summation would be exactly zero if it were not for rounding errors.

Although formula (12.5) defines $s_{y \cdot x}^2$ in the most meaningful manner, it does not directly provide a convenient computational procedure. The determination of each value of $\bar{y}_{x_i}$ and the computation of the resulting values of $(y_i - \bar{y}_{x_i})^2$, as is done in Table 12.4, is extremely laborious and the entire process is prone to error. However, algebraic manipulation of formula (12.5) leads to another form of $s_{y \cdot x}^2$ which is convenient for computation. This is

$$s_{y \cdot x}^2 = \frac{\Sigma y_i^2 - a\Sigma y_i - b\Sigma x_i y_i}{n - 2} \tag{12.6}$$

Table 12.3 contains all the quantities required by this formula, with the exception of Σy_i^2. This quantity is found to be 4,526.73. Substituting in (12.6),

$$s_{y\cdot x}^2 = \frac{4{,}526.73 - (41.85)(206.3) - (-.321)(12{,}845.7)}{8}$$
$$= 2.0681$$
$$s_{y\cdot x} = 1.44$$

This value of $s_{y\cdot x}^2$ agrees very closely with that given in Table 12.4, the difference being due to rounding. The value of $s_{y\cdot x}$ is the same in both instances.

12.2.4. *Some Problems of Inference in Linear Regression.* If the association between two quantitative variables can be described by the linear-regression model, then the components of the model can be used to provide answers to many questions of interest. Thus the equation of the population regression line immediately gives the mean value of y for all elements having any specified value of x, whether or not observations are taken from this x population; the slope of the regression line tells the amount by which this mean value changes when x is increased by one unit; and the variance about the regression line measures the amount of natural variation in the y values within any subpopulation defined by a fixed value of x. In practice, however, the exact values of the parameters of the regression model are unknown. Estimates of the y intercept and the slope of the regression line and of the variance about the regression line are obtained from a sample, and the sampling variability of these quantities must be taken into account.

This discussion of inference will be introduced by focusing attention on the slope of the population regression line β. The algebraic sign and the magnitude of β are of particular importance in any regression problem. Thus the linear regression of money value of food consumed per week on the total yearly family income, pictured in Fig. 12.3 for families having a total yearly income of less than \$7,500, shows a positive slope having a numerical value of approximately \$3.3. It is not surprising that food expenditures increase as family income increases, but it is important to know that an increase of \$1,000 a year in family income is accompanied by an average increase of approximately \$3.30 per week in food expenditures. On the other hand, the negative slope indicated in Fig. 12.4 for the regression of average speed on density of vehicles for traffic flow through a vehicular tunnel is not quite so obvious. One might argue that vehicles could be sent through such a tunnel at a constant speed, irrespective of the number of vehicles per mile. The existence of a negative slope means that one can possibly move a greater volume of traffic—volume of traffic being determined by the product of speed and density—by increasing the space between vehicles. The magnitude of β is important in determining the spacing that will move the greatest volume of traffic. This determination of a "best," or optimum, spacing was the most important goal of the study from which the data of Table 12.2 were taken [3].

Investigators are especially interested in detecting cases where the value of β is zero. If $\beta = 0$, as pictured in Fig. 12.6, the equation of the regression line is $\mu_{y \cdot x} = \alpha$, and each x population has exactly the same arithmetic mean. In this case, knowing that an element belongs to a specific x population does not assist one in predicting the value of y for that element. When β is equal to zero, it is customary to say that there

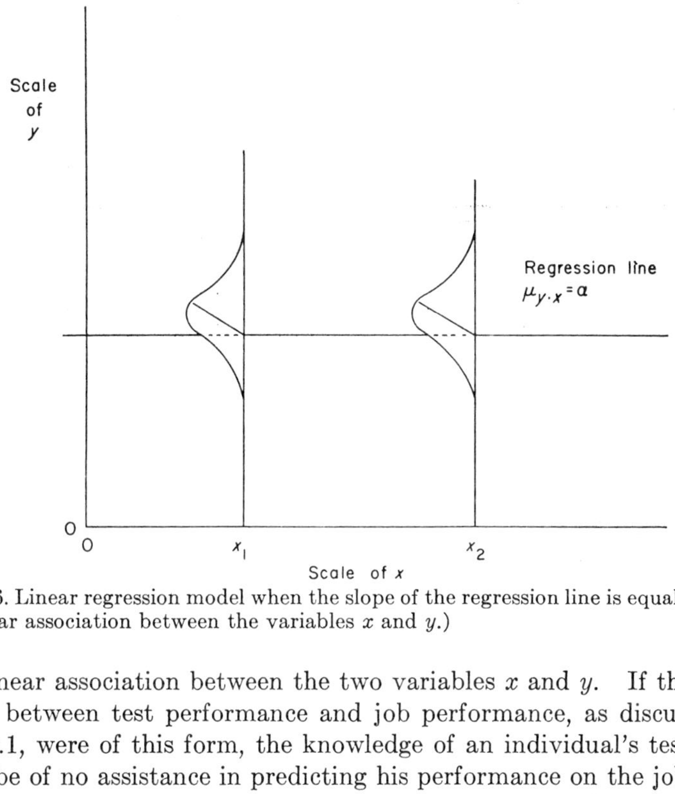

FIG. 12.6. Linear regression model when the slope of the regression line is equal to zero. (No linear association between the variables x and y.)

is no linear association between the two variables x and y. If the association between test performance and job performance, as discussed in Sec. 12.1, were of this form, the knowledge of an individual's test score would be of no assistance in predicting his performance on the job.

If repeated samples of n pairs of observations are drawn from a population satisfying the requirements of the linear-regression model, including the stated restrictions on sample selection, and if formula (12.3) is used to compute b for each sample, the value of b will vary from sample to sample. In order to evaluate the relation between an observed value of b and the unknown value of β, it is necessary to consider the sampling distribution of b. The following results can be proved concerning this sampling distribution:

If an indefinitely large number of samples of n pairs of observations are drawn from a population satisfying the requirements of the linear-regression model, and *if the n values of the independent variable, $x_1, x_2, \ldots, x_n$, are exactly the same for each of these samples*, then:

1. The long-run average of the values of b is equal to β. That is, $E(b) = \beta$.
2. The variance of b is given by

$$V(b) = \frac{\sigma^2}{\Sigma(x_i - \bar{x})^2} = \frac{\sigma^2}{\sum x_i^2 - \frac{(\Sigma x_i)^2}{n}} \tag{12.7}$$

If the sample size is large, then under the foregoing conditions the sampling distribution of b can be approximated by a normal distribution having a mean equal to β and a variance as given by (12.7). Note the requirement that each repeated sample have exactly the same n values of the independent variable. This means that an actual sample is drawn in accordance with either of the procedures set forth in Sec. 12.2.2, and the estimates obtained from this sample are evaluated in terms of an infinite population of samples that might be drawn, each sample in this infinite population having exactly the same n values of the independent variable.

The foregoing result concerning the sampling distribution of b can be used to determine confidence limits for β and to test hypotheses about the value of β. For example, 95 per cent confidence limits for β may be obtained in the following manner:

1. Draw a large sample of n pairs of observations, $(x_1,y_1), \ldots, (x_n,y_n)$.
2. Compute the value of b according to (12.3) and the value of $s_{y \cdot x}^2$ according to (12.6).
3. Determine the two quantities

$$b - 1.96 \sqrt{\frac{s_{y \cdot x}^2}{\Sigma(x_i - \bar{x})^2}}$$

and

$$b + 1.96 \sqrt{\frac{s_{y \cdot x}^2}{\Sigma(x_i - \bar{x})^2}}$$

4. State, with 95 per cent confidence, that the true value of β lies between the two numbers obtained in step 3.

If other confidence coefficients are desired, an appropriate multiplier can be selected from Table 9.3. Observe that the value of σ^2, which enters into formula (12.7), has been approximated by its sample estimate in computing the two confidence limits. This is possible because of the assumption that the sample size is large.*

As an application of these techniques, consider the 35 observations shown in Fig. 12.7. These data were obtained by recording, on each of 35 days, the number of customers x passing through a supermarket and

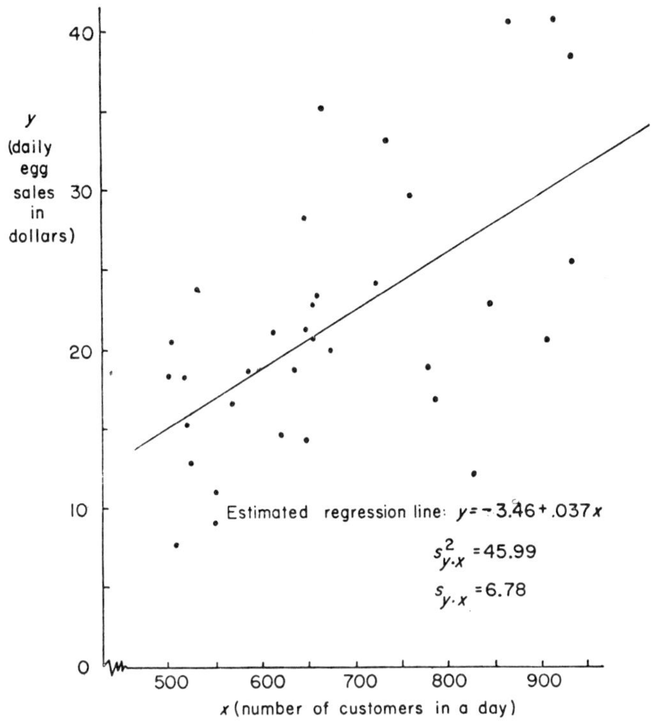

FIG. 12.7. Scatter diagram showing association between daily number of customers and value of eggs sold by a supermarket. SOURCE: Marketing study conducted by the Department of Agricultural Economics, Cornell University.

the value of eggs sold on that day, y. (The x scale has not been extended to zero, and it is therefore not possible to show the value of a on this figure.) The quantities required in the analysis have been computed from these data. Their values are:

* If the sample size is small, an additional condition must be imposed on the linear-regression model in order to obtain confidence limits for β. It is necessary to assume that the distribution of y, for each fixed value of x, is itself a normal distribution. Under these conditions, the sampling variability of $s^2_{y \cdot x}$ can be taken into account by obtaining the appropriate multiplier for a stated confidence coefficient from a Student's t distribution with $(n - 2)$ degrees of freedom.

Basic data:

$$\Sigma x_i = 23{,}858 \qquad \Sigma y_i = \$760.61$$
$$\Sigma x_i^2 = 16{,}924{,}770 \qquad \Sigma y_i^2 = 18{,}950.5165$$
$$\Sigma x_i y_i = 542{,}930.54$$

*Least-squares regression line:**

$$y = -3.46 + .037x$$

Variance about regression line:

$$s_{y \cdot x}^2 = 45.9864$$
$$s_{y \cdot x} = 6.78$$

Squared deviations of x values about their mean:

$$\Sigma(x_i - \bar{x})^2 = 661{,}794$$

Therefore the 95 per cent confidence limits for β are

$$\begin{aligned} b \pm 1.96 \sqrt{\frac{s_{y \cdot x}^2}{\Sigma(x_i - \bar{x})^2}} &= .037 \pm 1.96 \sqrt{\frac{45.9864}{661{,}794}} \\ &= .037 \pm 1.96(.0083) \\ &= .037 \pm .016 \\ &= .021 \text{ and } .053 \end{aligned}$$

The slope of the regression line has a particularly simple interpretation in this example. It is the amount, in dollars, by which the average expenditure for eggs increases for each additional customer passing through the store.

Procedures for testing hypotheses concerning the value of β also follow directly from the stated characteristics of the sampling distribution of b.

* It might be argued that this regression line should be *forced* to go through the origin since no customers obviously means no egg sales. However, the observed values of x are so far from zero that we have not wished to restrict our analysis in this manner. (There is no evidence that the regression would be linear for x close to zero, and the variance about the regression line would undoubtedly decrease as x approached closer and closer to zero.) If a regression line is forced through the origin, that is, the value of α is taken to be zero, it can be shown that the following are the appropriate formulas for estimating β and σ^2:

$$b = \frac{\Sigma x_i y_i}{\Sigma x_i^2}$$

and

$$\begin{aligned} s_{y \cdot x}^2 &= \frac{\Sigma(y_i - bx_i)^2}{n - 1} \\ &= \frac{\Sigma y_i^2 - b\Sigma x_i y_i}{n - 1} \end{aligned}$$

Ordinarily one has no a priori basis for hypothesizing some particular value for β and merely wishes to ascertain that β differs from zero, that is, to ascertain that the "no linear association" situation pictured in Fig. 12.6 does not hold. A large sample procedure for testing the hypothesis that $\beta = 0$ against the alternative that $\beta \neq 0$, at the 1 per cent level of significance, is the following:

1. Draw a large sample of n pairs of observations, $(x_1,y_1), \ldots, (x_n,y_n)$.
2. Compute the value of b according to (12.3) and the value of $s^2_{y \cdot x}$ according to (12.6).
3. If b is less than

$$-2.58 \sqrt{\frac{s^2_{y \cdot x}}{\Sigma(x_i - \bar{x})^2}}$$

or greater than

$$+2.58 \sqrt{\frac{s^2_{y \cdot x}}{\Sigma(x_i - \bar{x})^2}}$$

reject the hypothesis that $\beta = 0$ in favor of the alternative that $\beta \neq 0$.

4. Otherwise, *accept* the hypothesis that $\beta = 0$.

Other levels of significance are obtained by choosing an appropriate multiplier from Table 9.3.

The y intercept of the population regression line must also be estimated from a sample. The following results can be proved concerning the sampling distribution of a:

> If an indefinitely large number of samples of n pairs of observations are drawn from a population satisfying the requirements of the linear-regression model and if the n values of the independent variable, $x_1, x_2, \ldots, x_n$, are exactly the same for each of these samples, then:
>
> 1. The long-run average of the values of a is equal to α. That is, $E(a) = \alpha$.
> 2. The variance of a is given by
>
> $$V(a) = \frac{\sigma^2 \Sigma x_i^2}{n\Sigma(x_i - \bar{x})^2} \tag{12.8}$$

If the sample size n is large, then under the foregoing conditions the sampling distribution of a can be approximated by a normal distribution having a mean equal to α and a variance as given by (12.8).

These results can be used to obtain confidence limits for α and to test hypotheses concerning the value of α. As in the case of b, the value of σ^2 is estimated by $s^2_{y \cdot x}$. Thus the 95 per cent confidence limits for α in the case of the data of Fig. 12.7 are

$$\begin{aligned} a \pm 1.96\sqrt{\frac{s_{y\cdot x}^2 \Sigma x_i^2}{n\Sigma(x_i - \bar{x})^2}} &= -3.46 \pm 1.96\sqrt{\frac{(45.9864)(16{,}924{,}770)}{(35)(661{,}794)}} \\ &= -3.46 \pm 1.96(5.80) \\ &= -3.46 \pm 11.37 \\ &= -14.83 \text{ and } +7.91 \end{aligned}$$

These confidence limits appear to be extremely far apart. This happens because α is the value on the line at $x = 0$ and the observed values of x are all much greater than zero. This would be more apparent if the x scale in Fig. 12.7 had been continued all the way to $x = 0$.

One of the principal advantages associated with linear-regression analysis is that it is possible to estimate the mean value of the variable y for any fixed value of the variable x, even though one has few, if any, observations from the specified x population. For example, let us focus attention on a value of x, say x_0. Then we know that the mean value of y for all elements identified by x_0 is given by

$$\mu_{y\cdot x_0} = \alpha + \beta x_0$$

The values of α and β are unknown, but they can be estimated by a and b as computed from a sample. Therefore, $\mu_{y\cdot x_0}$ is estimated by

$$\bar{y}_{x_0} = a + bx_0$$

Thus with the data of Fig. 12.7, the average daily egg sales for all days in which exactly 600 customers passed through the store would be estimated by

$$\begin{aligned} \bar{y}_{600} &= -3.46 + (.037)(600) \\ &= \$18.74 \end{aligned}$$

But this is only an estimate of the true average, and we would like to obtain a measure of sampling variability to attach to this estimate.

The sampling variability of $\bar{y}_{x_0}$ is clearly influenced by the sampling variability of both a and b. The combined influence of these two sources of sampling variability is mirrored in the following:

> If an indefinitely large number of samples of n pairs of observations are drawn from a population satisfying the requirements of the linear-regression model, and if the n values of the independent variable, $x_1, x_2, \ldots, x_n$, are exactly the same for each of these samples, then:
>
> 1. The long-run average of the values of $\bar{y}_{x_0} = a + bx_0$ is equal to $\alpha + \beta x_0$. That is, $E(\bar{y}_{x_0}) = \mu_{y\cdot x_0}$.

2. The variance of $\bar{y}_{x_0}$ is

$$V(\bar{y}_{x_0}) = \left[\frac{1}{n} + \frac{(x_0 - \bar{x})^2}{\Sigma(x_i - \bar{x})^2}\right]\sigma^2 \tag{12.9}$$

If the sample size n is large, then under the foregoing conditions the sampling distribution of $\bar{y}_{x_0}$ can be approximated by a normal distribution. Applying these results to obtain 95 per cent confidence limits for $\mu_{y \cdot 600}$ from the data of Fig. 12.7, we have

$$\begin{aligned} a + bx_0 \pm 1.96\sqrt{\left[\frac{1}{n} + \frac{(x_0 - \bar{x})^2}{\Sigma(x_i - \bar{x})^2}\right]s^2_{y\cdot x}} &= 18.74 \pm 1.96\sqrt{\left[\frac{1}{35} + \frac{(600 - 681.66)^2}{(661{,}794)}\right](45.9864)} \\ &= 18.74 \pm 1.96(1.33) \\ &= 18.74 \pm 2.61 \\ &= 16.13 \text{ and } 21.35 \end{aligned}$$

Thus we state, with 95 per cent confidence, that the true value of $\mu_{y \cdot 600}$ lies between \$16.13 and \$21.35.

Notice that the variance of $\bar{y}_{x_0}$, as given by formula (12.9), depends directly upon the quantity $(x_0 - \bar{x})^2$. The farther x_0 is from the arithmetic mean of the observed values of the x_i's the greater is the variance of the sampling distribution of $\bar{y}_{x_0}$ and the greater is the length of the confidence interval for $\mu_{y \cdot x_0}$. This is illustrated in Fig. 12.8 where the 95 per cent confidence limits for $\mu_{y \cdot x_0}$ have been plotted with solid curves for values of x_0 ranging from 400 to 1,000, the computations having been performed on the data of Fig. 12.7. It can be seen that these confidence limits are closest together when $x_0 = \bar{x}$ and that they become farther and farther apart as x_0 departs from $\bar{x}$ in either direction. (When $x_0 = 0$, we would have the limits -14.83 and $+7.91$, and these are the 95 per cent limits that were previously obtained for the value of α.) The region in which the observed points of Fig. 12.7 fall is roughly indicated by the shaded area in Fig. 12.8, and the limits become quite far apart when x_0 lies outside this area. It is also true that if x_0 lies outside the area of observed values of x, then the population regression may depart from linearity and one has no evidence to indicate whether or not this is happening. Thus inferences about $\mu_{y \cdot x_0}$, where x_0 falls outside the area in which the observed values of x fall, must be made with great care.

As a final problem in inference for the linear-regression model consider the following situation. A large sample of n pairs of observations, $(x_1,y_1), \ldots, (x_n,y_n)$ is available from which to compute estimates of α, β, and σ^2. A *single additional* observation is to be obtained from a known x population. Let us again designate this known value of x as x_0.

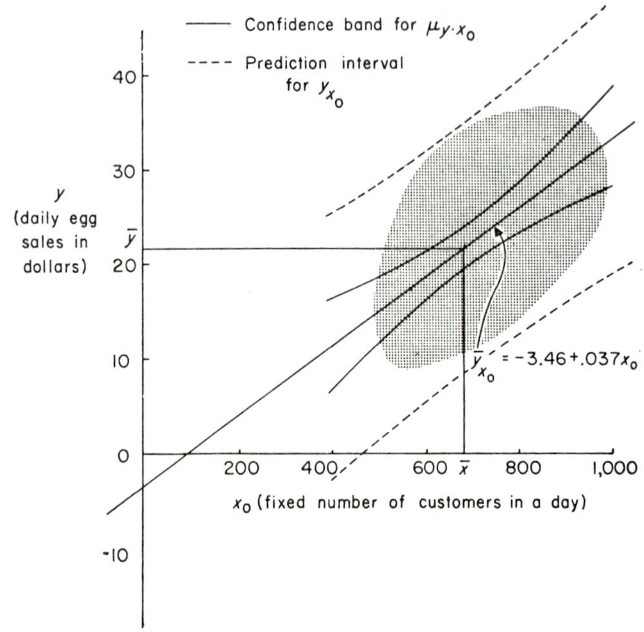

FIG. 12.8. 95 per cent confidence intervals for $\mu_{y \cdot x_0}$ plotted as a function of x_0. (Data taken from Fig. 12.7; shaded area roughly indicates region in which observed data fall.)

We desire an objective statement or prediction, on the basis of the n observations, concerning the value of y that will be associated with this element. This single value of y will be designated as y_{x_0}. Thus for the supermarket example of Fig. 12.7, the manager might know that there will be approximately 700 customers on a given day. He wishes to predict the range within which he can expect the egg sales to fall on that day. Note that this is quite a different problem from that of estimating $\mu_{y \cdot 700}$, the average daily sale of eggs for an indefinitely large number of days on which exactly 700 customers pass through the store on each day.

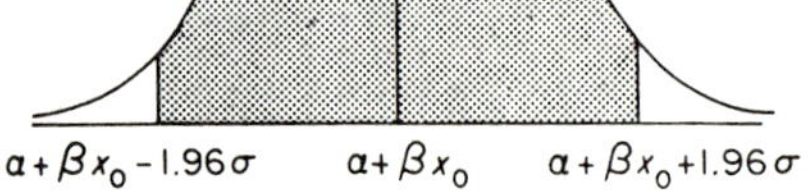

FIG. 12.9. 95 per cent prediction interval for an individual observation with x known to be x_0. (Values of α, β, and σ assumed known.)

If the exact values of α, β, and σ^2 were available, the desired goal could be easily attained, provided that the distribution of individual y values in the x_0 population were known to follow a normal distribution. The situation would then be as pictured in Fig. 12.9. Since the distribution of y_{x_0} is assumed to be normal with mean $\alpha + \beta x_0$ and variance σ^2, we can make such probability statements as

$$\Pr[\alpha + \beta x_0 - 1.96\sigma \leq y_{x_0} \leq \alpha + \beta x_0 + 1.96\sigma] = .95$$

Hence, if in advance of obtaining the single observation we always made the statement "The value of y will fall in the interval $\alpha + \beta x_0 - 1.96\sigma$ to $\alpha + \beta x_0 + 1.96\sigma$," this statement would be correct 95 per cent of the time and incorrect 5 per cent of the time. We shall speak of the interval $\alpha + \beta x_0 - 1.96\sigma$ to $\alpha + \beta x_0 + 1.96\sigma$ as a 95 per cent prediction interval.

When a prediction interval is to be determined from a sample it is necessary to take into account not only the variability of the individual y_{x_0}'s about their mean $\mu_{y \cdot x_0}$ but also the fact that this mean is estimated from a sample. It can be shown that the following is a 95 per cent prediction interval for an individual y_{x_0}:

$$a + bx_0 - 1.96\sqrt{\left[\frac{n+1}{n} + \frac{(x_0 - \bar{x})^2}{\Sigma(x_i - \bar{x})^2}\right] s^2_{y \cdot x}}$$

to

$$a + bx_0 + 1.96\sqrt{\left[\frac{n+1}{n} + \frac{(x_0 - \bar{x})^2}{\Sigma(x_i - \bar{x})^2}\right] s^2_{y \cdot x}}$$

The use of the multiplier 1.96 assumes that the sample size is large so that the sampling variability of $s^2_{y \cdot x}$ can be ignored. [If the sample size is small, the multiplier can be obtained from Student's t distribution with $(n - 2)$ degrees of freedom.]

Returning to the data of Fig. 12.7, we have as a 95 per cent prediction interval for an individual y_{700}:

$$-3.46 + (.037)(700) \pm 1.96\sqrt{\left[\frac{36}{35} + \frac{(700 - 681.66)^2}{661{,}794}\right] (45.9864)}$$
$$= 22.44 \pm 1.96(6.88)$$
$$= 22.44 \pm 13.48$$
$$= 8.96 \text{ to } 35.92$$

This is an extremely wide interval, but this is simply a result of the fact that the observations in Fig. 12.7 show considerable variability about the regression line. The 95 per cent prediction intervals for other values of x_0 are indicated by the dashed lines in Fig. 12.8. Here also, as in the case of confidence intervals for $\mu_{y \cdot x_0}$, the interval is narrowest when $x_0 = \bar{x}$, but the increase in length is not very marked when x_0 deviates from $\bar{x}$.

12.2.5. *Additional Comments on the Linear-regression Model and the Least-squares Criterion.* The n pairs of observations required for making a regression analysis can be obtained in either of the ways pictured in Fig. 12.1, just so long as the deviations of the individual y values from their respective means are independent of one another. If it is possible to select arbitrarily the x populations from which observations are to be drawn and if a particular goal is to be achieved with the regression analysis, then a judicious choice of the x populations will often permit one to achieve "the best possible results" for given expenditure of effort, that is, for a given sample size. For example, if interest is centered primarily on obtaining an estimate of β, an examination of formula (12.7) indicates that $V(b)$ will be small when $\Sigma(x_i - \bar{x})^2$ is large. It can be shown that $\Sigma(x_i - \bar{x})^2$ will be large, for fixed sample size, when all observations are taken from two x populations, say x_1 and x_2, where x_1 and x_2 are as far apart as possible. Furthermore, Fig. 12.8 shows that the confidence interval for $\mu_{y \cdot x_0}$ is narrowest when $x_0 = \bar{x}$. Therefore, if interest is to be centered on some particular value of x_0, the x populations should be chosen so that $\bar{x}$ is close to x_0.

Many different problems of inference were treated in Sec. 12.2.4. Thus confidence intervals were presented for α, β, and $\mu_{y \cdot x_0}$, and a prediction interval was constructed for y_{x_0}. If a *single* sample is used to compute a large number of these intervals, then the probability relations underlying the construction of the intervals will not hold exactly. The estimated values of these quantities depend upon one another. Thus an "odd" value of b will lead to an "odd" value of a and to "odd" values of $\bar{y}_{x_0}$. A strict interpretation of theory requires that a new sample be drawn for each problem of inference. This is rarely possible in practice, but an investigator should at least be aware of the fact that he may be working with confidence coefficients which differ from those that appear on the surface to be correct.

The general approach outlined in the preceding sections can be extended to cover two-variable situations where a curvilinear- rather than linear-regression model is appropriate. The steps required in this type of analysis are:

1. One attempts to find an equation expressing the approximate relationship between $\mu_{y \cdot x}$ and x over the range of values of x in which interest is centered. Thus the equation $\mu_{y \cdot x} = \alpha + \beta x + \gamma x^2$ is one equation which produces a curvilinear relation.

2. The manner in which $\sigma^2_{y.x}$ depends upon the value of x must be specified. As for the linear-regression model, the simplest case occurs when $\sigma^2_{y.x} = \sigma^2$ for all values of x.

3. Procedures must be developed for estimating the parameters of the model from a sample. For example, if $\mu_{y \cdot x} = \alpha + \beta x + \gamma x^2$, it would be necessary to estimate the values of α, β, γ, and σ^2. The criterion of least squares can often be used to determine these estimation procedures.

4. Finally, the estimation procedures developed in step 3 must be such

that it is possible to solve stated problems of inference. A discussion of many of these problems will be found in Anderson and Bancroft [4, chaps. 13–15].

Even though the relation between $\mu_{y \cdot x}$ and x is curvilinear, it is sometimes possible to make changes, or *transformations*, in the original variables y and x so that the techniques developed for the linear-regression model can be applied directly. For example, it was noted in the discussion of Fig. 12.3 that the regression of money value of food consumed per week per family on total yearly family income is definitely curvilinear when the entire range of possible values of yearly family income is considered. The cited source for these data [1] performed linear-regression analyses, but these analyses were made principally on the logarithms of the weekly food expense and the yearly family income rather than on the original observations. The use of *logarithmic transformations*, on one or both of the variables y and x, is treated in Yule and Kendall [5, chap. 15].

The criterion of least squares can be employed in "fitting" lines or curves to a group of observed points even though there is apparently only a single value of y, rather than a distribution of y's, corresponding to each fixed value of x. If a straight line is to be fitted to such a group of points, formulas (12.3) and (12.4) can still be used to obtain the slope and y intercept of the best-fitting line. This situation arises frequently in the study of economic time series. The points plotted in Fig. 12.10 provide one example of a time series. Time is taken as the independent variable in an analysis of this type, and the horizontal scale shows the years from 1920 to 1955. The successive values of x are taken as 0, 1, 2, . . . , 35, where $x = 0$ corresponds to 1920, and these numbers are used in the analysis, thus simplifying computations. The values of y in this instance are the per-person (14 years of age and over in the United States) consumption of cigarette tobacco as measured in pounds of unstemmed-processing weight.

The points of Fig. 12.10 show somewhat irregular year-to-year fluctuations, probably due to such factors as economic conditions or number of individuals in the Armed Forces, and in recent years to publicity concerning the association between smoking and lung cancer. However, the general trend in cigarette consumption from 1920 to 1955 has been clearly upward. It is possible to fit a *linear-trend* line to these points by computing a and b from formulas (12.3) and (12.4), and this line is shown in Fig. 12.10. The line does not appear to fit too well since all points below 1930 are above the line, all points from 1931 to 1941 are below the line, and all points (except the last) beginning with 1943 are above the line. This situation is improved somewhat by fitting a curve of the form $y = ab^x$, which is shown with the dashes in Fig. 12.10, but there are still large groups of successive points which are either all above or all below the curve. The use of the curve $y = ab^x$ assumes that there is a *constant percentage increase* in cigarette consumption from year to year, and the techniques* for fitting such a curve follow easily from the least-squares

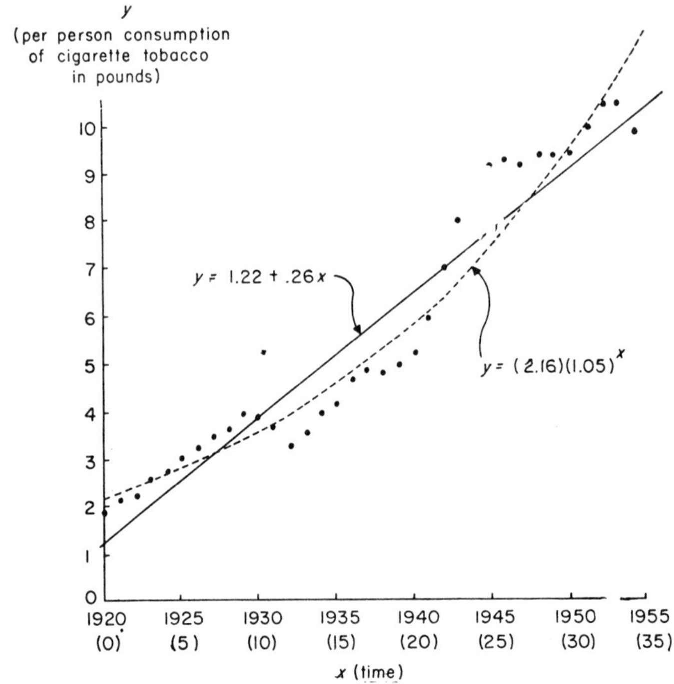

FIG. 12.10. Consumption of cigarette tobacco in the United States, 1920 to 1954. (Per person, 14 years of age and over, consumption of cigarette tobacco as measured in pounds of unstemmed-processing weight.) SOURCE: Benno K. Milmore and Arthur G. Conover, "Tobacco Consumption in the United States, 1880 to 1954," *Agr. Econ. Res.*, **8**:9–13 (1956).

procedures developed for estimating a linear-regression line. The equation of the fitted curve, $y = (2.16)(1.05)^x$, indicates a 5 per cent increase in cigarette consumption per year.

Under some circumstances it may be possible to argue that a time series follows either a linear-regression model or some related regression model, even though the value of the dependent variable at first sight appears to be a fixed quantity for any given year. Thus the data of Fig. 12.10 can be viewed in the following manner. Cigarette consump-

* Formulas (12.3) and (12.4) are used to obtain the least-squares line for the n pairs of observations $(x_1, \log y_1)$, $(x_2, \log y_2)$, . . . , $(x_n, \log y_n)$. Let the slope of this line be b' and the y intercept a'. Then a is the number whose logarithm is a', and b is the number whose logarithm is b'.

tion, over the time period covered, is increasing on the average at a constant percentage rate as described by an equation of the form $y = ab^x$. There is a fixed amount of consumption for any given year, but this consumption varies from the regression curve because of the net effect of many chance factors—some people decide to give up smoking, others change from cigars to cigarettes, economic conditions force individuals to curtail their amount of smoking, and so on. If any of these conditions were changed slightly, a different value would have been observed for that year. However, even with this view of the situation, special problems arise in trying to apply the techniques of inference that were developed in Sec. 12.2.4. In particular, the basic requirement of independence of observations is violated. The level of cigarette consumption for a year to some extent determines the level of cigarette consumption for the following year. Habits and general economic conditions will tend to persist from one year to the next. Thus if a point is above the regression curve for one year, it will tend also to be above the regression curve the next year. This dependence of observations from one time period to the next time period raises very difficult problems in studying time series. An introductory account of some of the approaches to this problem will be found in Yule and Kendall [5, chaps. 26, 27].

As a final point, we observe that the criterion of least squares is not the only criterion that can be used to obtain estimation procedures for the parameters of the linear-regression model. However, a discussion of these other criteria would require methods beyond the scope of this book, and it can be shown that the least-squares estimates of α and β satisfy most reasonable requirements that one may choose to impose.

12.3. Measuring the Degree of Linear Association between Two Quantitative Variables

12.3.1. *The Correlation Coefficient.* When two quantitative variables are related in accordance with the linear-regression model, where the slope of the regression line is different from zero, the two variables are said to be linearly associated. The parameters of the regression model (α, β, and σ^2), or sample estimates of them, describe this association, but they do not directly provide a measure of the degree or strength of the association. This section will be devoted to the presentation and discussion of one quantity which is often used as a measure of degree of linear association.

Intuitive notions concerning degree of association can be developed by examining and comparing a number of different scatter diagrams. Consider, for example, the data of Figs. 12.4 and 12.7. The observed points in Fig. 12.4—average density in vehicles per mile and average speed in miles per hour for 5-minute intervals of traffic flow through the fast lane of the Lincoln Tunnel—appear to be closely grouped about the least-squares regression line. On the other hand, the observed points in Fig. 12.7—number of customers and value of egg sales for the daily operation of a supermarket—appear to be widely scattered about the regression line. This leads to a feeling that the data of Fig. 12.4 indicate a higher degree of association than do those of Fig. 12.7 and suggests that a measure of association can be based in some fashion on the variation about the regression line. It is quite clear, however, that the variances themselves will not serve this purpose since different units of measurement (feet, dollars, miles, pounds, and so on) will be involved in different regression analyses, and a direct comparison of quantities having different units would be meaningless. Hence a measure is desired which depends on the variance about the regression line and which is a pure number, that is, a number having no units of measurement.

It was shown in Fig. 12.6 that the absence of linear association is indicated by a regression line whose slope β is equal to zero. This fact can be used in the following manner to obtain a measure of association:

1. We first act as if there were no association between the two variables x and y and fit a line whose slope is forced to be zero. That is, we fit a regression line of the form $\mu_{y \cdot x} = \alpha$. It was shown in the discussion of Form A of Fig. 12.4 that the least-squares *horizontal* line is the line $y = \bar{y}$ and the variation about this line, as measured by the sum of squared deviations, is $\Sigma(y_i - \bar{y})^2$. We may, if we wish, view this line as the one we would have to use if the x populations of the individual sample elements were not identified.

2. We next proceed as if there were linear association between the two variables x and y and fit the general regression line $\mu_{y \cdot x} = \alpha + \beta x$. The variation of the observed points about this line, as measured by the sum of squared deviations, is $\Sigma[y_i - (a + bx_i)]^2$. It can be shown that this sum of squared deviations can never be greater than $\Sigma(y_i - \bar{y})^2$. That is, the worst that can happen is that the sample estimate of β turns out to be zero. In this case $\Sigma[y_i - (a + bx_i)]^2$ will be exactly equal to $\Sigma(y_i - \bar{y})^2$. Otherwise it will be smaller.

3. We compute the amount by which we have reduced $\Sigma(y_i - \bar{y})^2$ by letting the value of β be different from zero, namely, $\Sigma(y_i - \bar{y})^2 - \Sigma[y_i - (a + bx_i)]^2$, and express this as a fraction of the "no association" variation. This fraction, designated by r^2, will be taken as a measure of linear association. Thus

$$r^2 = \frac{\Sigma(y_i - \bar{y})^2 - \Sigma[y_i - (a + bx_i)]^2}{\Sigma(y_i - \bar{y})^2}$$

$$= 1 - \frac{\Sigma[y_i - (a + bx_i)]^2}{\Sigma(y_i - \bar{y})^2} \qquad (12.10)$$

This formula serves to define r^2, but it does not provide a convenient means of computation. Formula (12.11) will ordinarily be used for computation.

The quantity r^2 is known as *the square of the linear, or product-moment, correlation coefficient.* Its value for the data of Fig. 12.4 is .94 and for the data of Fig. 12.7 is .37, and these values are at least in line with the intuitive observation that the degree of association between density and speed is greater than that between number of customers and value of egg sales. Some of the properties of r^2 will now be considered.

The following conclusions concerning r^2 can be drawn from an examination of the defining formula (12.10):

1. Both the numerator and denominator of r^2 are in squared units of the variable y. These units of measurement cancel one another, and the quantity r^2 is therefore a pure number. Hence values of r^2 can be compared for any two situations, no matter how different may be the variables involved.

2. The sum of squared deviations about the fitted regression line, $\Sigma[y_i - (a + bx_i)]^2$ can never be less than zero since it is the sum of positive terms. This sum is exactly equal to zero only if each observed point falls on the fitted regression line, as pictured in Fig. 12.11. The quantity r^2 is then equal to $1 - 0/\Sigma(y_i - \bar{y})^2 = 1$, and this is the greatest possible value of r^2. If this is the case, we say that the sample indicates *perfect linear association* between x and y.

3. It can be shown that the greatest possible value of $\Sigma[y_i - (a + bx_i)]^2$ is $\Sigma(y_i - \bar{y})^2$, and this occurs when the slope of the regression line is equal to zero (see Fig. 12.6). If the slope is zero, formula (12.10) shows that $r^2 = 0$, and this is taken to indicate the absence of linear association between the two variables.

4. A value of r^2 falling between the limits 0 and 1 gives the fraction of the variability in the observed y values which is explained or accounted for by employing the regression of y on x. Thus the regression of speed on density (Fig. 12.4) accounts for 94 per cent of the variability in the observations in speed, while the regression of egg sales on number of customers accounts for only 37 per cent of the variability in egg sales (Fig. 12.7).

Although the quantity r^2 has a direct quantitative meaning as the fraction of the observed variability which is explained or accounted for by employing the regression of y on x, the square root of r^2 is most commonly reported as a measure of degree of association. The value of r, which may be assigned a plus or minus sign, is called *the linear, or product-moment, correlation coefficient* between x and y. The sign of r is taken to be positive if the slope of the regression line is positive, and negative if the slope of the regression line is negative. Thus for the data of Fig. 12.4, $r = -\sqrt{.94} = -.97$, and for the data of Fig. 12.7,

$$r = +\sqrt{.37} = +.61$$

The value of r can range from -1 to $+1$, corresponding to the range of values of r^2 from 0 to 1. A value of -1 indicates perfect negative linear association (as in Fig. 12.11, but with the line having a negative slope); a value of 0 indicates the absence of linear association (as in Fig. 12.6); and a value of $+1$ indicates perfect positive linear association (as in Fig. 12.11).

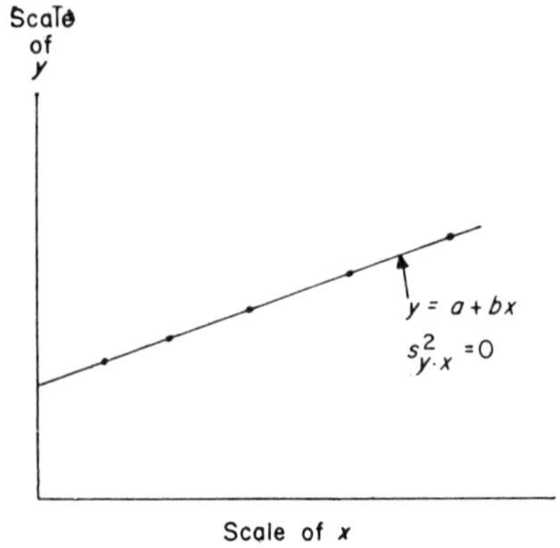

FIG. 12.11. A sample indicating perfect linear association between two quantitative variables. (Observed points fall exactly on the fitted regression line.)

If the values of $\Sigma[y_i - (a + bx_i)]^2$ and $\Sigma(y_i - \bar{y})^2$ are available, r^2 can be computed directly from formula (12.10). It often happens, however, that the value of the correlation coefficient is to be obtained directly from the basic data. Under these circumstances, it is not convenient to use (12.10) as a computational formula. It can be shown by algebraic manipulation that the value of r is given directly by the formula

$$r = \frac{n\Sigma x_i y_i - (\Sigma x_i)(\Sigma y_i)}{\sqrt{[n\Sigma x_i^2 - (\Sigma x_i)^2][n\Sigma y_i^2 - (\Sigma y_i)^2]}} \tag{12.11}$$

The use of this formula in computation also results in the appropriate sign (plus or minus) for r. Its application to the data of Fig. 12.4 (see Tables 12.2 and 12.3) is as follows:

Basic quantities:

$$n = 10$$
$$\Sigma x_i = 661 \qquad \Sigma y_i = 206.3$$
$$\Sigma x_i^2 = 46{,}153 \qquad \Sigma y_i^2 = 4{,}526.73$$
$$\Sigma x_i y_i = 12{,}845.7$$

Computation of r:

$$r = \frac{(10)(12{,}845.7) - (661)(206.3)}{\sqrt{[(10)(46{,}153) - (661)^2][(10)(4{,}526.73) - (206.3)^2]}}$$
$$= \frac{-7{,}907.3}{\sqrt{[(24{,}609)(2{,}707.61)]}}$$
$$= \frac{-7{,}907.3}{(156.9)(52.0)} = \frac{-7{,}907.3}{8{,}158.8}$$
$$= -.97$$

The reader may have observed that the foregoing discussion of the correlation coefficient has violated one of the conventions which has been followed throughout this book. Only the sample correlation coefficient r has been introduced, and at no point has the population correlation coefficient been mentioned. The reason for this is very simple. The value of the correlation coefficient, as defined by formula (12.10) or (12.11), depends directly upon the x populations chosen for study, even though these particular x populations are completely enumerated. This

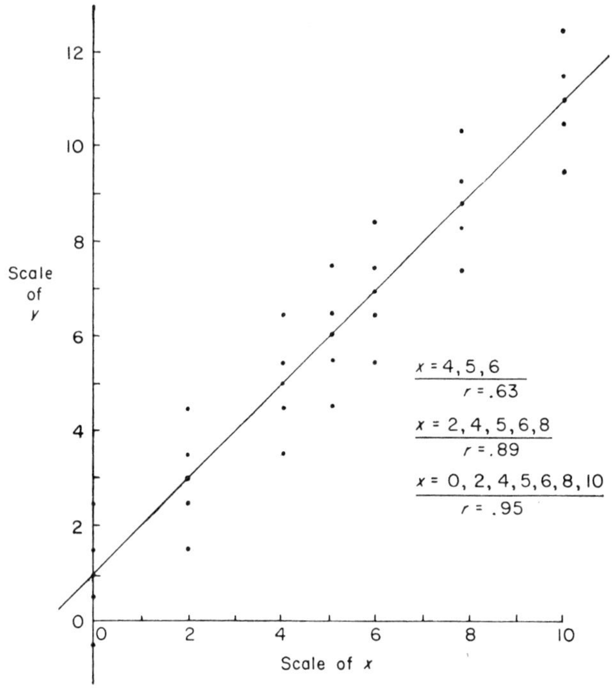

FIG. 12.12. Hypothetical data, following the linear-regression model, used to indicate the effect of choice of x populations on the value of the correlation coefficient.

is illustrated by the hypothetical data of Fig. 12.12. The mean of each group of y values, defined by a fixed value of x, falls exactly on the regression line $y = 1 + x$, and the variance about the mean is the same for each group. In accordance with the treatment of the linear-regression model in Sec. 12.2, we would obtain the same estimate of α and β and almost the same estimate of σ^2 no matter which two or more of these seven groups of points were chosen for study. (All five points in each of the chosen groups are used in the analysis.) But this is not true for the correlation coefficient! If we take the three x populations for $x = 4$, 5, and 6, the value of r is found to be .63; for the five x populations for

$x = 2, 4, 5, 6$, and 8, r is .89; and for the seven x populations for $x = 0, 2, 4, 5, 6, 8$, and 10, r is .95. Speaking roughly, the variance about the regression line remains the same no matter which of the x populations is used. However, the variance of the y values, with which this is compared in obtaining the correlation coefficient, increases as the x values become more and more spread apart. This leads to increases in the value of r. Thus it is difficult to speak meaningfully about the correlation coefficient in the case of the linear-regression model unless conditions are added to those stated in Sec. 12.2. If it is convenient or necessary to use the correlation coefficient in conjunction with the linear-regression model, as described in Sec. 12.2, one should clearly recognize that its value will depend on the x populations used in the analysis.

This difficulty in interpretation can be removed if the use of the correlation coefficient is restricted to situations where the linear association between two variables can be viewed in a symmetric fashion. Not only must we consider the distribution of y for fixed values of x, but we must also be able to speak of the distribution of x values for a fixed value of y. More specifically, it is most convenient to work with the correlation coefficient when the following conditions hold:

1. There exists an indefinitely large population in which each element can be characterized with respect to a pair of measurements (x,y). If one examines the *entire* population on the variable x alone, then it is assumed that the distribution of x is normal. Similarly, it is assumed that the variable y has a normal distribution in the entire population.

2. For each fixed value of x, there is a distribution of y values. This distribution will be taken to be a normal distribution for all values of x. Designating its mean as $\mu_{y \cdot x}$ and its variance as $\sigma^2_{y \cdot x}$, we assume that

$$\mu_{y \cdot x} = \alpha + \beta x$$

and that $\sigma^2_{y \cdot x}$ is the same for each fixed value of x. The line $\mu_{y \cdot x} = \alpha + \beta x$ is the regression line of y on x.

3. For each fixed value of y, there is a normal distribution of x values whose mean is $\mu_{x \cdot y}$ and whose variance is $\sigma^2_{x \cdot y}$. We assume that

$$\mu_{x \cdot y} = \alpha' + \beta' y$$

and that $\sigma^2_{x \cdot y}$ is the same for all values of y. The line $\mu_{x \cdot y} = \alpha' + \beta' y$ is called the regression line of x on y, and this will not ordinarily be the same as the regression line of y on x.

4. A sample of n pairs of observations, (x_1,y_1), (x_2,y_2), . . . , (x_n,y_n) is obtained by drawing a random sample from the *entire* population. The x value for any one pair has no influence on the x value for any other pair, and the y value for any one pair has no influence on the y value for any other pair. Among other things, this means that we can no longer con-

sider a sampling scheme such as is illustrated in Form B of Fig. 12.1 or in Fig. 12.12. Under these circumstances, it is meaningful to talk about the sample correlation coefficient—as computed from formula (12.10) or (12.11)—as an estimate of a population correlation coefficient ρ. Furthermore, procedures can be developed for drawing inferences concerning ρ from sample correlation coefficients r.

The scatter diagram of n observations obtained from a population with these characteristics will ordinarily have the elliptical appearance

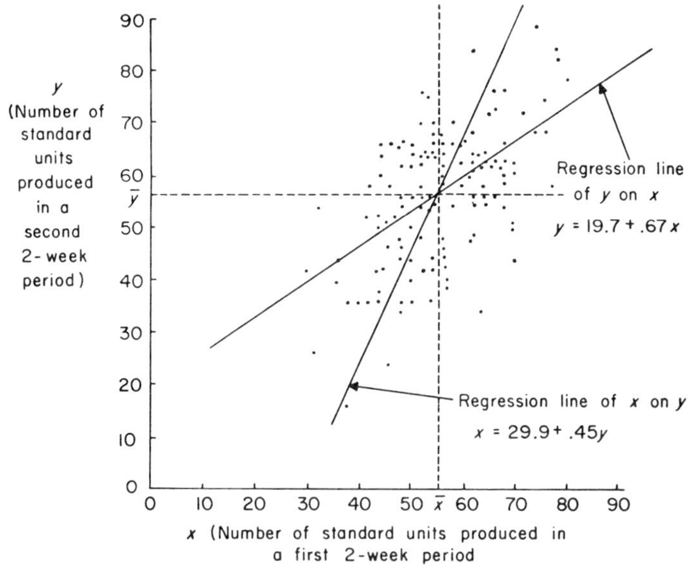

FIG. 12.13. Scatter diagram showing number of standard units produced in each of two production periods. (Production period of 2 weeks; data obtained from records of 130 regular employees in a machine shop.) SOURCE: Harold F. Rothe, "Output Rates among Machine Operators. I. Distributions and Their Reliability," *Jour. Appl. Psych.*, **31**:484–489 (1947).

of the data shown in Fig. 12.13. These data were obtained by examining the production records of a sample of 130 regular employees in a machine shop. A pair of observations was obtained for each of these employees. The x value of the pair gives the number of standard production units achieved in a first 2-week period and the y value gives the number of standard production units achieved in a second 2-week period. Since there is no reason for singling out one of these two production periods for special consideration, the variables x and y can be viewed in the symmetric fashion that has just been described.

A summary of the basic data and of the quantities that one would compute from these data in a linear-regression and correlation analysis is as follows:

*Basic data:**

$$n = 130$$
$$\Sigma x_i = 7{,}216 \qquad \Sigma y_i = 7{,}410$$
$$\Sigma x_i^2 = 414{,}984 \qquad \Sigma y_i^2 = 443{,}926$$
$$\Sigma x_i y_i = 421{,}012$$

Estimated regression line of y on x:

$$y = a + bx$$
$$b = \frac{n\Sigma x_i y_i - (\Sigma x_i)(\Sigma y_i)}{n\Sigma x_i^2 - (\Sigma x_i)^2} = .67$$
$$a = \bar{y} - b\bar{x} = 19.7$$
$$y = 19.7 + .67x$$

Estimated regression line of x on y:

$$x = a' + b'y$$
$$b' = \frac{n\Sigma x_i y_i - (\Sigma x_i)(\Sigma y_i)}{n\Sigma y_i^2 - (\Sigma y_i)^2} = .45$$
$$a' = \bar{x} - b'\bar{y} = 29.9$$
$$x = 29.9 + .45y$$

Correlation coefficient between x and y:

$$r = \frac{n\Sigma x_i y_i - (\Sigma x_i)(\Sigma y_i)}{\sqrt{[n\Sigma x_i^2 - (\Sigma x_i)^2][n\Sigma y_i^2 - (\Sigma y_i)^2]}}$$
$$= .55$$

Notice particularly that the regression line of y on x is quite distinct from the regression line of x on y. These two lines are drawn on Fig. 12.13. Formulas (12.3) and (12.4) can be used in computing the slope and intercept of the estimated regression line of x on y if we merely interchange the roles of x and y. The value of r is found to be .55 for this example. This means that $(.55)^2 = 30$ per cent of the variation in the y values is accounted for by the regression of y on x *and* that 30 per cent of the variation in the x values is accounted for by the regression of x on y.

12.3.2. *Inferences Concerning the Value of the Population Correlation Coefficient.* If repeated samples are drawn from a population in which the linear association between the two variables x and y can be viewed in a symmetric fashion, the sample correlation coefficient can be used as an estimate of a population correlation coefficient. Procedures for

* The cited source gave only the scatter diagram. The observations (x_i, y_i) were read from the scatter diagram, and so these quantities may differ slightly from the actual records.

obtaining confidence limits for ρ and for testing hypotheses about ρ will follow from the sampling distribution of r.

A discussion of the sampling distribution of r is beyond the scope of this book, but Fig. 12.14 presents a set of curves which lead to confidence

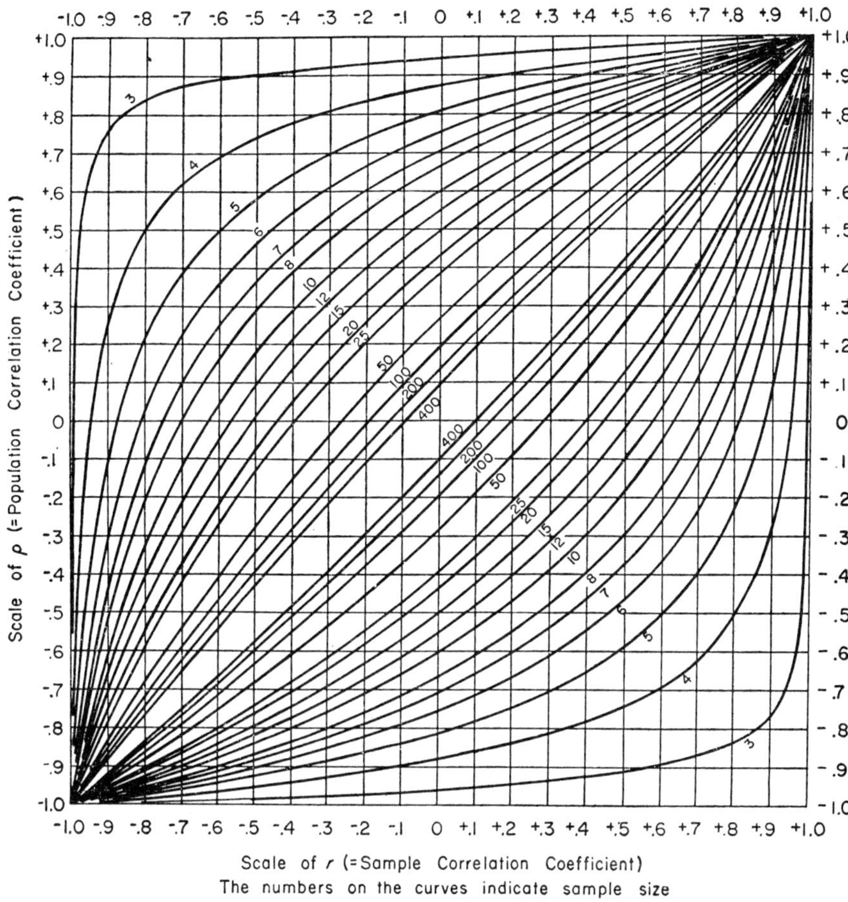

FIG. 12.14. 95 per cent confidence intervals for the correlation coefficient. (This chart is reproduced with the permission of Professor E. S. Pearson, from F. N. David, *Tables of the Ordinates and Probability Integral of the Distribution of the Correlation Coefficient in Small Samples*, Cambridge University Press for the Biometrika Trustees, 1938.)

limits for ρ and to tests of hypotheses concerning the value of ρ. These curves are interpreted and used in exactly the same manner as were the curves appearing in Figs. 8.5, 8.6, and 8.7 for obtaining confidence limits for the fraction of a population possessing some stated characteristic. For example, suppose we desire 95 per cent confidence limits for ρ in the population from which the data of Fig. 12.13 were drawn. The sample size was 130, and the value of r was found to be .55. Entering the chart of Fig. 12.14 on the horizontal scale—which includes the possible values of r from -1 to $+1$—at $r = .55$, we read up vertically to the two curves indicating a sample size of 130. (There are no curves for $n = 130$, and we must therefore interpolate between $n = 100$ and $n = 200$.) We read the two values of ρ on the vertical scale—which includes the possible values of ρ from -1 to $+1$—at which the vertical line through $r = .55$ intersects the two curves for $n = 130$. These two values are approximately $\rho_L = +.43$ and $\rho_U = +.66$, and they are 95 per cent confidence limits for the true value of ρ. That is, if repeated samples of size 130 were drawn from the population, the value of r would vary from sample to sample and therefore the pair of values, ρ_L and ρ_U, would vary from sample to sample. However, the curves in Fig. 12.14 have been drawn so that 95 per cent of these pairs of values will contain the true value of ρ between them.

The curves of Fig. 12.14 may also be used to test hypotheses concerning the unknown value of ρ. Suppose a sample of $n = 20$ is to be drawn from a population satisfying the conditions stated at the end of the preceding section for the purpose of testing the hypothesis that there is no linear association between the two variables ($\rho = 0$) against the alternative that there is association ($\rho \neq 0$). Entering the *vertical* scale at $\rho = 0$ and reading across, we find that the horizontal line intersects the two curves for $n = 20$ at $r = -.44$ and $+.44$. This means that

$$\Pr(-.44 \leq r \leq +.44 \mid \rho = 0) = .95$$
$$\Pr(r < -.44 \mid \rho = 0) = .025$$
$$\Pr(r > +.44 \mid \rho = 0) = .025$$

Therefore a two-sided test procedure at the 5 per cent level of significance is:

1. Obtain 20 independent pairs of observations (x_i,y_i) from the population.
2. If r is less than $-.44$ or greater than $+.44$, *reject* the hypothesis that $\rho = 0$ in favor of the alternative that $\rho \neq 0$.
3. If r is between $-.44$ and $+.44$, *accept* the hypothesis that $\rho = 0$.

If the sample size is large, the results given by this test will agree exactly with those that would be obtained by using the techniques of Sec. 12.2.4 to test the hypothesis that $\beta = 0$.

12.3.3. *The Use of the Correlation Coefficient in Selection Problems.* It was noted in Sec. 12.1 that problems of personnel selection may be usefully viewed in terms of the relation between two quantitative variables. Individuals can often be assigned scores x with respect to one or more aptitude tests and can also be assigned scores y with respect to a scale of performance rating. If positive linear association exists between the variables x and y, selecting individuals with "high" test scores will tend to select individuals who will achieve "high" or satisfactory performance ratings on the job. The value of the correlation coefficient ρ plays an extremely important role in determining the characteristics of such a selection scheme.

Figure 12.15 provides a diagram of the most useful way of viewing this selection problem. Consider an extremely large population of individuals constituting the pool of talent available for filling a particular type of

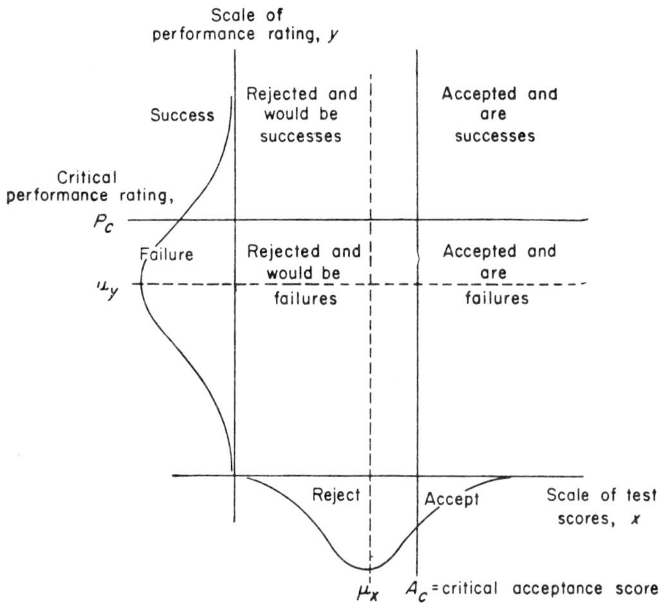

FIG. 12.15. Schematic representation of the performance of a selection procedure.

position. Associated with each individual will be a value of the variable x, the score he can attain on a specified aptitude test. The distribution of the variable x in this population is taken to be normal with mean μ_x and variance σ_x^2. This distribution is pictured at the foot of the diagram. A selection procedure then consists of choosing a critical acceptance score A_c. If an individual's test score is greater than A_c, he is accepted for the position; if it is less than A_c, he is rejected. One possible value of A_c is indicated by the solid vertical line in the middle of the diagram.

Each individual in the pool of available talent will also have a value of the variable y which is viewed as a performance rating. The distribution of the variable y in this population is taken to be normal with mean μ_y and variance σ_y^2. This distribution is pictured at the left of the diagram. We shall consider that it is possible to specify a critical performance rating P_c such that individuals with ratings higher than this are viewed as "successes" and individuals with ratings lower than P_c are classed as "failures." One possible value of P_c is indicated by the solid horizontal line in the middle of the diagram.

Individuals can now be classified according to their scores on both variables (x,y), and each individual will therefore be represented by a point in the interior of the diagram. Notice that the lines for A_c and P_c divide this area into four parts. These are:

1. Points falling to the right of A_c and above P_c represent individuals who will be accepted and who will be successes.
2. Points falling to the right of A_c and below P_c represent individuals who will be accepted and who will be failures.
3. Points falling to the left of A_c and above P_c represent individuals who will be rejected and who would have been successes.
4. Points falling to the left of A_c and below P_c represent individuals who will be rejected and who would have been failures.

The performance of the selection procedure depends upon the fractions of the total pool of available talent falling in these four classifications. These in turn depend upon the value of A_c, the value of P_c, *and* upon the value of the correlation coefficient ρ between x and y.

Extensive tables of these four fractions, for various values of A_c, P_c, and ρ, have been prepared by Arbous [6], and selections from these tables are reproduced in Table 12.5. Consider, for example, the first set of values where A_c is set at μ_x and P_c at μ_y. Because the distributions of x and y are normal and therefore symmetric, setting P_c at μ_y means that half the available pool of talent will be successes and half will be failures; furthermore, setting A_c at μ_x means that half of the applicants will be accepted and half will be rejected. If the value of ρ is zero, the table shows that half of those accepted are successful and half are failures. The selection scheme is consequently worthless since one could do as well by choosing applicants by the flip of a coin. As the value of ρ increases, the selection scheme does better and better in relation to pure random selection. Thus when $\rho = .5$:

33.3	per cent are accepted and are successes	50 per cent are accepted
16.7	per cent are accepted and are failures	
16.7	per cent are rejected and would be successes	50 per cent are rejected
33.3	per cent are rejected and would be failures	

$33.3/(33.3 + 16.7) = 66.7$ per cent of those accepted are successful
$16.7/(33.3 + 16.7) = 33.3$ per cent of those accepted are failures

The effects of changing the values of A_c and P_c can be seen by examining and comparing the other entries in this table. For fixed values of P_c and ρ, the effect of increasing A_c is to decrease the fraction selected but to increase the fraction of successes among those who are selected.

The determination of A_c in a practical situation will depend upon the factors set forth in Exercise 8.7 relating to the size of the pool of talent and to the investment in training that is necessary before an individual can be classed as a success or as a failure.

An empirical verification of the fact that actual data do conform roughly to the values given in Table 12.5 can be obtained from the data of Fig. 12.13. The mean of the x observations $\bar{x}$ and the mean of the y observations $\bar{y}$ are shown respectively by a dashed vertical line and a dashed horizontal line in Fig. 12.13. We find that 45, or 35 per cent, of the points are to the right of $\bar{x}$ and above $\bar{y}$; 18, or 14 per cent, of the points are to the right of $\bar{x}$ and below $\bar{y}$; 25, or 19 per cent, of the points are to the left of $\bar{x}$ and above $\bar{y}$; and 42, or 32 per cent, of the points are to the left of $\bar{x}$ and below $\bar{y}$. The value of the correlation coefficient for these data was .55. Reading in Table 12.5 for $A_c = \mu_x$, $P_c = \mu_y$, and $\rho = .5$, we find corresponding values of 33 per cent, 17 per cent, 17 per cent, and 33 per cent.

TABLE 12.5. PERFORMANCE CHARACTERISTICS OF A NUMBER OF SELECTION PROCEDURES

Critical acceptance score A_c set at:	Critical performance rating P_c set at:	Correlation coefficient ρ	Fraction of population which are:			
			Accepted and are successes	Accepted and are failures	Rejected and would be successes	Rejected and would be failures
Population mean of test scores μ_x	Population mean of performance ratings μ_y	0	.250	.250	.250	.250
		.2	.282	.218	.218	.282
		.5	.333	.167	.167	.333
		.9	.428	.072	.072	.428
One standard deviation above population mean test score $\mu_x + \sigma_x$	Population mean of performance ratings μ_y	0	.079	.079	.421	.421
		.2	.099	.060	.401	.440
		.5	.127	.031	.373	.469
		.9	.158	.001	.342	.499
Population mean of test scores μ_x	One standard deviation above the population mean performance rating $\mu_y + \sigma_y$	0	.079	.421	.079	.421
		.2	.099	.401	.060	.440
		.5	.127	.373	.031	.469
		.9	.158	.342	.001	.499

SOURCE: Ref. 6.

The numerical values of the quantities required for determining a selection procedure—μ_x, σ_x, μ_y, σ_y, and ρ—must be estimated from a random sample drawn from the entire pool of available applicants. If one first *selects*, on any basis other than random sampling, the group of individuals in which tests scores and performance ratings are to be compared, the resulting estimates will not apply to the entire pool. They will apply only to that part of the pool which satisfies the original selection criterion.

12.3.4. *Some Special Uses of the Correlation Coefficient.* A number of fields of investigation have developed their own particular applications of the correlation coefficient and have employed terminology which is descriptive of these applications. For example, the brief discussion of economic time series in Sec. 12.2.5 noted that observations made in successive periods of time are apt to be dependent upon one another. The extent of this dependence may be investigated by computing the correlation coefficient between successive observations.

As an illustration of this approach, consider the data of Fig. 12.10. They show the consumption of cigarette tobacco for each year from 1920 to 1954. A few of the values are: 1920—1.89 pounds per person, 1921—2.07, 1922—2.14, 1923—2.51, 1924—2.69, . . . , 1953—10.46, and 1954—9.84. Pairs of observations can be obtained by considering observations for all pairs of consecutive years. This gives the pairs (1.89, 2.07), (2.07, 2.14), (2.14, 2.51), (2.51, 2.69), . . . , (10.46, 9.84). Each observation, except the first and last, is used once as a y value (with the preceding year as the x) and once as an x value (with the succeeding year as the y). A correlation coefficient obtained in this manner is called a *serial correlation coefficient of order 1.* The value of this correlation coefficient for the data of Fig. 12.10 is +.99, thus indicating almost perfect linear association between the observations for consecutive years. If a correlation coefficient were computed by forming pairs of observations 2 years apart, we would speak of a *serial correlation of order 2,* and so on.

Special applications of correlation coefficients may also be found in the development and application of mental tests. These tests are used for measuring various types of mental abilities (general intelligence or aptitude for special pursuits), for measuring achievement in a defined course of study, and for predicting degree of success in carrying out stated tasks. Many of these applications are centered around the general concepts of *reliability* and *validity.* Reliability refers to a single test and is often viewed in the following manner. If an individual is given a specific test, his observed score is viewed as the sum of two components. The first component is his true score on the test. The second component is the amount by which the observed score deviates from the true score because of random *errors of measurement,* arising from such factors as the way in which he feels at the time he takes the test, the manner in which he reacts to the specific wording of questions, and so on. The *index of reliability* is defined to be the correlation between observed scores and true scores. This correlation coefficient cannot be estimated from the scores of individuals on a single test, and special approaches are necessary to estimate its value. Thus one may administer two parallel (or equivalent) forms of a test and correlate the two scores or may correlate the scores on the two halves of the same test. An extensive treatment of these matters is given by Gulliksen [7], and the interested reader should consult this reference.

The *validity* of a test is the correlation of test scores with some criterion and is in a sense a measure of the predictive value of the test. Great care must be exercised in talking about the validity of a test, and Gulliksen [7, p. 88] says:

> Reliability has been regarded as the correlation of a given test with a parallel form. Correspondingly, the validity of a test is the correlation of the test with some criterion. In this sense a test has a great many different "validities." For example, the ACE Psychological Examination has a validity for predicting grades in English and a different validity for predicting grades in Latin. It is also found in studying various validity coefficients for a given test that they may vary from school to school, and from time to time. In other words, validity cannot be regarded as a fixed or a unitary characteristic of a test. As new uses for a test are contemplated, new validity coefficients must be determined; and, when use of a test is continued, the validity coefficients must be redetermined at intervals.

In concluding this treatment of the correlation coefficient we should like to emphasize that it measures degree of association only when that association is linear. There may be a very strong degree of curvilinear association between two variables, and yet the correlation coefficient may be close to zero. If the association between two variables is curvilinear, then it is necessary to find an appropriate equation which gives $\mu_{y \cdot x}$ as a function of x and to measure association with respect to this regression curve. A measure of association, related to the correlation coefficient, that is often used under these circumstances is the *square of the correlation ratio.* The square of the correlation ratio is defined to be [(sum of squared deviations of the y values about $\bar{y}$) minus (sum of squared deviations of the y values about the regression curve)] divided by (sum of squared deviations of the y values about $\bar{y}$). That is, the square of the correlation ratio is the fraction of the variability in the y values which is explained by the curvilinear regression of y on x.

Finally, the interpretation of r must be a purely mathematical one and, unless one can bring to bear arguments peculiar to a specific field of application, must be completely devoid of any implications of *cause and effect.* The fact that two variables tend to increase or decrease together does not imply that one has any direct or indirect effect on the other. Both the variables may be influenced by other variables in such a manner as to give rise to a strong mathematical relationship. An interesting account of the approaches that one may employ in searching for cause-and-effect relationships between qualitative variables of interest in social-science research is given by Zeisel [8, chaps. 8–10]. A more formal approach to this problem has been outlined by Simon [9].

12.4. The Rank Correlation Coefficient

Social-science research often deals with qualitative variables which have an implied *order* relation among the categories of the variable. That is, even though it is not possible to assign a numerical value to each category of such a variable, it is possible to state that one category indicates the presence of some characteristic to a greater or lesser degree than does a second category. The response categories of a question which might be used in an employee-attitude survey were cited as an illustration in Sec. 2.2:

"How closely does your foreman check up on his men?"

________ Much too closely
________ Somewhat too closely
________ Just about right
________ Not quite closely enough
________ Not nearly closely enough

Although this question and its response categories define a qualitative variable, there is an implied order relation among the categories of the variable. That is, the "closeness" of supervision is greater for the "much too closely" category than it is for the "not quite closely enough" category. We shall now treat the problem of detecting association and measuring degree of association between two variables which are distinguished by this ordering of qualitative categories.

As an illustration of the type of problem which can be studied in this manner, consider the data of Table 12.6. These data were obtained from tape recordings of 12 mediation sessions involving small businesses, conducted under the auspices of a state mediation service [10]. (Mediation is the process by which agreements on labor-management disputes are reached with the assistance of a neutral party.) Although "success" was regarded as a qualitative variable, the author felt that he could place the 12 cases in order with respect to the extent to which the original issues were settled at the close of mediation. The "most successful" case was then assigned a *rank* of 1, the next "most successful" case was assigned a rank of 2, . . . , and the "least successful" case was assigned a rank of 12. The rank of the ith case, one of the numbers from 1 to 12, is designated by r_i, and these ranks appear in the second column of the table.

The major technique of analysis used in this study of mediation was that of going through the tape recordings of the sessions and identifying distinct units of verbal behavior. Each of these units was identified as to its originator (the actor), the party to which it was directed (the recipient), and the content of the unit. The third column in Table 12.6 refers to units which occurred in the early part of the mediation session, which originated with either the labor party or the management party and which were directed toward the opposing side. (Units originating with the mediator or directed toward the mediator were not used in this

TABLE 12.6. TWELVE MEDIATION CASES RANKED ACCORDING TO SUCCESS OF OUTCOME AND TO AMOUNT OF HOSTILITY SHOWN IN THE EARLY PART OF THE MEDIATION SESSION

Mediation case	Ranking* for success of outcome r_i	Per cent hostility†	Ranking on hostility r_i'	Difference in rankings d_i	Squared differences d_i^2
A	1	0.0	12	−11	121
B	2	10.2	7	− 5	25
C	3	6.4	10	− 7	49
D	4	4.0	11	− 7	49
E	5	30.6	4	+ 1	1
F	6	66.7	1	+ 5	25
G	7	9.0	8	− 1	1
H	8	7.8	9	− 1	1
I	9	33.3	3	+ 6	36
J	10	18.2	6	+ 4	16
K	11	40.0	2	+ 9	81
L	12	25.7	5	+ 7	49
					$\Sigma d_i^2 = 454$

* Most successful case is given rank 1.

† This is the per cent of interactions, in the early part of the session, between labor representatives and management representatives which indicated antagonism or hostility.

SOURCE: Ref. 10, pp. 71, 98.

particular analysis.) The per cents in the third column are the per cents of these interparty units of verbal behavior which indicated hostility. These 12 per cents are used to rank the 12 cases on the amount of hostility shown. The case with the largest per cent of hostile behavior was assigned a rank of 1, . . . , and the case with the smallest per cent of hostile behavior was assigned a rank of 12. The rank of the ith case, one of the numbers from 1 to 12, is designated by r_i', and these ranks appear in the fourth column of the table. We wish to know if these data indicate any association between success of outcome and amount of initial hostility between the parties. Although there is a quantitative basis for the hostility ranking, the succeeding analysis would still be appropriate if this hostility ranking had been obtained in the same manner as the success ranking.

The foregoing analysis provides 12 pairs of ranks: case A (1,12); case B (2,7); . . . ; and case L (12,5). If there were perfect agreement between the two rankings, we would have the pairs (1,1), (2,2), (3,3), . . . , (12,12); if there were a complete reversal of rankings, we would have the pairs (1,12), (2,11), (3,10), . . . , (12,1). A measure of the extent of agreement between the two rankings is desired, and one such measure can be obtained by computing a linear correlation coefficient from the 12 pairs of ranks. That is, formula (12.11) is used for computing a correlation coefficient, where the 12 pairs of observations (r_1,r'_1), (r_2,r'_2), . . . , (r_{12},r'_{12}) are used in place of the pairs (x_1,y_1), (x_2,y_2), . . . , (x_{12},y_{12}). The coefficient thus obtained is known as the Spearman *rank correlation coefficient*, and its value for the data of Table 12.6 is $-.587$. This coefficient is designated by r_s, and its interpretation will be treated shortly.

Actually, it is not necessary to use formula (12.11) for the computation of the rank correlation coefficient. Consider the n pairs of ranks, (r_1,r'_1), (r_2,r'_2), . . . , (r_n,r'_n). The fact that each of the sets $r_1, r_2, \ldots, r_n$ and $r'_1, r'_2, \ldots, r'_n$ consists of the integers from 1 up to n simplifies the computations. It can be shown that the formula for the Spearman rank correlation coefficient becomes

$$r_s = 1 - \frac{6\Sigma(r_i - r'_i)^2}{n(n^2 - 1)}$$

$$= 1 - \frac{6\Sigma d_i^2}{n(n^2 - 1)} \qquad (12.12)$$

where $d_i = (r_i - r'_i)$, the difference in ranks for the ith observation. These differences and their squares are given in the last two columns of Table 12.6, and

$$r_s = 1 - \frac{6(454)}{12(144 - 1)}$$

$$= 1 - \frac{2{,}724}{1{,}716} = 1 - 1.587$$

$$= -.587$$

Now let us return to the problem of interpreting an observed value of the rank correlation coefficient. It can be demonstrated that r_s will be $+1$ if there is perfect agreement between the two rankings, that is, if the pairs are (1,1), (2,2), . . . , $(n - 1, n - 1)$, (n,n). Similarly, r_s will be -1 if there is a complete reversal of rankings, that is, if the pairs are $(1,n)$, $(2, n - 1)$, . . . , $(n - 1, 2)$, $(n,1)$. Otherwise the value of r_s will lie between -1 and $+1$, a value close to zero indicating the absence of linear association between the rankings. The question immediately arises as to how much smaller or greater than zero must r_s be in order to provide evidence that there actually exists association between the two variables in the population from which the sample of n pairs of observations have been drawn. This question can be answered by considering the sampling distribution of r_s *under the hypothesis of independence.*

TABLE 12.7. CRITICAL VALUES OF THE RANK CORRELATION COEFFICIENT TO BE USED IN TESTING THE HYPOTHESIS OF INDEPENDENCE

(The column headings give the probability that an observed value of r_s will either be equal to or larger than the critical value with a plus sign *or* will be equal to or smaller than the critical value with a minus sign, provided that the sample is drawn from a population in which the two variables are independent.)

n	.05	.10	n	.05	.10
2			16	.506	.425
3			17	.490	.411
4		1.000	18	.475	.399
5	1.000	.900	19	.462	.388
			20	.450	.377
6	.886	.829			
7	.786	.714	21	.438	.368
8	.738	.643	22	.428	.359
9	.683	.600	23	.418	.351
10	.648	.564	24	.409	.343
			25	.400	.336
11	.620	.520			
12	.591	.496	26	.392	.329
13	.566	.475	27	.384	.323
14	.544	.456	28	.377	.317
15	.524	.440	29	.370	.311
			30	.364	.305

SOURCE: Edwin G. Olds, "The 5% Significance Levels for Sums of Squares of Rank Differences and a Correction," *Annals Math. Stat.*, **20**:117–118 (1949).

If the rankings of elements in a population on each of two variables are independent, then the rank of a single element on one variable has no influence on its rank on the second variable, and vice versa. Under these circumstances, observed pairs of ranks, such as are exhibited in Table 12.6, may be viewed as having arisen in the following manner. Consider the first ranking of the 12 cases, on the success of the outcome of mediation, as a standard or fixed ranking. Now consider a container in which are placed 12 identical chips numbered from 1 up to 12. These chips are thoroughly mixed and are drawn out of the container one by one without replacement. The number on the first chip drawn is assigned to case A as its rank on hostility, the number on the second chip drawn is assigned to case B as its rank on hostility, . . . , and the number on the last chip drawn is assigned to case L as its rank on hostility. The value of r_s is computed from the resulting pairs of ranks.

If the foregoing process of assigning ranks at random is repeated time after time, a possibly different value of r_s will be obtained each time. An indefinitely large number of repetitions will lead to a distribution of r_s, and this is the *sampling distribution of r_s under the hypothesis of independence.* The arithmetic mean of this distribution is zero. This sampling distribution of r_s can be determined exactly if the value of n is small, and reasonable approximations to it can be obtained for larger values of n. Critical values of r_s, for use in testing the hypothesis of independence, can be read from these distributions, and a selection of such critical values is given in Table 12.7. When n is outside the range of this table, the critical values of r, as read from the curves of Fig. 12.14, can be used as critical values for r_s.

The critical values of r_s are viewed in the following manner, where we take $n = 8$ for illustrative purposes. If samples of size 8 are drawn from a population in which two ordered variables are *independent,* then

$$\begin{aligned} \Pr(\text{either } r_s \leq -.738 \text{ or } r_s \geq +.738) &= .05 \\ \Pr(r_s \leq -.738) &= .025 \\ \Pr(r_s \geq +.738) &= .025 \end{aligned}$$

Thus we may test the hypothesis of independence, at the 5 per cent level of significance, against the alternative that there is association (either positive or negative) in the following manner:

1. Draw a random sample of observations from the population.
2. Compute the value of r_s.
3. If r_s is less than or equal to $-.738$ *or* greater than or equal to $+.738$, *reject* the hypothesis of independence.
4. Otherwise, *accept* the hypothesis of independence.

If we wish to test the hypothesis of independence, at the 2.5 per cent level of significance, against the alternative that there is negative association, then:

1. If r_s is less than or equal to $-.738$, *reject* the hypothesis.
2. Otherwise, *accept* the hypothesis.

A blank in the table indicates that it is not possible to reject the hypothesis for that particular sample size and significance level, no matter how large or small may be the value of r_s.

Let us now apply these considerations to the data of Table 12.6. It would seem reasonable that success of outcome would be associated in a negative fashion with amount of initial hostility. Therefore the hypothesis of independence is tested against the alternative of negative association. The critical value of r_s, at the 2.5 per cent level of significance, for $n = 12$, is $-.591$. Since the observed value of r_s, $-.587$, is not less than this value, the hypothesis of independence is not rejected at the 2.5 per cent level of significance. On the other hand, this hypothesis is rejected at the 5 per cent level of significance since the critical value is $-.496$. In an exploratory study of this kind, this would probably be taken as indication of negative association between the two variables, but this would have to be confirmed by further investigation.

The rank correlation coefficient can also be used to test the hypothesis that two quantitative variables are independent. Consider the n pairs

TABLE 12.8. COMPUTATION OF RANK CORRELATION COEFFICIENT FOR DATA ON SPEED AND DENSITY OF TRAFFIC FLOW THROUGH A VEHICULAR TUNNEL (Data from Table 12.2)

x_i	Rank on the x variable r_i	y_i	Rank on the y variable r_i'	d_i	d_i^2
69	5	16.8	7.5	−2.5	6.25
88	1	14.8	10	−9	81
65	6	19.1	6	0	0
84	2	14.9	9	−7	49
63	7	21.6	4	+3	9
53	8.5	25.7	2	+6.5	42.25
82	3	16.8	7.5	−4.5	20.25
53	8.5	25.1	3	+5.5	30.25
70	4	19.7	5	−1	1
34	10	31.8	1	+9	81
					$\Sigma d_i^2 = 320$

$$r_s = 1 - \frac{6(320)}{10(100 - 1)} = 1 - 1.939$$
$$= -.939$$

of observations (x_1,y_1), (x_2,y_2), . . . , (x_n,y_n). Replace the n x observations by their ranks r_i where a rank of 1 is assigned to the element having the largest x value, a rank of 2 to the element having the next largest x value, . . . , and a rank of n to the observation having the smallest x value.* In a similar fashion, the n y observations are replaced by their ranks r_i', thus leading to n pairs of ranks, (r_1,r_1'), (r_2,r_2'), . . . , (r_n,r_n'). The value of r_s is again computed from formula (12.12).

This entire process is illustrated in Table 12.8 on the data of Table 12.2. The only difficulty arises when *ties* occur, that is, when two or more

* The ranks can also be assigned in the opposite direction, from the smallest to largest. One merely takes this into account in interpreting the resulting value of r_s.

observations have the same numerical value. There are a number of different procedures which can be used in the event this situation arises, but we shall confine attention to the *mid-rank method.* Each of the tied observations is assigned the rank which would ordinarily be given to the middle observation if there were no ties. This is illustrated in Table 12.8 where two different observations have x values of 53. One of these would ordinarily be given a rank of 8, and the other a rank of 9. Both are therefore assigned the mid-rank of 8.5. The computation of r_s then proceeds as before, and its value is $-.939$. From Table 12.7 for $n = 10$, the probability is .05 of obtaining a value of r_s less than or equal to $-.648$ or greater than or equal to $+.648$ under the hypothesis of independence. Since the observed value is less than $-.648$, the hypothesis of independence between speed and density of traffic flow through a vehicular tunnel is rejected at the 5 per cent level of significance.

When dealing with quantitative variables, one has a choice in testing for linear association of using the correlation coefficient as computed from the original observations or of using the rank correlation coefficient. The rank correlation has the advantage that the distribution of r_s under the hypothesis of independence is the same no matter how the variables x and y are distributed in the population from which the sample is drawn. On the other hand, the distribution of the ordinary correlation coefficient, upon which the curves of Fig. 12.14 are based, depends upon the assumption that both x and y have normal distributions. Thus the hypothesis-testing procedure based on r_s is an example of the nonparametric, or distribution-free, procedures that were mentioned in Sec. 9.6. If the population does satisfy the conditions stated at the end of Sec. 12.3.1, then a test based on the correlation of the original observations will, for a fixed level of significance, have a somewhat smaller chance of accepting the hypothesis of independence when it is false than will the test based on the rank correlation coefficient.

In conclusion, we observe that:

1. The rank correlation coefficient, like the ordinary product-moment correlation coefficient, is most sensitive to departures from independence that are characterized by linear association. There are other distribution-free tests of association which will detect departures from independence which are characterized by certain types of curvilinear association (see [11, chap. 3]).

2. The Spearman rank correlation coefficient discussed here is not the only rank correlation coefficient. Kendall has developed a rank correlation coefficient, denoted by τ, which is based on the number of interchanges between the two rankings. A discussion of this coefficient, as well as extensive treatment of other problems that can be studied by means of ranks, will be found in [12].

12.5. Multiple Linear Regression and Correlation

The simple linear-regression model, from which follow the results of Sec. 12.2, was defined in these terms: (1) there exists a family of populations, each population being identified by a fixed and known value of an independent variable x; (2) within a given x population the elements may vary with respect to the dependent variable y; (3) the mean values of the distributions of y for different values of x are associated by the linear equation $\mu_{y \cdot x} = \alpha + \beta x$; and (4) the variance of the distribution of y is the same for each fixed value of x.

Now it is quite clear from the examples cited in Sec. 12.1 that it is often possible to further subdivide the x populations according to a second independent variable, say x'. For example, the data of Table 12.1 and Fig. 12.3 refer to the population of urban housekeeping families of two or more persons in the United States, 1947. The independent variable x was taken to be "yearly income in dollars," and the dependent variable y was taken to be "money value of food consumed per week." If we think of the population of families having some stated amount of yearly income, then it is possible to subdivide these families with respect to the variable "number of persons in the household," which can be designated by the symbol x'. Thus any family would be characterized by a triplet of numbers (x,x',y) where x is yearly income, x' is number of persons in the household, and y is money value of food consumed per week. The variable x' is of obvious interest since the food consumption of a family will depend in some manner on the number of persons in the family.

These considerations lead to an immediate extension of the simple linear-regression model. This is:

1. There exists a family of populations, each population being identified by fixed and known values of *two* independent variables x and x'.
2. Within a given x-x' population the elements will vary with respect to the dependent variable y, thus defining a distribution of y's. Denote the mean and variance of this distribution by $\mu_{y \cdot xx'}$ and $\sigma^2_{y \cdot xx'}$, respectively.
3. The mean values of the distributions of y for different x-x' populations are related by the linear equation

$$\mu_{y \cdot xx'} = \alpha + \beta x + \gamma x' \tag{12.13}$$

This is known as the *regression plane*, or *regression surface*, of y on x and x'.

4. The variance of the distributions of y is the same for each fixed pair of values, x and x'. That is $\sigma^2_{y \cdot xx'} = \sigma^2$ for all values of x and x'.

These conditions define a *multiple-linear-regression* model, and they can be extended to cover any number of independent variables, x, x', x'', x''', and so on.

Although the details are beyond the scope of this book, the approach of Sec. 12.2 can be extended to cover the multiple-linear-regression model. Thus the criterion of least squares can be used to obtain estimates of α, β, γ, and σ^2 from a sample of n triplets of observations, (x_1,x'_1,y_1), (x_2,x'_2,y_2), . . . , (x_n,x'_n,y_n). Furthermore, procedures can be developed for obtaining a wide variety of confidence intervals and tests of hypotheses. The concept of correlation can also be used. Thus the *multiple correlation coefficient squared* R^2 is the fraction of the variability in the y values which is explained by the regression of y on x and x'. A treatment of multiple linear regression will be found in [4, chap. 14] and [5, chap. 12]. The study of the food consumption of urban families in the United States [1] considers, among other things, the multiple linear regression of family milk consumption on the independent variables, family income, number of persons in the family, number of children in the family, age of the homemaker (in years), and schooling of the homemaker (in years). Multiple-regression techniques are also very useful in selection problems where performance is the dependent variable and scores on a number of different tests are the independent variables.

EXERCISES

12.1. The data of Table 12.1 provide an estimate of the regression line of weekly food expenditure on total yearly income for families receiving less than $7,500 per year. This estimated regression line is $y = 12.26 + 3.21x$, where x is expressed in thousands of dollars. Use this equation to estimate the food expenditure of urban housekeeping families having incomes of exactly $5,400. What is your interpretation of the computed value?

12.2. The following table shows the age and strength of grip, as measured in kilograms (kg), of 552 male manual workers in industry.

Age group	Assumed mid-points of class intervals x	Number of workers in class interval	*Average* strength of grip for workers in an age interval	Variance of strength of grip for workers in an age interval
18–22	20.0	28	53.46	33.64
23–27	25.0	82	56.05	48.02
28–32	30.0	153	54.23	43.96
33–37	35.0	126	53.33	48.44
38–42	40.0	72	52.42	37.82
43–47	45.0	42	50.36	51.55
48–52	50.0	29	48.31	42.51
53–68	60.5	20	46.80	34.57

SOURCE: M. B. Fisher and J. E. Birren, "Age and Strength," *Jour. Appl. Psych.*, **31**:492 (1947).

The individual observations for workers in an age group were not reported in the source, but the necessary quantities for studying the regression of strength of grip y on age x can be computed from the reported quantities. These will be given as needed.

a. Plot the average strength of grip, column 4, against the mid-points of the age groups, column 2. This is an approximation to the regression of y on x. Although curvilinear regression would be required in studying the entire range of x, observe that this regression appears to be linear if the lowest age group is omitted from the analysis. This age group will not be considered in the following.

b. A random sample of nine observations has been drawn from the above distribution, three from the 23–27 age group, three from the 38–42 age group, and three from the 48–52 age group. These resulting pairs of observations (x_i,y_i) are (25,61), (25,45), (25,52), (40,56), (40,57), (40,54), (50,47), (50,39), and (50,52). Plot these points on the diagram prepared in *a*, compute from them the least-squares estimates of the slope and y intercept of the regression line of strength of grip on age, and draw this line on the diagram prepared in *a*.

c. With the omission of the 18–22-year age group, the original table contains 524 observations. These 524 observations provide the quantities that would ordinarily be obtained in a regression analysis: $\bar{x} = 35.27$, $\bar{y} = 53.13$, $a = 62.51$, $b = -.266$, $s^2_{y \cdot x} = 45.51$, and $\sum_{i=1}^{524} (x_i - \bar{x})^2 = 37{,}513$. Use these data in performing the following operations: (1) Plot the regression line on the diagram prepared in *a*. (2) Obtain 90 per cent confidence limits for the true slope of the regression line β. (3) Compute an estimate of the average strength of grip of manual workers 41 years of age, $\mu_{y \cdot 41}$, and obtain 99 per cent confidence limits for this average. (4) Suppose the strength of grip of an additional worker, 41 years of age, is to be determined. What is a 95 per cent prediction interval for his strength of grip?

d. Comment on the applicability of this analysis to the general population of United States males, particularly with reference to males in older age groups.

12.3. The table on page 390 shows the joint classification of 85 male white adolescent delinquents on school age scores x as measured by the Otis Classification Test and on reported scholastic achievement y in school years completed. These data were obtained from inmates, between 16 and 22 years of age, of the New York City Reformatory. The number in a cell is the number of individuals showing that particular combination of an x and a y value. Thus there were five individuals having an x score of 6.0 *and* a y score of 8.5.

a. A random sample of 10 cases has been drawn from this table, leading to the 10 pairs of observations (x_i,y_i): (6.5, 9.0), (7.0, 8.5), (7.0, 10.5), (5.0, 9.5), (7.5, 7.0), (4.5, 10.5), (4.0, 8.0), (10.0, 10.5), (9.0, 10.5), and (7.5, 7.0). Compute the value of the correlation coefficient from these 10 pairs of observations and obtain the 95 per cent confidence limits for ρ from the curves of Fig. 12.14.

b. The basic data for the entire 85 observations are $\Sigma x_i = 515$, $\Sigma x_i^2 = 3{,}322$, $\Sigma y_i = 712.5$, $\Sigma y_i^2 = 6{,}146.75$, and $\Sigma xy = 4{,}355.5$. (*Note:* Σx_i, Σx_i^2, Σy_i, Σy_i^2 are easily obtained from the two sets of marginal frequencies by using the computational format of Table 5.3. Σxy is obtained by computing the sum of x's *for each row*, multiplying a row sum by the value of y for that row, and adding these products for all rows.) The value of r is .21. Using the curves of Fig. 12.14, test the hypothesis that $\rho = 0$ in the population from which these data may be regarded as a sample.

c. Determine the fraction of these observations falling above $\bar{x}$ and above $\bar{y}$, above $\bar{x}$ and below $\bar{y}$, below $\bar{x}$ and above $\bar{y}$, and below $\bar{x}$ and below $\bar{y}$. Compare these with the fractions obtained from Table 12.5 for $A_c = \mu_x$, $P_c = \mu_y$, and $\rho = .20$.

12.4. If there is a tendency for the absenteeism behavior of workers to persist from one time period to the next, then it is customary to speak of workers with consistently high absence rates as "absence-prone" individuals. An empirical investigation of absenteeism behavior thus requires the comparison of absence records among two or

Data for Exercise 12.3

y \ x	3.0	3.5	4.0	4.5	5.0	5.5	6.0	6.5	7.0	7.5	8.0	8.5	9.0	9.5	10.0	Total
12.5						1										1
12.0																0
11.5																0
11.0							1									1
10.5				1	1			1	1				1			5
10.0					1		1	1		1					1	5
9.5				1	1	1	1	1	1	2	1	1	1			11
9.0			1	1			2	2	1		1		1			9
8.5			1	2	2	3	5	1	4			1				19
8.0			4		1	4	1	1		1	1	1				14
7.5				1	1		1									3
7.0		1		1	1				1	2						6
6.5				1		1										2
6.0		1		2		1					1					5
5.5															1	1
5.0	1										1					2
4.5				1												1
Total	1	2	6	11	8	11	12	7	8	6	5	3	3	0	2	85

x = school-age scores, rounded to nearest half year, as measured by Otis Classification Test, pt. I

y = reported scholastic achievement in school years completed

Source: Philip Ash, "The discrepancy between Reported Schooling and Tested Scholastic Ability among Adolescent Delinquents," *Jour. Appl. Psych.*, **31**:323–328 (1947).

more consecutive time periods. The table below shows one such set of data, obtained by comparing the number of absences from work in each of two consecutive 6-month periods for 318 industrial workers.

Data for Exercise 12.4

y \ x	0	1	2	3	4	5	6	7	8	9	10	11	12	Total
0	31	16	10	2	1	1				1				62
1	30	26	16	12	2	2								88
2	6	18	12	6	8	1	2	1						54
3	8	14	5	2	1	2	5	2	3					42
4	2	4	2	8	1	4	1	1	1					24
5	1	2	2	2	3	1	2	2		1		2		18
6			1	2		2	2	2						9
7						1	3	1						5
8			1	1	1	1	2		1					7
9				1	2	1		1						5
10				1										1
11									2					2
12									1					1
Total	78	80	49	37	19	16	17	10	8	2	0	2	0	318

x = number of absences from work, for all causes, during a first 6-month period
y = number of absences from work, for all causes, during a second 6-month period

Source: A. G. Arbous and H. S. Sichel, "New Techniques for the Analysis of Absenteeism Data," fig. 4, South African Council for Scientific and Industrial Research, 1952, mimeographed.

a. Compute the correlation coefficient between number of absences in the first 6-month period x and number of absences in the second 6-month period y.

b. Although the association between x and y can be viewed in a symmetric fashion and is apparently linear in form, the data indicate at least one very clear-cut reason why the contents of Fig. 12.14 and Table 12.5 are not entirely appropriate for interpreting and applying the value of r computed in *a*. What is the reason? The cited source used another theoretical probability distribution, the negative binomial distribution, in making these interpretations and applications.

12.5. In an investigation of the social integration of American cities, a comparison was made between the value of a welfare-effort index and a crime index for a group of 28 selected cities. The reported values of these indices were:

DATA FOR EXERCISE 12.5

City	Welfare index	Crime index	City	Welfare index	Crime index
Buffalo	266.7	9.14	San Francisco	101.7	29.81
Syracuse	235.6	11.99	Bridgeport	101.4	17.35
Springfield, Mass.	233.5	14.79	Dayton	93.2	26.15
Milwaukee	199.5	6.22	St. Louis	89.9	19.46
Cincinnati	191.5	42.23	Kansas City, Mo.	83.3	35.32
Baltimore	166.1	24.43	Columbus, Ohio	72.2	55.54
Richmond, Va.	153.6	54.25	New Orleans	71.7	15.07
Providence	148.2	10.20	Grand Rapids	65.7	21.73
Hartford	139.0	27.18	Wichita	65.2	19.34
Indianapolis	132.0	47.90	Atlanta	63.5	77.46
Cleveland	131.2	33.79	Houston	63.5	41.89
Los Angeles	115.5	43.36	Birmingham	62.8	48.54
Louisville	111.6	59.83	Dallas	50.7	45.86
Canton	105.0	32.91	Wilkes-Barre	50.6	14.18

SOURCE: Robert C. Angell, "The Social Integration of Selected American Cities," *Amer. Jour. Soc.*, **47**:575–592 (1942).

Compute a rank correlation coefficient from these data and test the hypothesis of independence between value of welfare index and value of crime index.

REFERENCES CITED

1. Clark, Faith, Janet Murray, Gertrude S. Weiss, and Evelyn Grossman: *Food Consumption in the United States*, Agriculture Information Bulletin 132, U.S. Department of Agriculture, 1954.
2. Arbous, A. G.: *Selection for Industrial Leadership*, Oxford University Press, New York, 1953.
3. Olcott, Edward S.: "The Influence of Vehicular Speed and Spacing on Tunnel Capacity," *Jour. Operations Res. Soc. America*, **3**:147–167 (1955).
4. Anderson, R. L., and T. A. Bancroft: *Statistical Theory in Research*, McGraw-Hill Book Company, Inc., New York, 1952.
5. Yule, G. Udny, and M. G. Kendall: *An Introduction to the Theory of Statistics*, 14th ed., Hafner Publishing Company, New York, 1950.
6. Arbous, A. G.: *Tables for Aptitude Testers*, South African Council for Scientific and Industrial Research, National Institute for Personnel Research, 1952.
7. Gulliksen, Harold: *Theory of Mental Tests*, John Wiley & Sons, Inc., New York, 1950.
8. Zeisel, Hans: *Say It with Figures*, 4th ed., Harper & Brothers, New York, 1957.
9. Simon, Herbert A.: "Spurious Correlation: A Causal Interpretation," *Jour. Amer. Stat. Assoc.*, **49**:467–479 (1954).
10. Landsberger, Henry A.: *A Study of Mediation through an Analysis of the Background of Disputes and the Activities of the Mediator*, unpublished Ph.D. thesis, Cornell University, 1954.
11. Quenouille, M. H.: *Associated Measurements*, Academic Press, Inc., New York, 1952.
12. Kendall, Maurice G.: *Rank Correlation Methods*, Charles Griffin & Co., Ltd., London, 1948.

INDEX